P9-CRX-871

Cardiovascular Molecular Morphogenesis

Series Editor
Roger R. Markwald
Medical University of South Carolina

Editorial Advisory Board

Paul Barton
National Heart and Lung Institute
London, United Kingdom

Clayton Buck
University of Pennsylvania

María V. de la Cruz
Hospital Infantil de Mexico
"Federico Gómez," Mexico

Mark Fishman
Massachusetts General Hospital

Adriana Gittenberger-de Groot
University of Leiden
The Netherlands

Julie Korenberg
Cedars-Sinai Medical Center
Los Angeles

Wout H. Lamers
University of Amsterdam
The Netherlands

Kersti Linask
University of Medicine and Dentistry
   of New Jersey

Charles D. Little
Medical University of South Carolina

John Lough
Medical College of Wisconsin

Takashi Mikawa
Cornell Medical College

Jeffrey Robbins
Children's Hospital Medical Center,
   Cincinnati

Thomas Rosenquist
University of Nebraska Medical
   School

Raymond B. Runyan
University of Arizona

Robert Schwartz
Baylor College of Medicine

Assembly of the Vasculature and Its Regulation

Robert J. Tomanek
Editor

With 55 Figures, Including 2 Color Figures

Birkhäuser
Boston • Basel • Berlin

Robert J. Tomanek
Department of Anatomy and Cell Biology
University of Iowa
I-402 Bowen Science Building
Iowa City, IA 52242-1109
USA
robert-tomanek@uniowa. edu

Cover illustration: Early formation of the coronary vascular arterial tree represented by a coronary cast formed by injection of a plastic polymer. The cast is from a stage HH 33 quail which is about 1 day after the two main coronary arteries attach to the aorta. The scanning electron microscopic micrograph is from the laboratory of Robert Tomanek.

Library of Congress Cataloging-in-Publication Data

Assembly of the vasculature and its regulation / edited by Robert J. Tomanek.
    p.   cm.—(Cardiovascular molecular morphogenesis)
    Includes bibliographical references and index.
    ISBN 0-8176-4229-3 (h/c : alk. paper)
    1. Neovascularization—Regulation.   2. Blood-vessels—Growth—Regulation.
  I. Tomanek, Robert J.   II. Series.
    [DNLM: 1. Blood Vessels—embryology.   2. Blood Vessels—physiology.
    3. Endothelium, Vascular—embryology.   4. Endothelium, Vascular—physiology.
    5. Morphogenesis. WG 101 A844 2001]
    QP106.6.A85 2001
    612.1'3—dc21   2001025378

Printed on acid-free paper.
© 2002 Birkhäuser Boston

All rights reserved. This work may not be translated or copied in whole or in part without the written permission of the publisher (Birkhäuser Boston, c/o Springer-Verlag New York, Inc., 175 Fifth Avenue, New York, NY 10010, USA), except for brief excerpts in connection with reviews or scholarly analysis. Use in connection with any form of information storage and retrieval, electronic adaptation, computer software, or by similar or dissimilar methodology now known or hereafter developed is forbidden.
The use of general descriptive names, trade names, trademarks, etc., in this publication, even if the former are not especially identified, is not to be taken as a sign that such names, as understood by the Trade Marks and Merchandise Marks Act, may accordingly be used freely by anyone.

ISBN 0-8176-4229-3
ISBN 3-7643-4229-3   SPIN 10838366

Production managed by Michael Koy, manufacturing supervised by Jeffrey Taub.
Typeset by SNP Best-set Typesetter Ltd., Hong Kong.
Printed and bound by Maple-Vail Book Manufacturing Group, York, PA.
Printed in the United States of America.

9 8 7 6 5 4 3 2 1

Birkhäuser Boston Basel Berlin
A member of BertelsmannSpringer Science+Business Media GmbH

Contents

Contributors

H. Scott Baldwin, G. Abramson Research Bldg., Children's Hospital of Philadelphia, 3516 Civic Center Blvd., Philadelphia, PA 19104-4318

Georg Breier, Abteilung Molekulare Zellbiologie Max-Planck-Institut für Physiologische und Klinische Forschung, W.G. Kerckhoff-Institut, Parkstrasse 1, D-61231 Bad Nauheim, Germany

Peter C. Brooks, New York University School of Medicine Depts. of Radiation Oncology and Cell Biology, Kaplan Cancer Center, 400 E. 34 Street, New York, NY 10016

Peter Carmeliet, Center for Transgene Technology and Gene Therapy, Katholieke Universiteit, Campus Gasthuisberg, O and N Herestradt 49, B-3000 Louvain, Belgium

Bodo Christ, Anatomisches Institut II, Universität Freiburg, Albertstr. 17, 79104 Freiburg, Germany

Daphne E. deMello, Department of Pathology, St. Louis University, Health Sciences Center and Cardinal Glennon Children's Hospital, 1465 South Grand Blvd., St. Louis, MO 63104

Xiu-Rong Dong, Molecular and Cellular Biology, Baylor College of Medicine, One Baylor Plaza, Houston, TX 77030

Ingo Flamme, Zentrum für Molekularbiologische Medizin, Universität Köln, Joseph-Stelzmann-Str. 9, D-50931 Cologne, Germany

R. Ariel Gómez, University of Virginia, School of Medicine, MR-4 Bldg., Room 2001, Charlottesville, VA 22908

Po-Tsan Ku, Medicine Athero and Lipo, Baylor College of Medicine, One Baylor Plaza, Houston, TX 77030

Haymo Kurz, Anatomisches Institut II, Universität Freiburg, Albertstr. 17, D-79104 Freiburg, Germany

María Luisa S. Sequeira López, University of Virginia, School of Medicine, Charlottesville, VA 22908

Mark W. Majesky, Department of Pathology, Baylor College of Medicine, One Baylor Plaza, Houston, TX 77030

Susan Pichla, The Joseph Stokes Research Institute, Children's Hospital of Philadelphia, Philadelphia, PA 19104-4318

Lynne M. Reid, Department of Pathology, Harvard Medical School, Children's Hospital, 300, Longwood Avenue, Boston, MA 02115

Paul Robson, The Joseph Stokes Research Institute, Children's Hospital of Philadelphia, Philadelphia, PA 19104-4318

Joanna Schwartz, College of Pharmacy and Health Sciences, 2507 University Avenue, Drake University, Des Moines, IA 50311-4505

Chitra Suri, Regeneron Pharmaceuticals, Inc., 777 Old Saw Mill River Road, Tarrytown, NY 10591

Robert J. Tomanek, Department of Anatomy and Cell Biology, Bowen Science Bldg., University of Iowa, Iowa City, IA 52242

Donald S. Torry, Department of Medical Microbiology and Immunology, Southern Illinois University School of Medicine, Springfield, IL 62794

Ronald J. Torry, College of Pharmacy and Health Sciences, 2507 University Avenue, Drake University, Des Moines, IA 50311-4505

Jingsong Xu, University of Southern California School of Medicine, Department of Biochemistry and Molecular Biology, Norris Cancer Center, Topping Tower Room 6409, 1441 Eastlake Ave., Los Angeles, CA 90033

George D. Yancopoulos, Regeneron Pharmaceuticals, Inc., 777 Old Saw Mill River Road, Tarrytown, NY 10591

Xinping Yue, Department of Anatomy and Cell Biology, Bowen Science Bldg., University of Iowa, Iowa City, IA 52242

Wei Zheng, Department of Anatomy and Cell Biology, Bowen Science Bldg., University of Iowa, Iowa City, IA 52242

Bin Zhou, The Joseph Stokes Research Institute, Children's Hospital of Philadelphia, Philadelphia, PA 19104-4318

Corresponding Authors

H. Scott Baldwin, Room 702, G. Abramson Research Bldg., Children's Hospital of Philadelphia, 3516 Civic Center Blvd., Philadelphia, PA 19104-4318, Phone: (215) 590-2938, Fax: (215) 590-5454, Email: *sbaldwin@mail.med.upenn.edu*

Peter C. Brooks, New York University School of Medicine, Depts. of Radiation Oncology and Cell Biology, Kaplan Cancer Center, 400 E. 34 Street, New York, NY 10016

Peter Carmeliet, Center for Transgene Technology and Gene Therapy, Katholieke Universiteit, Campus Gasthuisberg, O and N Herestradt 49, B-3000 Louvain, Belgium

Daphne E. deMello, Department of Pathology, St. Louis University, Health Sciences Center and Cardinal Glennon Children's Hospital, 1465 South Grand Blvd., St. Louis, MO 63104, Phone: (314) 577-5338, Email: *demellde@slu.edu* or ddemello@ssmhcs.com

Ingo Flamme, Zentrum für Molekularbiologische Medizin, Universität Köln, Joseph-Stelzmann-Str. 9, D-50931 Cologne, Germany

R. Ariel Gómez, University of Virginia, School of Medicine, MR-4 Bldg., Room 2001, Charlottesville, VA 22908, Phone: (804) 924-0078, Fax: (804) 982-4328, Email: *rg@virginia.edu*

Haymo Kurz, Anatomisches Institut II, Universität Freiburg, Albertstr. 17, 79104 Freiburg, Germany

Mark W. Majesky, Department of Pathology, Baylor College of Medicine, One Baylor Plaza, Houston, TX 77030, Phone: (713) 798-5837, Fax: (713) 798-8920, Email: *mmajesky@bcm.tcm.edu*

Chitra Suri, Regereron Pharmaceuticals, Inc, 777 Old Saw Mill River Road, Tarrytown, NY 10591

Robert J. Tomanek, Department of Anatomy and Cell Biology, Bowen Science Bldg., University of Iowa, Iowa City, IA 52242, Phone: (319) 335-7740, Fax: (319) 335-7198, Email: *robert-tomanek@uiowa.edu*

Ronald J. Torry, College of Pharmacy and Health Sciences, 2507 University Avenue, Drake University, Des Moines, IA 50311-4505, Phone: (515) 271-2750, Fax: (515) 271-4171, Email: *ron.torry@drake.edu*

Series Preface

The overall scope of this new series will be to evolve an understanding of the genetic basis of (1) how early mesoderm commits to cells of a heart lineage that progressively and irreversibly assemble into a segmented, primary heart tube that can be remodeled into a four-chambered organ, and (2) how blood vessels are derived and assembled both in the heart and in the body. Our central aim is to establish a four-dimensional, spatiotemporal foundation for the heart and blood vessels that can be genetically dissected for function and mechanism.

Since Robert DeHaan's seminal chapter "Morphogenesis of the Vertebrate Heart" published in *Organogenesis* (Holt Rinehart & Winston, NY) in 1965, there have been surprisingly few books devoted to the subject of cardiovascular morphogenesis, despite the enormous growth of interest that occurred nationally and internationally. Most writings on the subject have been scholarly compilations of the proceedings of major national or international symposia or multiauthored volumes, often without a specific theme. What is missing are the unifying concepts that can make sense out of a burgeoning database of facts. The Editorial Board of this new series believes the time has come for a book series dedicated to cardiovascular morphogenesis that will serve not only as an important archival and didactic reference source for those who have recently come into the field but also as a guide to the evolution of a field that is clearly coming of age. The advances in heart and vessel morphogenesis are not only serving to reveal general basic mechanisms of embryogenesis but are also now influencing clinical thinking in pediatric and adult cardiology.

Undoubtedly, the Human Genome Project and other genetic approaches will continue to reveal new genes or groups of genes that may be involved in heart development. A central goal of this series will be to extend the identification of these and other genes into their functional role at the molecular, cellular, and organ levels. The major issues in morphogenesis highlighted in the series will be the local (heart or vessel) regulation of cell growth and death, cell adhesion and migration, and gene expression responsible for the cardiovascular cellular phenotypes.

Specific topics will include the following:

- The roles of extracardiac populations of cells in heart development.
- Coronary angiogenesis.
- Vasculogenesis.
- Breaking symmetry, laterality genes, and patterning.
- Formation and integration of the conduction cell phenotypes.
- Growth factors and inductive interactions in cardiogenesis and vasculogenesis.
- Morphogenetic role of the extracellular matrix.
- Genetic regulation of heart development.
- Application of developmental principles to cardiovascular tissue engineering.

Cardiovascular Developmental   *Roger R. Markwald*
   Biology Center
Medical University of South Carolina
Charleston, South Carolina

Preface

In 1983 Ciba Foundation Symposium 100, entitled *Development of the Vascular System* (London: Pitman), was published, in which the chairman's introduction noted that this was "the first meeting to be recorded on the development of the cardiovascular system." The chairman also stated that it was "surprising that such an important subject has been so neglected." Since that time, considerable progress has been made regarding vascular formation and the molecular mechanisms that regulate the process. Although much of the work has focused on angiogenesis in disease states, such as tumors, retinopathy, and ischemic models, our understanding of vasculogenesis and angiogenesis during development has advanced. Progress in this field has been facilitated by the rapid advances in the field of growth factor biology, as well as new discoveries concerning adhesion molecules, various extracellular components, and signaling pathways.

Considering the expanding literature in this area, it is obvious that a comprehensive book addressing the major areas of vessel formation during development is long overdue. Accordingly, *Assembly of the Vasculature and Its Regulation* addresses key topics in the field of vessel formation during development. In the first section of the book major issues regarding vessel formation and its regulation (growth factors and their receptors, extracellular matrix, adhesion molecules, and smooth muscle differentiation) are detailed. The second section addresses vascular development in a number of organs (heart, central nervous system, kidney, lung, and placenta). The rationale for the book is to provide a reference treatise that covers the events and regulatory factors characterizing vessel formation during the prenatal and perinatal periods. In aggregate, this book addresses the proliferation, migration, and differentiation of angioblasts, their assembly into vascular tubes, the subsequent growth of these structures, and the addition of medial and adventitial components to form a functional network, i.e., the vascular system.

Assembly of the Vasculature and Its Regulation is dedicated to the memory of Werner Risau (1953–1998) whose work has had profound impact on the field of vascular development. His contributions are many and will long be quoted and appreciated. Drs. Breier and Flamme provide a tribute to Dr. Risau on the following pages.

University of Iowa   *Robert J. Tomanek*
Iowa City, Iowa, USA

Werner Risau (1953–1998)

Werner Risau, who died in December 1998, will be remembered for his outstanding work on the development of the vascular system in the vertebrate embryo and on the mechanisms of tumor angiogenesis.

Werner Risau was born in Rheine, Germany, in 1953, and educated in his native city. In 1973 he began to study chemistry in the nearby city of Münster; three years later he moved to Tübingen, Germany, and focused on biochemistry. His first appointment was at the Max Planck Institute for Developmental Biology in Tübingen, where he finished his Ph.D. thesis on ribonucleoprotein particles in *Drosophila* in the department of Prof. Friedrich Bonhoeffer in 1983. In parallel, Risau also studied medicine. During this time, his interest in biochemistry, developmental biology, and medicine amalgamated and he became fascinated by the question of how blood vessels develop and grow in the vertebrate embryo.

The defining period in Risau's scientific life was his postdoctoral time in Prof. Judah Folkman's laboratory at the Children's Hospital, Harvard University, in Boston from 1983 to 1984. There he learned endothelial cell biology and became familiar with angiogenic growth factors. Inspired by Folkman's brilliant concept of tumor angiogenesis, Risau had the vision that angiogenic growth factors not only might be involved in the vascularization of tumors, but also might regulate the development of the vascular system. This idea opened a new era in vascular biology and became a dominant research theme during the 1990s.

After having returned to the Max Planck Institute in Tübingen, Risau founded his own laboratory and initiated vigorous and creative research on the molecular mechanisms of endothelial cell growth and differentiation in the vertebrate embryo. He considered the brain to be an ideal organ in which to investigate the regulation of embryonic angiogenesis. By using biochemical methods, he showed that angiogenic growth factors of the fibroblast growth factor family are present in the embryonic chick brain. On the other hand, he put forward the hypothesis that the locally produced endogenous inhibitors of angiogenesis would counteract angiogenic growth factors and induce blood vessel regression in the embryo, for example in the limb bud. Risau's analyses of blood-brain barrier induction and function founded his reputation as an internationally recognized expert in endothelial cell biology. During his period, he also became interested in the question of how endothelial cells differentiate from their mesodermal precursor cells. He put forward the concept of vasculogenesis as the principal mechanism by which the primary vascular system is established, as a mechanism distinct from angiogenesis, the blood vessel formation from preexisting vessels.

In 1988, Risau's interest in the vasculature of the brain led him and his group to the Max Planck Institute of Neurobiology in Martinsried (the former Max Planck Institute of Psychiatry). There he increasingly utilized molecular approaches to address the problems of vascular biology. This period was characterized by studies on the function of vascular endothelial growth factor (VEGF). In several papers published in 1992 and 1993, he laid down the basis for the work of many other researchers. These studies included the first identification of VEGF and the high-affinity VEGF receptor, Flk-1, as a critical signaling system involved in the development of the vascular system, and the first identification of the VEGF as tumor angiogenesis factor in malignant glioma and its regulation by hypoxia. Of the many fruitful collaborations from this time, two stand out—that with Dr. Erwin Wagner (Vienna), using the polyoma middle T oncogene in the study of endothelial cell growth and transformation, and that with Dr. Axel Ullrich (Martinsried) in the functional analysis of the Flk-1 receptor tyrosine kinase.

In 1992, Risau was appointed director at the Max Planck Institute for Physiological and Clinical Research in Bad Nauheim. In his department, he combined expertise in various research areas, including vasculogenesis, endothelial cell differentiation and blood-brain barrier development, endothelial cell signaling, endothelial cell-specific transcriptional control, mechanisms of hypoxia-induced expression, and endothelial cell apoptosis. He extended the investigation of the molecular players that cooperate to regulate endothelial cell growth and differentiation. These studies included other endothelial receptor tyrosine kinases such as Tie2 and its role in vascular remodeling and angiogenesis, endothelial phosphatases, hypoxia-inducible factors, and other transcription factors involved in blood vessel growth and development.

Werner Risau was a generalist interested in many areas of vascular biology. He was not only an excellent researcher, but also had a rare sense of community. This is documented by the large number of scientific collaborations that were initiated on the basis of his enthusiastic will to understand the biology of the vascular system.

Georg Breier
Ingo Flamme

Werner Risau

CHAPTER 1

The Development of Blood Vessels: Cellular and Molecular Mechanisms

Peter Carmeliet

All blood vessels share a common function to deliver oxygen to and remove metabolites from peripheral tissues. However, we now know that not all vessels are alike. In fact, they significantly differ in many respects. For instance, some only consist of endothelial cells (capillaries), while others are surrounded by smooth muscle cells (arteries, veins). Arteries differ from veins by the number of smooth muscle cells, but even capillaries without smooth muscle cells express selective arterial or venous markers. Various arteries differ by the embryonic origin of their smooth muscle cells. But even arteries and veins in different organs express specific markers and have specialized functions to accommodate the distinct needs in each organ. Furthermore, vessels in an embryo are actively growing in an ordered pattern and are quiescent in a healthy adult, but they are completely chaotic and growing in an uncontrolled manner in a sick patient with a tumor. Genetic analysis over the last 10 years has provided stunning insights in the cellular and molecular mechanisms that define the formation and function of these distinct vessels. These insights are discussed in this overview.

CELLULAR INTERACTIONS DURING VASCULAR GROWTH

Blood vessels in the embryo initially form via vasculogenesis (Fig. 1.1). Endothelial precursors (angioblasts) differentiate *in situ* and assemble into a vascular labyrinth.[1] Subsequently, these primitive vessels expand, grow, and remodel into a complex mature network via angiogenesis. This process involves enlargement of venules to "mother" vessels that subsequently extend sprouts or become divided by intervascular pillars of periendothelial cells (intussusception) or by trans-endothelial cell bridges that then split into individual capillaries. Periendothelial cells (pericytes in small vessels, smooth muscle cells in large vessels) exhibit functions that are essential for vascular development and function.[2] Once endothelial cells become surrounded by smooth muscle cells (vascular myogenesis), they stop proliferating and migrating and become quiescent (Figs. 1.1 and 1.2). Mural cells also produce extracellular matrix, which further stabilizes the nascent vessels. Vessels that contain a smooth muscle coat are more resistant to traumatic rupture

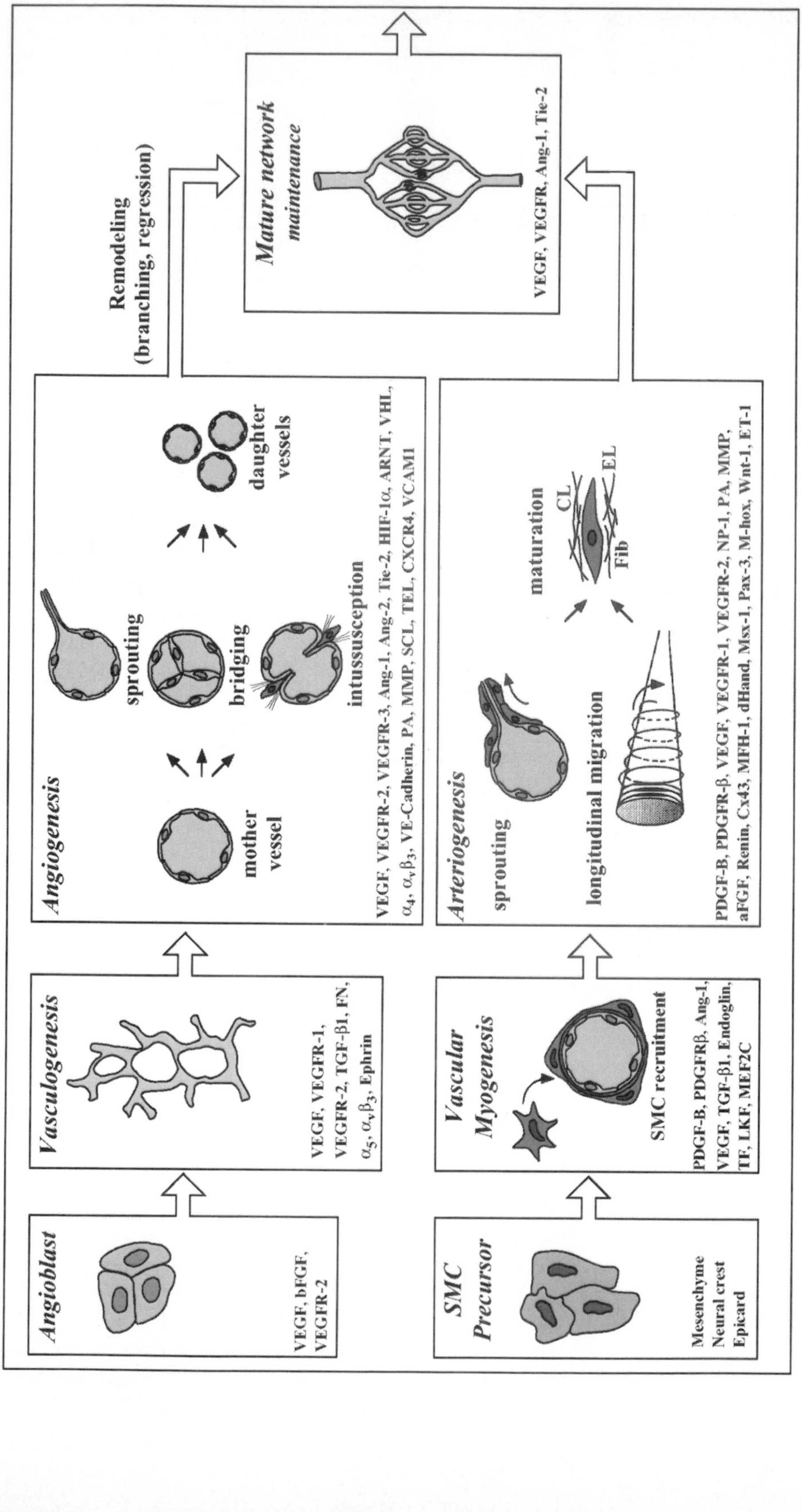

EMBRYO

Angioblast

VEGF, bFGF, VEGFR-2

Vasculogenesis

VEGF, VEGFR-1, VEGFR-2, TGF-β1, FN, α5, αvβ3, Ephrin

Angiogenesis

mother vessel

sprouting

bridging

intussusception

daughter vessels

VEGF, VEGFR-2, VEGFR-3, Ang-1, Ang-2, Tie-2, HIF-1α, ARNT, VHL, α4, αvβ3, VE-Cadherin, PA, MMP, SCL, TEL, CXCR4, VCAMl

Remodeling (branching, regression)

Mature network maintenance

VEGF, VEGFR, Ang-1, Tie-2

SMC Precursor

Mesenchyme
Neural crest
Epicard

Vascular Myogenesis

SMC recruitment

PDGF-B, PDGFRβ, Ang-1, VEGF, TGF-β1, Endoglin, TF, LKF, MEF2C

Arteriogenesis

sprouting

longitudinal migration

maturation

CL

EL

Fib

PDGF-B, PDGFR-β, VEGF, VEGFR-1, VEGFR-2, NP-1, PA, MMP, aFGF, Renin, Cx43, MFH-1, dHand, Msx-1, Pax-3, M-hox, Wnt-1, ET-1 or regression.[3] During subsequent arteriogenesis, vessels become extensively covered by a muscular coat. The blood vessels then acquire viscoelastic and vasomotor properties, which are necessary to accommodate the fluctuating needs in tissue perfusion. In addition, periendothelial cells assist endothelial cells in acquiring specialized functions in different vascular beds.[2]

MOLECULAR MECHANISMS OF VASCULOGENESIS AND ANGIOGENESIS

It is now widely accepted that vascular growth during development and pathologic conditions is tightly regulated by a complex interplay and balance of angiogenic stimulators and inhibitors.[4] The angiogenic switch is off when the effect of proangiogenic molecules is balanced by that of antiangiogenic molecules, and is on when the net balance is tipped in favor of angiogenesis. Various signals that trigger this switch have been discovered. These include metabolic stress (e.g., low pO_2, low pH, hypoglycemia), mechanical stress (e.g., pressure generated by proliferating cells), and immune/inflammatory response (e.g., immune/inflammatory cells that have infiltrated the tissue). How the interplay between environmental and genetic mechanisms influences angiogenesis and growth is a complex and largely unresolved matter.

The first blood vessels in an embryo consist only of endothelial cells. These cells share a common origin with hematopoietic cells (hemangioblast). We still know very little about the mechanisms governing endothelial cell fate; *Ets-1*,[5] *Hex*,[6] *Vezf1*,[7] *Hox-*,[8,9] and *GATA* family members,[10–13] and basic helix-loop-helix (bHLH) factors and their inhibitors of differentiation (*Id*-proteins)[14] have been implicated, but their function remains largely unknown. Another outstanding, though most intriguing, question is how endothelial cells "know" that they belong to an artery or a vein. Recent genetic studies have provided initial insights in the molecular mechanims of arteriovenous specification. The bHLH transcription factor gridlock appears to be the first switch driving angioblasts into arterial lineage,[15] while members of the ephrin family—signals that are also involved in guidance of axons and repulsion of neurons[16,17]—are essential at later stages.

FIGURE 1.1.   Vascular development in the embryo. Endothelial precursors (angioblasts) differentiate in endothelial cells that assemble in a primitive vascular labyrinth (vasculogenesis). This network subsequently expands and remodels via sprouting, bridging, and intussusception (angiogenesis). Smooth muscle cell (SMC) precursors—derived from the mesenchyme, neural crest, or epicard—differentiate and become recruited around endothelial cell channels (vascular myogenesis). During subsequent arteriogenesis, smooth muscle cells migrate over endothelial sprouts (sprouting) or longitudinally alongside preexisting endothelial channels. During further maturation, the smooth muscle cells produce matrix molecules like collagen (CL), elastin (EL), or fibrillin (Fib). Both endothelial and smooth muscle lineages essentially contribute to a mature vascular network. VEGF, vascular endothelial growth factor; bFGF, basic fibroblast growth factor; VEGFR-1, VEGF receptor-1; VEGFR-2, VEGF receptor-2; VEGFR-3, VEGF receptor-3; TGF-β1, transforming growth factor-β1; FN, fibronectin; Ang-1, angiopoietin-1; Ang-2, angiopoietin-2; HIF-1α, hypoxia-inducible factor-1α; ARNT, aryl hydrocarbon receptor nuclear translocator (HIF-1β); VHL, von Hippel–Lindau; PA, plasminogen activator; MMP, matrix metalloproteinase; SCL, stem cell leukemia factor; CXCR4, chemokine receptor 4; VCAM-1, vascular cell adhesion molecule-1; PDGF-B, platelet-derived growth factor-B; PDGFR-β, PDGF receptor-β; TF, tissue factor of coagulation; MEF2C, myocyte enhancer binding factor-2; NP-1, neuropilin-1; Cx43, connexin 43; ET-1, endothelin-1. (From Carmeliet,[127] with permission.)

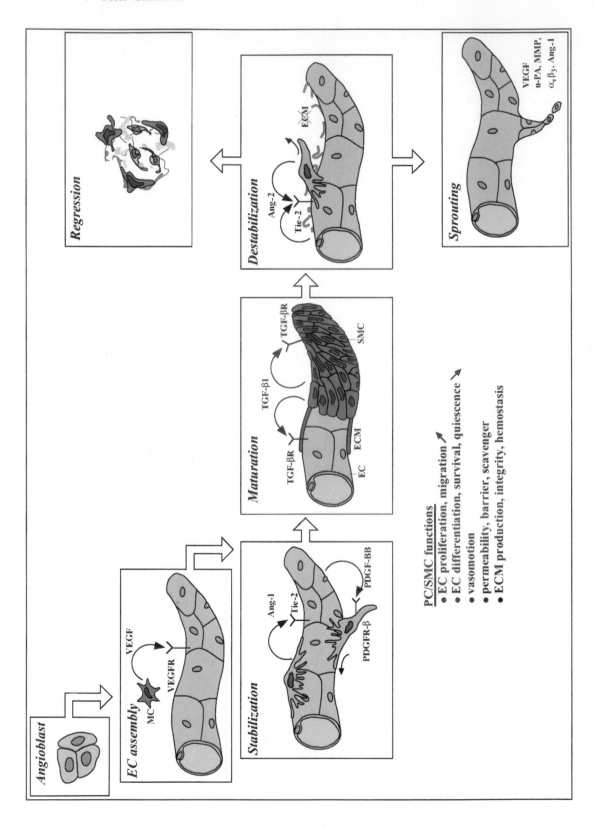

In the embryonic yolk sac, hemangioblasts form cellular aggregates in which the inner cells develop into hematopoietic precursors and the outer population into endothelial cells. The endothelial precursors or angioblasts may migrate extensively before *in situ* differentiation and plexus formation. Vascular endothelial growth factor (VEGF), VEGF receptor-2 (VEGFR-2), and basic fibroblast growth factor (bFGF) stimulate angioblast differentiation.[18–21] Other molecules such as VEGF receptor-1 (VEGFR-1) suppress hemangioblast commitment.[22] Transforming growth factor-β1 (TGF-β1) and TGF-β receptor-2,[23] matrix macromolecules (fibronectin), or matrix receptors (α_5 integrin, $\alpha_v\beta_3$ integrin) also affect vasculogenesis.[24]

Endothelial precursors not only exist during embryonic life, but also have been identified in bone marrow and in peripheral blood in adults.[25] Pathologic angiogenesis in adults has been shown to involve circulating endothelial precursors (angioblasts).[25] Circulating endothelial cells can be shed from a vessel wall or their precursors can be mobilized from bone marrow and peripheral blood. Unlike shed cells, circulating CD34/VEGFR-2/AC133-positive marrow-derived angioblasts have a high proliferation rate.[26] These precursor cells can contribute to angiogenesis in ischemic, inflamed, or neoplastic tissues. This can occur in response to tissue ischemia, VEGF, bFGF, granulocyte-macrophage colony-stimulating factor (GM-CSF), insulin-like growth factor (IGF-1), and angiopoietins.[27] To what extent these precursors contribute to pathologic angiogenesis remains to be determined.

Several molecules have been implicated in the sequential steps in angiogenesis (Table 1.1). Initially, vessels dilate—a process involving nitric oxide (NO). The increase in permeability in response to VEGF allows extravasation of plasma proteins that lay down a provisional scaffold for migrating endothelial cells. This involves formation of fenestrations and vesiculovacuolar organelles and opening of intercellular junctions via redistribution of platelet endothelial cell adhesion molecule-1 (PECAM-1)[28] and the adherens-type junctional VE-cadherin.[29] VEGF, in synergism with its homologue placental growth factor (PlGF) is 1,000-fold more potent than histamine, which might play a more selective role in pathologic conditions. Angiopoietin-1 (Ang-1), a ligand of the endothelial Tie-2 receptor, is a natural inhibitor of vascular permeability, "tightening" preexisting vessels.[30]

For endothelial cells to emigrate from their resident site, they need to loosen interendothelial cell contacts and to relieve periendothelial cell support; that is, mature vessels need to become destabilized. Angiopoietin-2 (Ang-2), another

---

FIGURE 1.2.   Molecular and cellular mechanisms of angiogenesis. Angioblasts differentiate into endothelial cells. VEGF, produced by mesenchymal cells (MCs) and binding on VEGF receptors (VEGFRs), mediates endothelial cell (EC; yellow cells) differentiation, growth, migration, and assembly. The fragile endothelial vessels become stabilized by recruitment of periendothelial cells (pericytes and smooth muscle cells; gray spiderlike cells) in response to PDGF-BB. Mural cells produce angiopoietin-1 (Ang-1) that tightens the interaction between endothelial and periendothelial cells. During maturation, TGF-β1 induces endothelial quiescence, smooth muscle cell (SMC) differentiation, and extracellular matrix (ECM) production. Before vessel sprouting can initiate, angiopoietin-2 (Ang-2) is implicated in destabilizing the interaction between endothelial and periendothelial cells and degrading the extracellular matrix. In the presence of angiogenic stimuli (VEGF, u-PA, MMP, $\alpha_v\beta_3$, Ang-1), endothelial cell sprouting proceeds; in contrast, in the absence of endothelial survival factors, the vessel regresses due to vascular cell apoptosis. (From Carmeliet,[127] with permission.)

TABLE 1.1. Endogenous activators of angiogenesis.

Name	Function	Reference
Growth factors/receptors[a]		
Vascular endothelial growth factor (VEGF)	Predominant regulator of physiologic and pathologic angiogenesis and permeability; used for therapeutic angiogenesis	18, 128, 129
Placental growth factor (PlGF)	Homologue of VEGF, redundant for physiologic but essential for pathologic angiogenesis via amplification of VEGF activity	38, 130
VEGF-C[b]	Homologue of VEGF; (lymph)-angiogenic factor	39, 131
VEGF-B,D[b]	Homologues of VEGF; role in pathologic angiogenesis to be defined	39
VEGFR-2[b]	Receptor for VEGF and VEGF-C, mediating the angiogenic response	132
VEGFR-1[b]	Receptor for VEGF, PlGF and VEGF-B	133
VEGFR-3[b]	Receptor for VEGF-C and VEGF-D, involved in (lymph)-angiogenesis	132
Neuropilin-1 (NP1)[b]	Receptor for the $VEGF_{165}$ isoform; co-receptor of VEGFR-2 in tumor angiogenesis	41
Angiopoietin-1 (Ang-1)[b]	Ligand of the Tie-2 receptor; stabilizes nascent vessels by tightening endothelial-periendothelial cell interactions; inhibits permeability	17
Angiopoietin-2 (Ang-2)[b]	Ligand of the Tie-2 receptor; angiogenic in the presence of VEGF via loosening of periendothelial cells	17
Platelet-derived growth factor-BB (PDGF-BB)	Recruits smooth muscle cells around nascent endothelial channels	134
Transforming growth factor-β1 (TGF-β1)[c]	Stabilizes nascent vessels by stimulating extracellular matrix production	52
Basic fibroblast growth factors (bFGF)	Angiogenic factor when administered; used for therapeutic angiogenesis; endogenous vascular role unclear; affects other cell types	128, 135
Hepatocyte growth factor (HGF)	Therapeutically angiogenic; endogenous role unclear; affects other cell types	136
Monocyte chemoattractant protein-1 (MCP-1)	Inflammatory cytokine; stimulates growth of preexisting collaterals (arteriogenesis)	137
Matrix receptors, junctional molecules, proteinases, and others[a]		
Integrins $\alpha_v\beta_3$, $\alpha_v\beta_5$	Cellular receptors for matrix components and proteinases (MMP-2)	138
VE-cadherin	Endothelial-specific cadherin; mediates adhesion and the VEGF endothelial survival effect	139
Platelet endothelial cell adhesion molecule (PECAM/CD31)	Cell surface glycoprotein on hemopoietic and endothelial cells, mediating homophilic adhesion	140
Ephrins	Regulate arterial/venous specification	16, 17
Urokinase-type plasminogen activators (u-PA)	Activator of plasminogen, involved in cellular migration and matrix remodeling	34
Plasminogen activator inhibitor-1 (PAI-1)	Inhibitor of u-PA; stabilizes nascent vessels via preventing proteolytic breakdown of the vessel matrix; activator of MMP's	34, 36
Matrix metalloproteinases (MMPs)	Proteinases degrading collagen (MMP-1, -8, -13), gelatin (MMP-2, -9), elastin (MMP-9, -12) and other vascular matrix components; involved in matrix remodeling and cellular migration	34, 141
Tissue-inhibitors of MMP (TIMPs)	MMP-inhibitors; synthetic inhibitors tested for suppression of pathologic angiogenesis	141
Nitric oxide synthase (NOS)	Generates nitric oxide (NO); stimulates angiogenesis and vasodilation	142
Cyclooxygenase-2 (Cox2)	Generates inflammatory prostaglandins; inhibited by nonsteroidal antinflammatory drugs	143
Chemokines[d]	Pleiotropic role in angiogenesis	144
AC133	Orphan receptor involved in angioblast differentiation	145

[a] Selected list.

[c] Suppresses angiogenesis in some contexts.

[d] Can be angiogenic in some contexts.

ligand of Tie-2, has been implicated in detachment of smooth muscle cells and loosening of endothelial extracellular matrix[17,31] (Fig. 1.2). Proteinases of the plasminogen activator (PA), matrix metalloproteinases (MMP), chymase, or heparanase families influence angiogenesis by degrading extracellular matrix molecules and by activating or liberating growth factors (bFGF, VEGF, IGF-1) that are sequestered within the extracellular matrix.[32] Urokinase-type PA (u-PA) is essential for revascularization of myocardial infarcts,[33] whereas u-PA receptor antagonists inhibit tumor angiogenesis (reviewed in ref. 34). MMP-3, MMP-7, and MMP-9 affect angiogenesis in neonatal bones[35] and tumors,[36] while tissue inhibitors of MMPs (TIMP-1, TIMP-3) or a naturally occurring fragment of MMP-2 (PEX)—preventing binding of MMP-2 to $\alpha_v\beta_3$—inhibit tumor angiogenesis.[37] Paradoxically, TIMP-1 and the PA inhibitor PAI-1 are risk factors for a poor prognosis in cancer patients and promote tumor angiogenesis in mice,[36] perhaps because they prevent excessive proteolysis or influence endothelial cell adhesion.

A variety of candidates have been identified to stimulate endothelial growth and migration (Table 1.1). VEGF,[18] PLGF,[38] VEGF-B, VEGF-C,[39] VEGF-D, and their receptors VEGFR-2, VEGF receptor-3 (VEGFR-3),[40] and neuropilin-1 (NP-1)[41] have distinct and specific roles. VEGF and its receptor VEGFR-2 affect embryonic, neonatal, and pathologic angiogenesis and are therapeutic targets, even though much remains to be learned about the role of the distinct VEGF isoforms or heterodimers of VEGF family members.[18–20,42] Expression of $VEGF_{120}$ alone could initiate—but not complete—angiogenesis.[43] VEGFR-3 plays a role in embryonic angiogenesis[44] and is expressed in angiogenic endothelial cells during pathologic angiogenesis, whereas VEGF-C (a ligand of VEGFR-3) is angiogenic in adult pathology.[45] Targeted truncation of VEGFR-1 at the tyrosine kinase domain did not impair angiogenesis, questioning the signaling activity of this receptor *in vivo*.[46] The role of VEGF-B and VEGF-D in angiogenesis remains unknown, even though they are angiogenic when administered exogenously. PlGF specifically affects pathologic angiogenesis. Ang-1 tyrosine-phosphorylates Tie-2, is chemotactic for endothelial cells, induces endothelial sprouting, and potentiates the action of VEGF, but fails to induce endothelial proliferation.[17] In contrast to VEGF, Ang-1 does not initiate endothelial network organization by itself, but stabilizes networks initiated by VEGF. This suggests that Ang-1 may act at later stages than VEGF.[17,47] Ang-2—at least in the presence of VEGF—is also angiogenic. Members of the fibroblast growth factor (FGF) and platelet-derived growth factor (PDGF) family are redundant during normal development,[48,49] but affect angiogenesis when administered, likely by recruiting mesenchymal (smooth muscle precursor) or inflammatory cells that produce angiogenic factors (see below). TGF-β1 and tumor necrosis factor-α (TNF-α) stimulate or inhibit endothelial growth *in vitro*, but are angiogenic *in vivo* via recruiting wound cells.[50–53] Molecules involved in cell-cell or cell-matrix interactions—such as the $\alpha_v\beta_3$ integrin, which localizes MMP-2 to the endothelial cell surface—mediate endothelial spreading and protrusions.[54]

During angiogenesis, endothelial cells not only proliferate but also need to survive. The importance of endothelial survival for angiogenesis is exemplified by findings that reduced endothelial survival, in the presence of normal proliferation, is sufficient to induce vascular regression.[55] Endothelial apoptosis is a natural mechanism of vessel regression after birth (exemplified in the retina and ovary). Endothelial apoptosis is also induced by deprivation of nutrients or survival signals

when the lumen is obstructed by spasms, thrombi[56] or shedding of dead endothelial cells,[57] or when a change in the angiogenic gene profile occurs.[42,43,58] Growth
and maintenance of the vasculature is regulated by a balance between stimulatory
and inhibitory angiogenic factors. The survival function of VEGF depends on the
interaction between VEGFR-2, β-catenin, and VE-cadherin.[55] Exposure of premature babies to hyperoxia leads to blindness because reduced VEGF levels induce
pathologic vessel regression in the retina.[59] Ang1 also promotes endothelial survival. In contrast, Ang2 may suppress endothelial survival—at least in the absence
of VEGF or other angiogenic stimuli—and contributes to regression of co-opted
tumor vessels.[17,47,60] Disruption of the interaction with matrix macromolecules—
using $\alpha_v\beta_3$ antagonists or the desintegrin accutin—also results in endothelial apoptosis (anoikis).[61] In addition, endothelial apoptosis is induced by TNF-α, NO,
reactive oxygen species, angiostatin,[62] thrombospondin-1 (TSP-1)[63] and its metal-
loproteinase/TSP-derivative METH, TNF-α or interferon-γ, tissue factor pathway
inhibitor (TFPI), and vascular endothelial growth inhibitor (VEGI).[64–66] Hemodynamic forces are essential for maintenance of blood vessels as physiologic
levels of shear stress reduce endothelial cell turnover *in vivo* and abrogate TNF-
α—mediated endothelial apoptosis.[67]

Angiogenesis inhibitors such as angiostatin (an internal fragment of plasminogen),[68] endostatin (a fragment of collagen XVIII),[69] antithrombin III, interferon-
β, leukemia inhibitory factor, and platelet factor 4 have been characterized in
pathologic angiogenesis, in particular in tumors (Table 1.2). Other angiogenesis
inhibitors include prothrombin kringle-1 and kringle-2,[70,71] TSP-2,[72] PECAM-1
antagonists,[73] interleukin-4 (IL-4), IL-12, interferon-α,[74] cyclooxygenase-2 (Cox2)
inhibitors,[75] 1,25-dihydroxyvitamin-D$_3$,[76] and the N-terminal fragment of prolactin.[77] In general, their role in embryonic or physiologic angiogenesis remains
largely unknown.

Gene-inactivation studies suggest that a number of molecules confer morphogenetic signals that are essential for proper remodeling and pruning of capillary-
like vessels with uniform size and irregular organization into a structured network
of branching vessels. These include the distinct VEGF-isoforms and VEGFR-
3,[19,20,43,44] the endothelial orphan receptor Tie-1,[78] T-cell leukemia protein SCL/
tal-1,[79] TEL,[80] the guanosine triphosphate (GTP)-binding protein Gα$_{13}$,[81] Jagged,[82]
the endothelial chemokine receptor CXCR4, and molecules involved in cell-to-
cell or cell-matrix communication such as vascular cell adhesion molecule-1
(VCAM-1),[83,84] its cognate α$_4$ integrin receptor,[85] and fibronectin.[86]

Endothelial growth can escape the intricate control and become derailed. Vascular tumors (hemangiomas) in children are quite common and invalidating, yet
we know little about their etiology.[87] The angiogenic switch in hemangiomas is
tilted by elevated levels of the proangiogenic molecules VEGF, bFGF, and monocyte chemoattractant protein-1 (MCP-1). Interferon-β (IFN-β) is an endogenous
inhibitor, and treatment with IFN-α accelerates involution of hemangiomas[88]
(Table 1.2). Hemangiomas could arise from intrinsic genetic alterations of endothelial cells. Alternatively, viral infections have been implicated as a pathogenic
mechanism. Thus, the Orf virus produces a novel VEGF homologue (VEGF-E)
in human hemangiomas.[89] Kaposi's sarcoma (KS) affects ~30% of AIDS patients
and is another angioproliferative tumor of viral origin. KS is characterized by the
presence of spindle-like lymphatic/vascular endothelial cells. Angiogenic factors
released by both KS and host cells, as well as human herpesvirus-8 (HHV-8) and

TABLE 1.2. Endogenous inhibitors of angiogenesis.

Name	Function	Reference
Thrombospondin-1 and internal fragments of thrombospondin-1	Thrombospondin is a 180-kd, large, modular extracellular matrix protein	146
Angiostatin	A 38-kd fragment of plasminogen involving either kringle domains 1–3, or kringle 5 fragments	68, 147, 148
Endostatin	A 20-kd zinc-binding fragment of type XVIII collagen	69
Vasostatin	An N-terminal fragment of calreticulin	149
Vascular endothelial growth factor inhibitor (VEGFI)	A 174 amino acid protein with 20% to 30% homology to tumor necrosis factor superfamily	150
Fragment of platelet factor 4 (PF-4)	An N-terminal fragment of PF-4	151
Derivative of prolactin	A 16-kd fragment of the hormone	152
Restin	NC10 domain of human collagen XV	153
Proliferin-related protein (PRP)	A protein related to the proangiogenic proliferin molecule	154
Meth-1 and Meth-2	Proteins containing metalloprotease, thrombospondin, and disintegrin domains	155
SPARC cleavage product	Fragments of secreted protein, acid and rich in cysteine (SPARC)	156
Osteopontin cleavage product	Thrombin-generated fragment containing an RGD sequence	155
Interferon-α, β, γ; IL-10, IL-4, IL-12	Cytokines and chemokines	144, 88, 157
Inhibitors of differentiation (Id1/Id3)	Inhibitory helix-loop-helix transcription factors	14
PEX	Proteolytic fragment of MMP-2, blocking binding of MMP-2 to $\alpha_v\beta_3$	37
Troponin-I	Subunit of the troponin complex, inhibiting actomyosin adenosine triphosphatase (ATPase)	158
Maspin	Tumor suppressor gene product, encoding a serpin member	159
Canstatin	Fragment of the α_2-chain of collagen type IV	160
Angiopoietin-2 (Ang-2)	Antagonist of angiopoietin-1 binding to Tie-2	161
Antithrombin III fragment	A fragment missing C-terminal loop of antithrombin III (member of the serpin family)	162

IL, interleukin.

human immunodeficiency virus type-1 (HIV-1) viral products, have been implicated in the pathogenesis.[90] Thus, HIV-1 Tat protein activates VEGFR-2, binds endothelial $\alpha_5\beta_1$ and $\alpha_v\beta_3$ integrins, and retrieves bFGF from the extracellular matrix.[91] Further, VEGF and VEGF-C are autocrine growth factors for AIDS-KS cells. The angiogenic balances is further tipped by underexpression of the endothelial inhibitor TSP-1.[92]

MOLECULAR MECHANISMS OF VASCULAR MYOGENESIS AND ARTERIOGENESIS

Smooth muscle cells are not a homogeneous population of cells, derived from a common precursor. Instead, they have a complex origin depending on their location in the embryo. Smooth muscle cells transdifferentiate from endothelial cells[93] or differentiate from mesenchymal cells *in situ* in response to (yet unidentified) endothelial-derived stimuli.[94,95] These cells can also differentiate from bone marrow precursors or macrophages. Mural cells of the coronary arteries are recruited from the epicardial layer, while smooth muscle cells of the coronary veins are derived from atrial myocardium.[96] The large thoracic blood vessels contain smooth muscle cells, derived from cardiac neural crest cells.[97] These vessels are often affected by congenital malformations, which may relate to their complex remodeling and patterning during development—a process involving apoptosis of medial smooth muscle cells. Genetic analysis has revealed that loss of endothelin,[98] mesenchyme forkhead-1 (MFH-1),[99] dHand and Msx-1,[100] Pax-3, Prx-1, retinoid acid receptors, the neurofibromatosis type-1 gene product Wnt-1, or the gap junctional protein connexin 43 induce severe aortic arch malformations. Loss of neuropilin-1—a receptor for the neurorepulsive semaphorins and for $VEGF_{165}$, PlGF-2, and VEGF-B isoforms—also induces abnormal patterning of the large thoracic vessels.[101] Other transcription factors involved in smooth muscle cell differentiation include the serum response factor, *Prx-1* and *Prx-2*,[102] *CRP2/ SmLIM*,[103] capsulin,[104] or members of the *Hox*,[105] myocite enhancor factor-2 (*MEF-2*),[106,107] and *GATA*[108,109] family.

PDGF-BB is a chemoattractant for smooth muscle cells.[49] Besides its predominant effects on endothelial cells, VEGF also promotes mural cell investment, presumably via release of PDGF-BB or via binding to VEGFR-1 on smooth muscle cells.[3,43] Ang-1 and Tie-2 affect growth and maintenance of blood vessels by stabilizing the interaction of mural cells with nascent endothelial tubes, and by inducing branching and remodeling.[17,31,47,110–112] Hereditary dysfunction of Tie-2 in humans induces venous malformations, characterized by vessels with fewer smooth muscle cells.[113] TGF-β1, TGF-βR2, endoglin (an endothelial TGF-β binding protein), and Smad-5 (a downstream TGF-β signal) are involved in vessel maturation in a pleiotropic manner: they inhibit endothelial proliferation and migration, induce smooth muscle differentiation, and stimulate extracellular matrix production, thereby "solidifying" the endothelial-mural cell interactions.[23,114] Patients lacking endoglin suffer hereditary hemorrhagic telangiectasia type-1.[115] N-cadherin appears to "glue" endothelial and mural cells in close apposition. Endothelin-1, produced by endothelial cells of the large thoracic blood vessels, is chemotactic for neural crest cells that transform into smooth muscle cells.[98] Tissue factor, the initiator of coagulation, promotes pericyte recruitment, possibly via generation of thrombin and/or a fibrin-rich scaffold.[116] Other candidates are heparin-binding epidermal growth factor—like factor (HB-EGF) and the transcription factors *LKLF, COUP-TFII,* and *MEF2C*.[117]

Once mural cells have been recruited to the nascent endothelial-lined vessels, they further "muscularize" the emerging vasculature by sprouting or by migrating in a spiral motion alongside preexisting vessels, using the latter as guiding cues (longitudinal migration) (Fig. 1.1). This has been elegantly visualized in the retina where pericytes cover the preexisting endothelial channels in a centrifugal direction, or in the heart where smooth muscle coverage proceeds in a epicardial-to-endocardial pattern. In mesenchyme-rich tissues, such as in the lung, *in situ* differentiation of mesenchymal cells is a continuous mechanism of progressive muscularization. Presumably, similar signals as those mediating smooth muscle cell recruitment and growth during initial vascular myogenesis are involved in arteriogenesis. FGFs may play a role in branching of coronary arteries, whereas the renin-angiotensin system has been implicated in initiation, branching, and elongation of the renal arterial tree.[118] Signals involved in neuronal patterning appear to play a role in vascular patterning. In the avian heart, there is a close spatial juxtaposition between coronary arteries and Purkinje cells of the myocardial conduction system. Endothelin-1, locally generated in the coronary artery, is an instructive cue for differentiation of cardiomyocytes into Purkinje cells.[119]

Mural cells acquire specialized differentiation characteristics during development, including contractile components. Loss of intermediate filament desmin results in smooth muscle hypoplasia and degeneration,[120] whereas deficiency of MEF2C results in impaired smooth muscle differentiation.[117] Interstitial matrix components are deposited during the late prenatal period, thereby providing the developing arteries with viscoelastic properties (elastin and fibrillin-2) and structural strength (collagen and fibrillin-1). Deficiency of these components—in gene-inactivated mice, or in humans with hereditary Marfan syndrome or atherosclerotic media destruction—results in weakening and aneurysmal dilatation of the arteries.[34,121] However, elastin also regulates smooth muscle cell proliferation and differentiation, as mice lacking elastin die because of obstructive intimal hyperplasia due to uncontrolled proliferation and migration of medial smooth muscle cells.[122] Hereditary elasting gene disruption in humans causes supravalvular aortic stenosis.[122]

Hypoxia is an important stimulus of the formation of new blood vessels in normal tissues, tumors, atherosclerotic plaques, diabetic retinas, etc. Hypoxia-inducible transcription factors (HIF-1β, HIF-1α, and HIF-2α) trigger a coordinated response of angiogenesis and arteriogenesis by inducing expression of VEGF, VEGFR-1, VEGFR-2, NP-1, Ang-2, NO synthase, TGF-β1, PDGF-BB, endothelin-1, IL-8, IGF-2, Tie-1, cyclooxygenase-2, etc.[123] The von Hippel—Lindau (VHL) tumor suppressor gene product suppresses expression of hypoxia-inducible target genes during normoxia. Gene-inactivation studies reveal that the initial steps of vascular development are not critically regulated by oxygen but that subsequent vascular remodeling is regulated by hypoxia.[123,124] Tumors lacking HIF-1β or HIF-1α fail to develop vascularized tumors and lack hypoxic induction of VEGF expression.[125] Stabilization of HIF-1α by the PR39 peptide induces therapeutic angiogenesis in the myocardium.[126]

CONCLUSION

Cells in the mammalian body require oxygen and nutrients for their survival and are therefore located within 100 to 200 µm of blood vessels—the diffusion limit for oxygen. For most multicellular organisms to grow beyond this size, they must recruit new blood vessels. This process needs to be tightly regulated by an intricate balance between pro- and antiangiogenic molecules. Several disorders such as cancer, inflammatory disorders, and tissue ischemia arise when this process is derailed. With advances in molecular genetics and the availability of molecular probes, imaging technologies, and therapeutic opportunities, we are now obtaining stunning insight into physiologic and pathologic angiogenesis and starting to realize the marked impact that blood vessels have on the global quality of life. It is an exciting time to study the molecular basis of vascular growth. These insights have facilitated the development of therapeutic strategies to stimulate or suppress angiogenesis.

ACKNOWLEDGMENTS

The authors are grateful to the members of the Center for Transgene Technology and Gene Therapy and to all the other collaborators who contributed to these studies.

REFERENCES

1. Risau, W. (1997). Mechanisms of angiogenesis. Nature 386:671–674.
2. Hirschi, K.K., d'Amore, P.A. (1996). Pericytes in the microvasculature. Cardiovasc Res 32:687–698.
3. Benjamin, L.E., Hemo, I., Keshet, E. (1998). A plasticity window for blood vessel remodelling is defined by pericyte coverage of the preformed endothelial network and is regulated by PDGF-B and VEGF. Development 125:1591–1598.
4. Folkman, J. (2000). Tumor angiogenesis. In: Holland, J.F., et al., eds. Cancer medicine, pp. 132–152. Ontario, Canada: B.C. Decker.
5. Vandenbunder, B., Pardanaud, L., Jaffredo, T., Mirabel, M.A., Stehelin, D. (1989). Complementary patterns of expression of c-ets 1, c-myb and c-myc in the blood-forming system of the chick embryo. Development 107:265–274.
6. Thomas, P.Q., Brown, A., Beddington, R.S. (1998). Hex: a homeobox gene revealing peri-implantation asymmetry in the mouse embryo and an early transient marker of endothelial cell precursors. Development 125:85–94.
7. Xiong, J.W., Leahy, A., Lee, H.H., Stuhlmann, H. (1999). Vezf1: a Zn finger transcription factor restricted to endothelial cells and their precursors. Dev Biol 206: 123–141.
8. Belotti, D., Clausse, N., Flagiello, D., et al. (1998). Expression and modulation of homeobox genes from cluster B in endothelial cells. Lab Invest 78:1291–1299.
9. Boudreau, N., Andrews, C., Srebrow, A., Ravanpay, A., Cheresh, D.A. (1997). Induction of the angiogenic phenotype by Hox D3. J Cell Biol 139:257–264.
10. Pan, J., Xia, L., McEver, R.P. (1998). Comparison of promoters for the murine and human P-selectin genes suggests species-specific and conserved mechanisms for transcriptional regulation in endothelial cells. J Biol Chem 273:10058–10067.
11. Hewett, P.W., Daft, E.L., Murray, J.C. (1998). Cloning and partial characterization of the human tie-2 receptor tyrosine kinase gene promoter [in process citation]. Biochem Biophys Res Commun 252:546–551.

12. Cowan, P.J., Tsang, D., Pedic, C.M., et al. (1998). The human ICAM-2 promoter is endothelial cell-specific in vitro and in vivo and contains critical Sp1 and GATA binding sites. J Biol Chem 273:11737–11744.

13. Almendro, N., Bellon, T., Rius, C., et al. (1996). Cloning of the human platelet endothelial cell adhesion molecule-1 promoter and its tissue-specific expression. Structural and functional characterization. J Immunol 157:5411–5421.

14. Lyden, D., Young, A.Z., Zagzag, D., et al. (1999). Id1 and Id3 are required for neurogenesis, angiogenesis and vascularization of tumour xenografts. Nature 401:670–677.

15. Zhong, T.P., Rosenberg, M., Mohideen, M.A., Weinstein, B., Fishman, M.C. (2000). Gridlock, an HLH gene required for assembly of the aorta in zebrafish. Science 287:1820–1824.

16. Wang, H.U., Chen, Z.F., Anderson, D.J. (1998). Molecular distinction and angiogenic interaction between embryonic arteries and veins revealed by ephrin-B2 and its receptor Eph-B4. Cell 93:741–753.

17. Gale, N.W., Yancopoulos, G.D. (1999). Growth factors acting via endothelial cell-specific receptor tyrosine kinases: VEGFs, angiopoietins, and ephrins in vascular development. Genes Dev 13:1055–1066.

18. Ferrara, N. (1999). Role of vascular endothelial growth factor in the regulation of angiogenesis. Kidney Int 56:794–814.

19. Carmeliet, P., Ferreira, V., Breier, G., et al. (1996). Abnormal blood vessel development and lethality in embryos lacking a single vascular endothelial growth factor allele. Nature 380:435–439.

20. Ferrara, N., et al. (1996). Heterozygous embryonic lethality induced by targeted inactivation of the VEGF gene. Nature 380:439–442.

21. Shalaby, F., Ho, J., Stanford, W.L., et al. (1997). A requirement for Flk-1 in primitive and definitive hematopoiesis and vasculogenesis. Cell 89:981–990.

22. Fong, G.H., Zhang, L., Bryce, D.M., Peng, J. (1999). Increased hemangioblast commitment, not vascular disorganization, is the primary defect in flt-1 knock-out mice. Development 126:3015–3025.

23. Dickson, M.C., Martin, J.S., Cousins, F.M., et al. (1995). Defective haematopoiesis and vasculogenesis in transforming growth factor-beta 1 knock out mice. Development 121:1845–1854.

24. Bader, B.L., Rayburn, H., Crowley, D., Hynes, R.O. (1998). Extensive vasculogenesis, angiogenesis, and organogenesis precede lethality in mice lacking all alpha v integrins. Cell 95:507–519.

25. Asahara, T., Murohara, T., Sullivan, A., et al. (1997). Isolation of putative progenitor endothelial cells for angiogenesis. Science 275:964–967.

26. Lin, Y., Weisdorf, D.J., Solovey, A., Hebbel, R.P. (2000). Origins of circulating endothelial cells and endothelial outgrowth from blood. J Clin Invest 105:71–77.

27. Peichev, M., Naiyer, A.J., Pereira, D., et al. (2000). Expression of VEGFR-2 and AC133 by circulating human CD34(+) cells identifies a population of functional endothelial precursors. Blood 95:952–958.

28. Roberts, W.G., Palade, G.E. (1995). Increased microvascular permeability and endothelial fenestration induced by vascular endothelial growth factor. J Cell Sci 108:2369–2379.

29. Esser, S., Wolburg, K., Wolburg, H., et al. (1998). Vascular endothelial growth factor induces endothelial fenestrations in vitro. J Cell Biol 140:947–959.

30. Thurston, G., Rudge, J.S., Ioffe, E., et al. (2000). Angiopoietin-1 protects the adult vasculature against plasma leakage. Nat Med 6:1–4.

31. Maisonpierre, P.C., Suri, C., Jones, P.F., et al. (1997). Angiopoietin-2, a natural antagonist for Tie2 that disrupts in vivo angiogenesis. Science 277:55–60.

32. Coussens, L.M., Raymond, W.W., Bergers, G., et al. (1999). Inflammatory mast cells up-regulate angiogenesis during squamous epithelial carcinogenesis. Genes Dev 13: 1382–1397.

33. Heymans, S., Luttun, A., Nuyens, D., et al. (1999). Inhibition of plasminogen activators or matrix metalloproteinases prevents cardiac rupture but impairs therapeutic angiogenesis and causes cardiac failure. Nat Med 5:1135–1142.

34. Carmeliet, P., Collen, D. (1998). Development and disease in proteinase-deficient mice: role of the plasminogen, matrix metalloproteinase and coagulation system. Thromb Res 91:255–285.

35. Vu, T.H., Shipley, J.M., Bergers, G., et al. (1998). MMP-9/gelatinase B is a key regulator of growth plate angiogenesis and apoptosis of hypertrophic chondrocytes. Cell 93:411–422.

36. Bajou, K., Noel, A., Gerard, R.D., et al. (1998). Absence of host plasminogen activator inhibitor 1 prevents cancer invasion and vascularization. Nat Med 4:923–928.

37. Brooks, P.C., Silletti, S., von Schalscha, T.L., Friedlander, M., Cheresh, D.A. (1998). Disruption of angiogenesis by PEX, a noncatalytic metalloproteinase fragment with integrin binding activity. Cell 92:391–400.

38. Persico, M.G., Vincenti, V., DiPalma, T. (1999). Structure, expression and receptor-binding properties of placenta growth factor (PlGF). Curr Top Microbiol Immunol 237:31–40.

39. Olofsson, B., Jeltsch, M., Eriksson, U., Alitalo, K. (1999). Current biology of VEGF-B and VEGF-C. Curr Opin Biotechnol 10:528–535.

40. Taipale, J., Makinen, T., Arighi, E., et al. (1999). Vascular endothelial growth factor receptor-3. Curr Top Microbiol Immunol 237:85–96.

41. Soker, S., Takashima, S., Miao, H.Q., Neufeld, G., Klagsbrun, M. (1998). Neuropilin-1 is expressed by endothelial and tumor cells as an isoform-specific receptor for vascular endothelial growth factor. Cell 92:735–745.

42. Gerber, H.P., Hillan, K.J., Ryan, A.M., et al. (1999). VEGF is required for growth and survival in neonatal mice. Development 126:1149–1159.

43. Carmeliet, P., Ng, Y.S., Nuyens, D., et al. (1999). Impaired myocardial angiogenesis and ischemic cardiomyopathy in mice lacking the vascular endothelial growth factor isoforms VEGF164 and VEGF188. Nat Med 5:495–502.

44. Dumont, D.J., Jussila, L., Taipale, J., et al. (1998). Cardiovascular failure in mouse embryos deficient in VEGF receptor-3. Science 282:946–949.

45. Cao, Y., Linden, P., Farnebo, J., et al. (1998). Vascular endothelial growth factor C induces angiogenesis in vivo. Proc Natl Acad Sci USA 95:14389–14394.

46. Hiratsuka, S., Minowa, O., Kuno, J., Noda, T., Shibuya, M. (1998). Flt-1 lacking the tyrosine kinase domain is sufficient for normal development and angiogenesis in mice. Proc Natl Acad Sci USA 95:9349–9354.

47. Suri, C., McClain, J., Thurston, G., et al. (1998). Increased vascularization in mice overexpressing angiopoietin-1. Science 282:468–471.

48. Zhou, M., Sutliff, R.L., Paul, R.J., et al. (1998). Fibroblast growth factor 2 control of vascular tone. Nat Med 4:201–207.

49. Lindahl, P., Hellstrom, M., Kalen, M., Betsholtz, C. (1998). Endothelial-perivascular cell signaling in vascular development: lessons from knockout mice. Curr Opin Lipidol 9:407–411.

50. Battegay, E.J. (1995). Angiogenesis: mechanistic insights, neovascular diseases, and therapeutic prospects. J Mol Med 73:333–346.

51. Carmeliet, P., Collen, D. (1998). Vascular development and disorders: molecular analysis and pathogenetic insights. Kidney Int 53:1519–1549.

52. Pepper, M.S. (1997). Transforming growth factor-beta: vasculogenesis, angiogenesis and vessel wall integrity. Cytokine Growth Factor Res 8:21–43.

EMBRYO

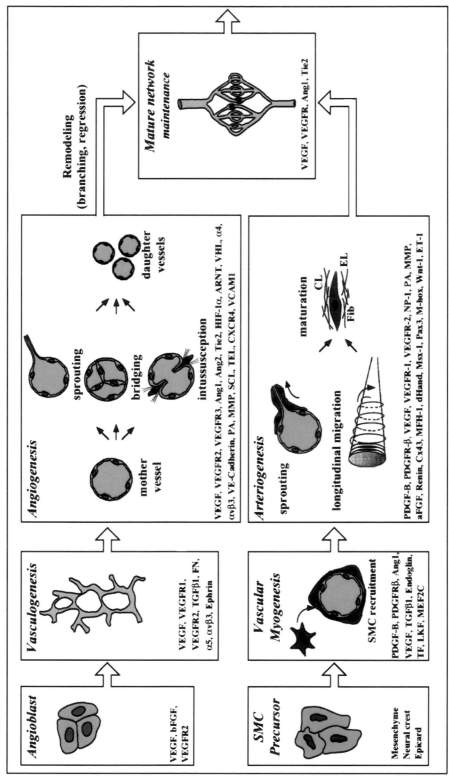

Angioblast

VEGF, bFGF, VEGFR2

Vasculogenesis

VEGF, VEGFR1, VEGFR2, TGFβ1, FN, α5, αvβ3, Ephrin

Angiogenesis

mother vessel

sprouting

bridging

intussusception

daughter vessels

VEGF, VEGFR2, VEGFR3, Ang1, Ang2, Tie2, HIF-1α, ARNT, VHL, α4, αvβ3, VE-Cadherin, PA, MMP, SCL, TEL, CXCR4, VCAM1

SMC Precursor

Mesenchyme
Neural crest
Epicard

Vascular Myogenesis

SMC recruitment

PDGF-B, PDGFRβ, Ang1, VEGF, TGFβ1, Endoglin, TF, LKF, MEF2C

Arteriogenesis

sprouting

longitudinal migration

maturation

CL
EL
Fib

PDGF-B, PDGFR-β, VEGF, VEGFR-1, VEGFR-2, NP-1, PA, MMP, aFGF, Renin, Cx43, MFH-1, dHand, Msx-1, Pax3, M-box, Wnt-1, ET-1

Remodeling (branching, regression)

Mature network maintenance

VEGF, VEGFR, Ang1, Tie2

Plate 1

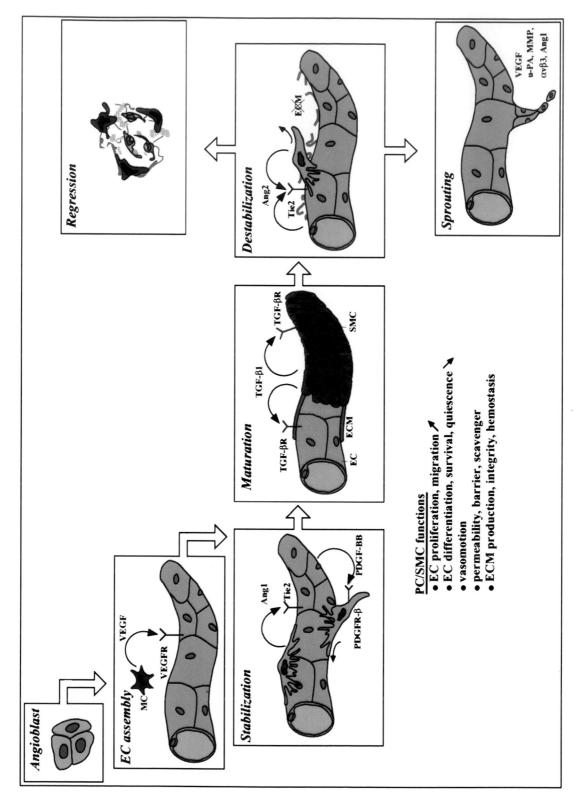

Plate 2

53. Fräter-Schröder, M., Risau, W., Hallmann, P., Gautschi, P., Böhlen, P. (1987). Tumor necrosis factor type alpha, a potent inhibitor of endothelial growth in vitro, is angiogenic in vivo. Proc Natl Acad Sci USA 84:5277–5281.

54. Varner, J.A., Brooks, P.C., Cheresh, D.A. (1995). Review: the integrin αvβ3: angiogenesis and apoptosis. Cell Adhesion Commun 3:367–374.

55. Carmeliet, P., Lampugnani, M.G., Moons, L., et al. (1999). Targeted deficiency or cytosolic truncation of the VE-cadherin gene in mice impairs VEGF-mediated endothelial survival and angiogenesis. Cell 98:147–157.

56. Huang, X., Molema, G., King, S., et al. (1997). Tumor infarction in mice by antibody-directed targeting of tissue factor to tumor vasculature [see comments]. Science 275: 547–550.

57. Meeson, A.P., Argilla, M., Ko, K., Witte, L., Lang, R.A. (1999). VEGF deprivation-induced apoptosis is a component of programmed capillary regression. Development 126:1407–1415.

58. Jain, R.K., Safabakhsh, N., Sckell, A., et al. (1998). Endothelial cell death, angiogenesis, and microvascular function after castration in an androgen-dependent tumor: role of vascular endothelial growth factor. Proc Natl Acad Sci USA 95:10820–10825.

59. Alon, T., Hemo, I., Itin, A., et al. (1995). Vascular endothelial growth factor acts as a survival factor for newly formed retinal vessels and has implications for retinopathy of prematurity. Nat Med 1:1024–1028.

60. Holash, J., Maisonpierre, P.C., Compton, D., et al. (1999). Vessel co-option, regression, and growth in tumors mediated by angiopoietins and VEGF. Science 284:1994–1998.

61. Brooks, P.C., Montgomery, A.M., Rosenfeld M., et al. (1994). Integrin alpha v beta 3 antagonists promote tumor regression by inducing apoptosis of angiogenic blood vessels. Cell 79:1157–1164.

62. Lucas, R., Holmgren, L., Garcia, I., et al. (1998). Multiple forms of angiostatin induce apoptosis in endothelial cells. Blood 92:4730–4741.

63. Guo, N., Krutzsch, H.C., Inman, J.K., Roberts, D.D. (1997). Thrombospondin 1 and type I repeat peptides of thrombospondin 1 specifically induce apoptosis of endothelial cells. Cancer Res 57:1735–1742.

64. Lang, R., Lustig, M., Francois, F., Sellinger, M., Plesken, H. (1994). Apoptosis during macrophage-dependent ocular tissue remodelling. Development 120:3395–3403.

65. Lang, R.A., Bishop, J.M. (1993). Macrophages are required for cell death and tissue remodeling in the developing mouse eye. Cell 74:453–462.

66. Meeson, A., Palmer, M., Calfon, M., Lang, R. (1996). A relationship between apoptosis and flow during programmed capillary regression is revealed by vital analysis. Development 122:3929–3938.

67. Dimmeler, S., Haendeler, J., Rippmann, V., Nehls, M., Zeiher, A.M. (1996). Shear stress inhibits apoptosis of human endothelial cells. FEBS Lett 399:71–74.

68. O'Reilly, M.S., Boehm, T., Shing, Y., et al. (1994). Angiostatin: a novel angiogenesis inhibitor that mediates the suppresion of metastases by a Lewis lung carcinoma. Cell 79:315–328.

69. O'Reilly, M.S., Boehm, T., Shing, Y., et al. (1997). Endostatin: an endogenous inhibitor of angiogenesis and tumor growth. Cell 88:277–285.

70. Lee, T.H., Rhim, T., Kim, S.S. (1998). Prothrombin kringle-2 domain has a growth inhibitory activity against basic fibroblast growth factor-stimulated capillary endothelial cells. J Biol Chem 273:28805–28812.

71. Rhim, T.Y., Park, C.S., Kim, E., Kim, S.S. (1998). Human prothrombin fragment 1 and 2 inhibit bFGF-induced BCE cell growth [in process citation]. Biochem Biophys Res Commun 252:513–516.

72. Volpert, O.V., Tolsma, S.S., Pellerin, S., et al. (1995). Inhibition of angiogenesis by thrombospondin-2. Biochem Biophys Res Commun 217:326–332.

73. DeLisser, H.M., Christofidou-Solomidou, M., Strieter, R.M., et al. (1997). Involvement of endothelial PECAM-1/CD31 in angiogenesis. Am J Pathol 151:671–677.
74. Mitsuyasu, R.T. (1991). Interferon alpha in the treatment of AIDS-related Kaposi's sarcoma. Br J Haematol 79 (suppl 1):69–73.
75. Majima, M., Isono, M., Ikeda Y., et al. (1997). Significant roles of inducible cyclooxygenase (COX)-2 in angiogenesis in rat sponge implants. Jpn J Pharmacol 75:105–114.
76. Majewski, S., Skopinska, M., Marczak, M., et al. (1996). Vitamin D3 is a potent inhibitor of tumor cell-induced angiogenesis. J Invest Dermatol Symp Proc 1:97–101.
77. Lee, H., Struman, I., Clapp, C., Martial, J., Weiner, R.I. (1998). Inhibition of urokinase activity by the antiangiogenic factor 16K prolactin: activation of plasminogen activator inhibitor 1 expression. Endocrinology 139:3696–3703.
78. Patan, S. (1998). TIE1 and TIE2 receptor tyrosine kinases inversely regulate embryonic angiogenesis by the mechanism of intussusceptive microvascular growth. Microvasc Res 56:1–21.
79. Visvader, J.E., Fujiwara, Y., Orkin, S.H. (1998). Unsuspected role for the T-cell leukemia protein SCL/tal-1 in vascular development. Genes Dev 12:473–479.
80. Wang, L.C., Kuo, F., Fujiwara, Y., et al. (1997). Yolk sac angiogenic defect and intra-embryonic apoptosis in mice lacking the Ets-related factor TEL. EMBO J 16:4374–4383.
81. Offermans, S., Mancino, V., Revel, J.-P., Simon, M.I. (1997). Vascular system defects and impaired cell chemokinesis as a result of G alpha 13 deficiency. Science 275:533–536.
82. Shirayoshi, Y., Yuasa, Y., Suzuki, T., et al. (1997). Proto-oncogene of int-3, a mouse Notch homologue, is expressed in endothelial cells during early embryogenesis. Genes Cells 2:213–224.
83. Gurtner, G.C., Davis, V., Li, H., et al. (1995). Targeted disruption of the murine VCAM1 gene: essential role of VCAM-1 in chorioallantoic fusion and placentation. Genes Dev 9:1–14.
84. Kwee, L., Baldwin, H.S., Shen H.M., et al. (1995). Defective development of the embryonic and extraembryonic circulatory systems in vascular cell adhesion molecule (VCAM-1) deficient mice. Development 121:489–503.
85. Yang, J.T., Rayburn, H., Hynes, R.O. (1995). Cell adhesion events mediated by alpha 4 integrins are essential in placental and cardiac development. Development 121:549–560.
86. George, E.L., Georges Labouesse, E.N., Patel King, R.S., Rayburn, H., Hynes, R.O. (1993). Defects in mesoderm, neural tube and vascular development in mouse embryos lacking fibronectin. Development 119:1079–1091.
87. Vikkula, M., Boon, L., Mulliken, J.B., Olsen, B.R. (1998). Molecular basis of vascular anomalies. Trends Cardiovasc Med 8:281–292.
88. Ezekowitz, R.A., Mulliken, J.B., Folkman, J. (1992). Interferon alfa-2a therapy for life-threatening hemangiomas of infancy [see comments] [published errata appear in N Engl J Med 1994; 330(4):300 and 1995; 31;333(9):595–596]. N Engl J Med 326:1456–1463.
89. Ogawa, S., Oku, A., Sawano, A., et al. (1998). A novel type of vascular endothelial growth factor, VEGF-E (NZ-7 VEGF), preferentially utilizes KDR/Flk-1 receptor and carries a potent mitotic activity without heparin-binding domain. J Biol Chem 273:31273–31282.
90. Flore, O., Rafii, S., Ely, S., et al. (1998). Transformation of primary human endothelial cells by Kaposi's sarcoma-associated herpesvirus. Nature 394:588–592.
91. Albini, A., Soldi, R., Giunciuglio, D., et al. (1996). The angiogenesis induced by HIV-1 tat protein is mediated by the Flk-1/KDR receptor on vascular endothelial cells. Nat Med 2:1371–1375.

92. Taraboletti, G., Benelli, R., Borsotti, P., et al. (1999). Thrombospondin-1 inhibits Kaposi's sarcoma (KS) cell and HIV-1 Tat-induced angiogenesis and is poorly expressed in KS lesions. J Pathol 188:76–81.

93. DeRuiter, M.C., Poelmann, R.E., VanMunsteren, J.C., et al. (1997). Embryonic endothelial cells transdifferentiate into mesenchymal cells expressing smooth muscle actins in vivo and in vitro [see comments]. Circ Res 80:444–451.

94. Nakamura, H. (1988). Electron microscopic study of the prenatal development of the thoracic aorta in the rat. Am J Anat 181:406–418.

95. de Ruiter, M.C., Poelmann, R.E., van Iperen, L., Gittenberger-de Groot, A.C. (1990). The early development of the tunica media in the vascular system of rat embryos. Anat Embryol 181:341–349.

96. Dettman, R.W., Denetclaw, W., Jr., Ordahl, C.P., Bristow, J. (1998). Common epicardial origin of coronary vascular smooth muscle, perivascular fibroblasts, and intermyocardial fibroblasts in the avian heart. Dev Biol 193:169–181.

97. Creazzo, T.L., Godt, R.E., Leatherbury, L., Conway, S.J., Kirby, M.L. (1998). Role of cardiac neural crest cells in cardiovascular development. Annu Rev Physiol 60: 267–286.

98. Yanagisawa, H., Hammer, R.E., Richardson, J.A., et al. (1998). Role of endothelin-1/endothelin-A receptor-mediated signaling pathway in the aortic arch patterning in mice. J Clin Invest 102:22–33.

99. Iida, K., Koseki, H., Kakinuma, H., et al. (1997). Essential roles of the winged helix transcription factor MFH-1 in aortic arch patterning and skeletogenesis. Development 124:4627–4638.

100. Thomas, T., Kurihara, H., Yamagishi, H., et al. (1998). A signaling cascade involving endothelin-1, dHAND and msx1 regulates development of neural-crest-derived branchial arch mesenchyme. Development 125:3005–3014.

101. Kawasaki, T., Kitsukawa, T., Bekku, Y., et al. (1999). A requirement for neuropilin-1 in embryonic vessel formation. Development 126:4895–4902.

102. Bergwerff, M., Gittenberger-de Groot, A.C., DeRuiter, M.C., et al. (1998). Patterns of paired-related homeobox genes PRX1 and PRX2 suggest involvement in matrix modulation in the developing chick vascular system [in process citation]. Dev Dyn 213:59–70.

103. Jain, M.K., Kashiki, S., Hsieh C.M., et al. (1998). Embryonic expression suggests an important role for CRP2/SmLIM in the developing cardiovascular system. Circ Res 83:980–985.

104. Hidai, H., Bardales, R., Goodwin, R., Quertermous, T., Quertermous, E.E. (1998). Cloning of capsulin, a basic helix-loop-helix factor expressed in progenitor cells of the pericardium and the coronary arteries. Mech Dev 73:33–43.

105. Miano, J.M., Firulli, A.B., Olson, E.N., et al. (1996). Restricted expression of homeobox genes distinguishes fetal from adult human smooth muscle cells. Proc Natl Acad Sci USA 93:900–905.

106. Firulli, A.B., Miano, J.M., Bi, W., et al. (1996). Myocyte enhancer binding factor-2 expression and activity in vascular smooth muscle cells. Association with the activated phenotype. Circ Res 78:196–204.

107. Katoh, Y., Molkentin, J.D., Dave, V., Olson, E.N., Periasamy, M. (1998). MEF2B is a component of a smooth muscle-specific complex that binds an A/T-rich element important for smooth muscle myosin heavy chain gene expression. J Biol Chem 273:1511–1518.

108. Perlman, H., Suzuki, E., Simonson, M., Smith, R.C., Walsh, K. (1998). GATA-6 induces p21(Cip1) expression and G1 cell cycle arrest. J Biol Chem 273:13713–13718.

109. Morrisey, E.E., Ip, H.S., Tang, Z., Lu, M.M., Parmacek, M.S. (1997). GATA-5: a transcriptional activator expressed in a novel temporally and spatially-restricted pattern during embryonic development. Dev Biol 183:21–36.

110. Sato, T.N., Tozawa, Y., Deutsch, U., et al. (1995). Distinct roles of the receptor tyrosine kinases Tie-1 and Tie-2 in blood vessel formation. Nature 376:70–74.
111. Puri, M.C., Rossant, J., Alitalo, K., Bernstein, A., Partanen, J. (1995). The receptor tyrosine kinase TIE is required for integrity and survival of vascular endothelial cells. EMBO J 14:5884–5891.
112. Suri, C., Jones, P.F., Patan, S., et al. (1996). Requisite role of angiopoietin-1, a ligand for the TIE2 receptor, during embryonic angiogenesis. Cell 87:1171–1180.
113. Vikkula, M., Boon, L.M., Carraway, K.L., 3rd, et al. (1996). Vascular dysmorphogenesis caused by an activating mutation in the receptor tyrosine kinase TIE2 [see comments]. Cell 87:1181–1190.
114. Li, D.Y., Sorensen, L.K., Brooke, B.S., et al. (1999). Defective angiogenesis in mice lacking endoglin. Science 284:1534–1537.
115. McAllister, K.A., Grogg, K.M., Johnson, D.W., et al. (1994). Endoglin, a TGF-beta binding protein of endothelial cells, is the gene for hereditary haemorrhagic telangiectasia type 1. Nat Genet 8:345–351.
116. Carmeliet, P., Mackman, N., Moons, L., et al. (1996). Role of tissue factor in embryonic blood vessel development. Nature 383:73–75.
117. Lin, Q., Lu, J., Yanagisawa, H., et al. (1998). Requirement of the MADS-box transcription factor MEF2C for vascular development. Development 125:4565–4574.
118. Reddi, V., Zaglul, A., Pentz, E.S., Gomez, R.A. (1998). Renin-expressing cells are associated with branching of the developing kidney vasculature. J Am Soc Nephrol 9:63–71.
119. Gourdie, R.G., Wei, Y., Kim, D., Klatt, S.C., Mikawa, T. (1998). Endothelin-induced conversion of embryonic heart muscle cells into impulse-conducting Purkinje fibers. Proc Natl Acad Sci USA 95:6815–6818.
120. Capetanaki, Y., Milner, D.J., Weitzer, G. (1997). Desmin in muscle formation and maintenance: knockouts and consequences. Cell Struct Funct 22:103–116.
121. Pereira, L., Andrikopoulos, K., Tian, J., et al. (1997). Targetting of the gene encoding fibrillin-1 recapitulates the vascular aspect of Marfan syndrome. Nat Genet 17: 218–222.
122. Li, D.Y., Brooke, B., Davis, E.C., et al. (1998). Elastin is an essential determinant of arterial morphogenesis. Nature 393:276–280.
123. Semenza, G.L. (1998). Hypoxia-inducible factor 1: master regulator of O2 homeostasis. Curr Opin Genet Dev 8:588–594.
124. Maltepe, E., Schmidt, J.V., Baunoch, D., Bradfield, C.A., Simon, C.M. (1997). Abnormal angiogenesis and responses to glucose and oxygen deprivation in mice lacking the protein ARNT. Nature 386:403–407.
125. Carmeliet, P., Dor, Y., Herbert, J.M., et al. (1998). Role of HIF-1alpha in hypoxia-mediated apoptosis, cell proliferation and tumour angiogenesis [published erratum appears in Nature 1998;395(6701):525]. Nature 394:485–490.
126. Li, J., Post, M., Volk, R., et al. (2000). PR39, a peptide regulator of angiogenesis. Nat Med 6:49–55.
127. Carmeliet, P. (2000). Mechanisms of angiogenesis and arteriogenesis. Nat Med 6: 389–395.
128. Isner, J.M., Asahara, T. (1999). Angiogenesis and vasculogenesis as therapeutic strategies for postnatal neovascularization. J Clin Invest 103:1231–1236.
129. Dvorak, H.F. (2000). VPF/VEGF and the angiogenic response. Semin Perinatol 24: 75–78.
130. Carmeliet, P., Stalmans, I., DiPalma, T., et al. (2000). Placental growth factor (PLGF): a regulator of vascular endothelial growth factor during pathological angiogenesis and arteriogenesis. (Submitted).
131. Ferrara, N., Alitalo, K. (1999). Clinical applications of angiogenic growth factors and their inhibitors. Nat Med 5:1359–1364.

132. Veikkola, T., Alitalo, K. (1999). VEGFs, receptors and angiogenesis. Semin Cancer Biol 9:211–220.
133. Shibuya, M., Ito, N., Claesson-Welsh, L. (1999). Structure and function of vascular endothelial growth factor receptor-1 and -2. Curr Top Microbiol Immunol 237: 59–83.
134. Lindahl, P., Bostrom, H., Karlsson, L., et al. (1999). Role of platelet-derived growth factors in angiogenesis and alveogenesis. Curr Top Pathol 93:27–33.
135. Bikfalvi, A., Klein, S., Pintucci, G., Rifkin, D.B. (1997). Biological roles of fibroblast growth factor-2. Endocr Rev 18:26–45.
136. Van Belle, E., Witzenbichler, B., Chen, D., et al. (1998). Potentiated angiogenic effect of scatter factor/hepatocyte growth factor via induction of vascular endothelial growth factor: the case for paracrine amplification of angiogenesis. Circulation 97:381–390.
137. Ito, W.D., et al. (1997). Monocyte chemotactic protein-1 increases collateral and peripheral conductance after femoral artery occlusion. Circ Res 80:829–837.
138. Eliceiri, B.P., Cheresh, D.A. (1999). The role of alpha-v integrins during angiogenesis: insights into potential mechanisms of action and clinical development. J Clin Invest 103:1227–1230.
139. Bazzoni, G., Martinez Estrada, O., Dejana, E. (1999). Molecular structure and functional role of vascular tight junctions. Trends Cardiovasc Med 9:147–152.
140. Newman, P.J. (1997). The biology of PECAM-1. J Clin Invest 100:S25–29.
141. Stetler-Stevenson, W.G. (1999). Matrix metalloproteinases in angiogenesis: a moving target for therapeutic intervention. J Clin Invest 103:1237–1241.
142. Fukumura, D., Jain, R.K. (1998). Role of nitric oxide in angiogenesis and microcirculation in tumors. Cancer Metastasis Rev 17:77–89.
143. Tsujii, M., Kawano, S., Tsuji, S., et al. (1998). Cyclooxygenase regulates angiogenesis induced by colon cancer cells [published erratum appears in Cell 1998;94(2):following 271]. Cell 93:705–716.
144. Murdoch, C., Finn, A. (2000). Chemokine receptors and their role in vascular biology [in process citation]. J Vasc Res 37:1–7.
145. Rafii, S. (2000). Circulating endothelial precursors: mystery, reality, and promise [comment]. J Clin Invest 105:17–19.
146. Tolsma, S.S., Volpert, O.V., Good, D.J., et al. (1993). Peptides derived from two separate domains of the matrix protein thrombospondin-1 have anti-angiogenic activity. J Cell Biol 122:497–511.
147. Ji, W.R., Barrientos, L.G., Llinas, M., et al. (1998). Selective inhibition by kringle 5 of human plasminogen on endothelial cell migration, an important process in angiogenesis. Biochem Biophys Res Commun 247:414–419.
148. Cao, Y., Chen, A., An, S.S.A., et al. (1997). Kringle 5 of plasminogen is a novel inhibitor of endothelial cell growth. J Biol Chem 272:22924–22928.
149. Pike, S.E., Yao, L., Jones, K.D., et al. (1998). Vasostatin, a calreticulin fragment, inhibits angiogenesis and suppresses tumor growth. J Exp Med 188:2349–2356.
150. Zhai, Y., Ni, J., Jiang, G.W., et al. (1999). VEGI, a novel cytokine of the tumor necrosis factor family, is an angiogenesis inhibitor that suppresses the growth of colon carcinomas in vivo. FASEB J 13:181–189.
151. Gupta, S.K., Hassel, T., Singh, J.P. (1995). A potent inhibitor of endothelial cell proliferation is generated by proteolytic cleavage of the chemokine platelet factor 4. Proc Natl Acad Sci USA 92:7799–7803.
152. Clapp, C., Martial, J.A., Guzman, R.C., Rentier-Delure, F., Weiner, R.I. (1993). The 16-kilodalton N-terminal fragment of human prolactin is a potent inhibitor of angiogenesis. Endocrinology 133:1292–1299.
153. Ramchandran, R., Dhanabal, M., Volk, R., et al. (1999). Antiangiogenic activity of restin, NC10 domain of human collagen XV: comparison to endostatin. Biochem Biophys Res Commun 255:735–739.

154. Jackson, D., Volpert, O.V., Bouck, N., Linzer, D.I. (1994). Stimulation and inhibition of angiogenesis by placental proliferin and proliferin-related protein. Science 266: 1581–1584.
155. Sage, E.H. (1999). Pieces of eight: bioactive fragments of extracellular proteins as regulators of angiogenesis. Trends Bio Sci 7:182.
156. Vazquez, F., Hastings, G., Ortega, M.A., et al. (1999). METH-1, a human ortholog of ADAMTS-1, and METH-2 are members of a new family of proteins with angio-inhibitory activity. J Biol Chem 274:23349–23357.
157. Volpert, O.V., Fong, T., Koch, A.E., et al. (1998). Inhibition of angiogenesis by interleukin 4. J Exp Med 188:1039–1046.
158. Moses, M.A., Wiederschain, D., Wu, I., et al. (1999). Troponin I is present in human cartilage and inhibits angiogenesis. Proc Natl Acad Sci USA 96:2645–2650.
159. Zhang, M., Volpert, O., Shi, Y.H., Bouck, N. (2000). Maspin is an angiogenesis inhibitor. Nat Med 6:196–199.
160. Kamphaus, G.D., Colorado, P.C., Panka, D.J., et al. (2000). Canstatin, a novel matrix-derived inhibitor of angiogenesis and tumor growth. J Biol Chem 275:1209–1215.
161. Davis, S., Yancopoulos, G.D. (1999). The angiopoietins: Yin and Yang in angiogenesis. Curr Top Microbiol Immunol 237:173–185.
162. O'Reilly, M.S., Pirie-Shepherd, S., Lane, W.S., Folkman, J. (1999). Antiangiogenic activity of the cleaved conformation of the serpin antithrombin [see comments]. Science 285:1926–1928.

CHAPTER 2

The Role of Vascular Endothelial Growth Factors and Their Receptors During Embryonic Vascular Development

Ingo Flamme and Georg Breier

The cardiovascular system of vertebrates emanates from the mesodermal layer of the primitive embryo. Angioblasts giving rise to endothelial cells and hematoblasts giving rise to blood cells differentiate from their fibroblast-like precursors shortly after having migrated through the primitive streak during gastrulation (Gonzalez Crussi, 1971). Nascent angioblasts and their assembly into the primordial vascular plexus were made visible for the first time in the quail embryo by means of monoclonal antibodies, which recognize epitopes on endothelial and hematopoietic cells (MB-1 and QH-1) (Peault et al., 1983; Pardanaud et al., 1987). The first angioblasts originate at the periphery of the extraembryonic mesoderm, but a little later (when the head fold is formed at the one-somite stage) in the embryo proper (Pardanaud et al., 1987; Coffin and Poole, 1988; Poole and Coffin, 1989). Along the anterior intestinal portal they establish the primordia of the endocardium, and along the lateral edges of the somites they establish the primordia of large body vessels, which become interconnected to the extraembryonic vasculature at the two-somite stage. The morphogenesis of the early vasculature has been described in a series of comprehensive articles (His, 1900; Evans, 1909; Sabin, 1917, 1920; Gonzalez Crussi, 1971; Haar and Ackerman, 1971; Lanot, 1980; Hirakow and Hiruma, 1981; Pardanaud et al., 1987; Poole and Coffin, 1989; De Ruiter et al., 1991, 1993).

In the extraembryonic mesoderm, angioblasts originate in close proximity to hematoblasts and together they give rise to the so-called blood islands of the yolk sac, while in the embryo proper angioblasts originate without accompanying hematoblasts. The reason for this difference between intra- and extraembryonic angioblasts is not clear. The origin of hematoblasts and angioblasts from blood islands has led to the assumption that they are derived from a common mesodermal population of precursor cells for which the term *hemangioblasts* was coined (Murray, 1932). The existence of hemangioblastic cells *in vivo* is still elusive. However, the term *hemangioblastic cells* is widely used in the literature for the constituents of the early vasculature (Pardanaud et al., 1989). The hemangioblast hypothesis is supported by the fact that several cell surface markers and receptor molecules are shared by hematopoietic and endothelial cells (Labastie et al., 1986; Pardanaud et al., 1987; Baumhueter et al., 1994; De Lisser et al., 1994; McGann et al., 1997; Yano et al., 1997; Sato et al., 1998), and that, at least *in vitro*, both hematopoietic and endothelial cells were formed from single precursor cells having differentiated from embryonic stem cells (Choi et al., 1998; Nishikawa et al., 1998).

The early steps of vascular development follow each other very rapidly, and most remarkably a complete capillary vascular plexus is established and ready for perfusion before the onset of the heart beat. This implies that perfusion-independent mechanisms are involved in the early *in situ* development of the vascular plexus from its angioblastic precursors and shows that formation of the primordial vasculature is independent of such important factors as shear stress and metabolical demands that are inductive of further growth and adaptation of the vascular system. This is in sharp contrast to the need of perfusion for the maintenance of any blood vessel later on. Hence, it appears justified to discriminate between the early precirculatory phase of vascular development and the later phase, in which new blood vessels originate from preexisting ones. Therefore, the terms *vasculogenesis* and *angiogenesis*, respectively, were introduced to make that distinction and are commonly accepted (Risau and Flamme, 1995). However, recent discoveries have suggested that vasculogenesis-related processes may still be occurring in the adult organism, when circulating angioblastic cells from the bone marrow contribute to newly formed vascular beds, e.g., during wound repair and tumor growth (Asahara et al., 1997, 1999).

Upon the onset of the heart beat and perfusion, expansive growth of the primordial vasculature is initiated. While the structure of the primordial vascular plexus is quite uniform and resembles a honeycomb at least in the yolk sac and the perineural plexus, now the primordial vasculature is expanded by interstitial growth, intussusception, and sprouting of new capillaries into avascular areas such as the neural epithelium (Evans, 1909). Large vessels are established, conducting the bloodstream from and to the heart, and pericytes and smooth muscle cells are recruited to form the walls of these large vessels. Eventually arteries and veins are differentiated (Hughes, 1935; Murphy and Carlson, 1978). Vessels excluded from perfusion regress and in response to increasing perfusion others are enlarged (Noden, 1989). Taken together these processes perform an extensive remodeling of the primordial vasculature. The hemodynamic forces generated by the heart are one of the most important factors underlying the initiation of angiogenesis and the appropriate morphogenesis of the entire vasculature. This was demonstrated more than 80 years ago by very simple experiments (Chapman, 1918). Extirpation of the embryo from chicken blastoderms resulted in growth retardation of the yolk sac. Surprisingly, a vascular plexus was formed and continuously expanded, nevertheless. This plexus closely resembled the primordial vasculature laid down during vasculogenesis. Interestingly, as we will discuss in more detail later, inactivation of genes in mice encoding such diverse proteins as growth factors, growth factor receptors, and transcription factors resulted in a similar loss of vascular remodeling as in the cardiac phenotype. Since most if not all of these vascular knockout phenotypes may be due to disturbance of distinct endothelial functions rather than to a reduction in cardiac output, it can be concluded that a complex interplay of mechanical, metabolic, and molecular factors is necessary to perform vascular remodeling subsequent to vasculogenesis. Of crucial importance for our understanding of angiogenesis are the signal transduction systems by which endothelial cells in response to hemodynamic factors and metabolical demands become instructed to divide, to sprout, to migrate or to produce cytokines to recruit pericytes. Endothelial receptor tyrosine and serine/threonine kinases and their soluble or membrane-bound ligands have been encountered to execute most of these functions. As a result of the gene knockout technology, it is now possible to elucidate endothelial signal transduction systems that are indispensable for the transition from vasculogenesis to angiogenesis. For example, it was found that the endothelial RTK Tie-2 and its ligand angiopoietin-1 play a pivotal role in the onset of angiogenesis (Dumont et al., 1994; Sato et al., 1995; Suri et al., 1996). Also, endothelial Eph receptors and their ephrin ligands apparently play an important role during early angiogenesis and, moreover, are likely to be involved in the determination of venous versus arterial endothelium (Wang et al., 1998; Adams et al., 1999). Platelet-derived growth factor-B (PDGF-B), transforming growth factor-β (TGF-β), and their corresponding receptors are of importance for the crosstalk between endothelial cells and surrounding mesenchyme and are involved in pericyte recruitment and maturation of capillary blood vessels (Dickson et al., 1995; Lindahl et al., 1997). Conversely, the targeted deletion of vascular endothelial growth factor (VEGF) and the VEGF receptors caused embryonic lethality during vasculogenesis (Fong et al., 1995; Shalaby et al., 1995; Carmeliet et al., 1996; Ferrara et al., 1996) and hence their role during angiogenesis was not assessed, although it is well established for normal and pathologic angiogenesis. Nevertheless, the specific interplay of endothelial signaling systems, the mechanisms of their regulation, and their impact for vascular morphogenesis are poorly understood. This chapter reviews the role of VEGF and VEGF receptors in embryonic vascular development.

THE VEGF FAMILY OF GROWTH FACTORS AND VEGF RECEPTORS

In 1989 the cloning and characterization of a secreted homodimeric protein purified from tissue culture cells was reported that turned out to be a mitogen specific for endothelial cells and to be angiogenic *in vivo* and therefore was named vascular endothelial growth factor (VEGF) (Leung et al., 1989; Tischer et al., 1989). VEGF was also found to be a potent chemoattractant for endothelial cells (Yoshida et al., 1996) and induced the expression of matrix-degrading enzymes, such as interstitial collagenase in endothelial cells (Unemori et al., 1992; Lamoreaux et al., 1998). The same protein was found to be a strong inducer of vascular permeability in the Miles assay after intradermal injection into guinea pig skin. Therefore, the molecule was also named vascular permeability factor (VPF) (Connolly et al., 1989). How permeability is induced by VEGF and whether it depends on induction of endothelial fenestrae or other mechanisms involving the interendothelial junctions are still matters of debate, and a role for VEGF as a permeability factor during development is questionable. Molecular cloning and sequencing revealed that VEGF/VPF was a PDGF-related protein; in mice, three isoforms exist, and in humans five isoforms exist, which after characterization of the genomic structure clearly could be attributed to distinct splice variants of a single messenger RNA (mRNA) that alternatively include exons VI and VII (Houck et al., 1991; Ferrara, 2000). The isoforms largely differ in their affinity for heparin, which was originally used for purification of the protein. The different biophysical properties of the isoforms and their role in angiogenesis has been extensively examined and described (reviewed in Ferrara 2000). However, it was shown that all the isoforms are active *in vivo* in inducing angiogenesis (Takeshita et al., 1996; Schmidt and Flamme, 1998), and it is due only to the elegant exon VI and VII specific knockout that distinct functions of the larger isoforms could be elaborated for postnatal cardiac vascularization (Carmeliet et al., 1999b). Induction of angiogenesis by VEGF *in vivo* has been demonstrated in a series of experimental models [including the chicken chorioallantoic membrane (Wilting et al., 1992), the rabbit and mouse cornea assay (Phillips et al., 1994; Kenyon et al., 1996), and the ischemic hindlimb of rabbits (Takeshita et al., 1994)], and was finally proven by transgenic approaches in which the gene was overexpressed in mice and chicken (Flamme et al., 1995b; Okamoto et al., 1997; Detmar et al., 1998; Larcher et al., 1998). The first clinical trials using VEGF as a drug for the therapy of ischemic diseases have been initiated (Isner and Asahara, 2000).

By exposure of tissues and cells with iodinated VEGF it was shown that high-affinity receptors for VEGF are present on endothelial cells (Jakeman et al., 1992). Later the cognate receptor tyrosine kinases flk-1 (KDR in humans, VEGFR-2) and flt-1 (VEGFR-1) were identified as the high-affinity receptors (De Vries et al., 1992; Terman et al., 1992; Millauer et al., 1993; Quinn et al., 1993). Both were found to be expressed with high specificity on endothelial cells in the mouse embryo. Like other tyrosine kinase receptors, VEGF receptors homodimerize and become autophosphorylated upon ligand binding on intracellular tyrosine residues, and thus signal transduction is activated. Several intracellular downstream molecules, including Shc, Grb2, Nck, and mitogen-activated protein (MAP) kinases, have been described to associate with the activated receptors and to be their immediate or more distant targets in the signaling cascade of gene activation (for comprehensive review, see Petrova et al., 1999). The expression pattern of VEGF and its receptors was highly indicative of a role of this signaling system in angiogenesis (Breier et al., 1995). Moreover, VEGF is discussed as an important survival factor for immature blood vessels. Dissection of the signaling cascade has revealed that activation of the phosphatidylinositol (PI_3) kinase-Akt/PKB signaling pathway is important for this function. The induction in endothelial cells of antiapoptotic proteins survivin, XIAP, and Bcl-2 by VEGF is in support of the survival factor concept (Gerber et al., 1998; Fujio and Walsh, 1999; Nor et al., 1999; Tran et al., 1999). Beside its function as an angiogenesis factor, VEGF is also a procoagulatory and proinflammatory cytokine that stimulates endothelial cells to express tissue factor and the intercellular adhesion molecule-I (ICAM-I) (Clauss et al., 1990, 1996a; Lu et al., 1999).

A third member of the VEGF receptor family (flt-4/VEGFR-3) was predominantly found on lymphatics during later embryogenesis, but during early stages its pattern of expression was similar to those of its relatives (Kaipainen et al., 1995). This receptor did not bind VEGF, and therefore a search for VEGF homologues that potentially could bind flt-4 was started. Five factors have been identified as being close relatives of VEGF, and thus a family of VEGFs was discovered, the members of which are now systematically termed VEGF-A to -E, with VEGF-A as the founding member. VEGF-C and -D have been identified as the ligands for flt-4, but are also ligands for flk-1 (but not flt-1) and are angiogenic *in vivo* (Joukov et al., 1996; Kukk et al., 1996; Achen et al., 1998; Eichmann et al., 1998). By overexpression in transgenic mice, VEGF-C was characterized as a lymphangiogenesis factor, which is in line with the restricted expression of flt-4 (Jeltsch et al., 1997). VEGF-D is highly and almost exclusively expressed in the lung, and its expression is upregulated prior to birth. A relevance of this factor for vascularization of the lung was therefore postulated (Farnebo et al., 1999).

While VEGF binds to both flk-1 and flt-1, VEGF-B, which is predominantly expressed in the cardiac muscle during development (Aase et al., 1999), and placenta growth factor (PlGF), which was cloned by chance as the first homologue of VEGF (Maglione et al., 1991), bind only flt-1 with high affinity but not flk-1 (Terman et al., 1994; Olofsson et al., 1998). PlGF is expressed almost exclusively in throphoblast cells in mice and humans (Achen et al., 1997; Clark et al., 1998). Since neither the targeted deletion of the PlGF gene nor overexpression of this factor showed any developmental phenotype, its function during embryogenesis remains puzzling (Carmeliet and Collen, 1999). The formation of heterodimers with VEGF has been described, which in contrast to PlGF homodimers activate flk-1 signaling (Cao et al., 1996). Similar to VEGF-A, -B, and -C, splice variants of this factor exist. VEGF-E was identified as a viral protein encoded by the parapoxvirus Orf virus genome, and most likely is the pathogenic agent, which is responsible for the hemorrhage caused by Orf virus infection. VEGF-E is angiogenic in *in vivo* assays, binds selectively to flk-1, and stimulates autophosphorylation of the receptor (Ogawa et al., 1998; Meyer et al., 1999).

Signal transduction in the vascular endothelium by VEGF receptors has mainly been shown for flk-1 (Waltenberger et al., 1994). The role of flt-1 is still controversial. As will be discussed later, signaling by flt-1 is dispensable for vascular development, whereas the presence of its extracellular ligand-binding domain is essential (Hiratsuka et al., 1998). Flk-1 and flt-1 are colocalized in the developing vasculature (Breier et al., 1995), but their expression, at least in part, may be differentially regulated (Gerber et al., 1997). Moreover, a soluble secreted splice variant of flt-1 exists, which is strongly inhibitory for signaling by VEGF (Aiello et al., 1995; Kendall et al., 1996; Ferrara et al., 1998; Gerber et al., 1999a). Both flk-1 and flt-1 also have extraendothelial sites of expression. In monocytes flt-1 is an important signaling receptor that activates migration. Consistently PlGF is a stimulator of monocyte migration, and mice lacking a signaling flt-1 have monocyte function defects (Barleon et al., 1996; Clauss et al., 1996b; Hiratsuka et al., 1998). Conversely, in humans flk-1 was also found to be expressed on circulating CD34 positive hematopoietic cells and most likely marks the subset of circulating pluripotent blood stem cells (Ziegler et al., 1999). The role of flk-1 expression in the developing neuroepithelium of the retina is still not clear (Yang and Cepko, 1996).

Beside the high-affinity VEGF receptors, neuropilin-1 recently was identified as a low-affinity receptor for VEGFs (Migdal et al., 1998; Soker et al., 1998; Makinen et al., 1999; Miao et al., 1999). Neuropilin was known as a receptor (or component of the receptor complex) for class 3 semaphorins, which are inhibitory axon guidance signals. Interestingly, targeted disruption of neuropilin resulted in a complex vascular phenotype in which pattern formation of the heart outflow tract was disturbed, yolk sac angiogenesis was reduced, and, most strikingly, vascularization of the neuroepithelium was impaired (Kawasaki et al., 1999). Conversely, overexpression of neuropilin-1 led to hypervascularization and to other complex malformations (Kitsukawa et al., 1995). These data strongly point to a

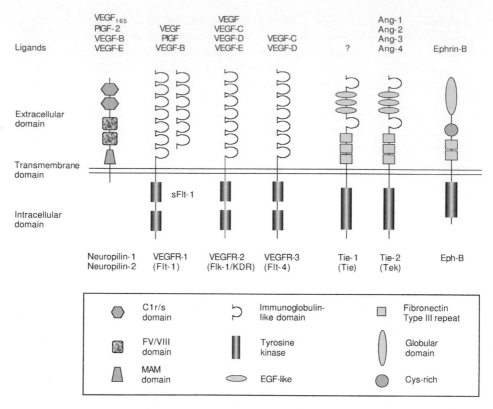

FIGURE 2.1.   Endothelial growth factor receptors and their ligands, which are involved in vasculogenesis, angiogenesis, and lymphangiogenesis.

role for this low-affinity VEGF receptor in embryonic angiogenesis and vascular pattern formation.

In conclusion, the families of VEGFs and VEGF receptors apparently constitute a network of endothelial paracrine signal transduction systems (Fig. 2.1). However, the experimental data, based on targeted deletion or overexpression of individual compounds of the network, clearly demonstrate that the functions of the individual receptors are distinct and that VEGF-A is probably the most important member of the VEGF family, being indispensable and sufficient for growth of new blood vessels.

THE ROLE OF VEGF SIGNALING IN VASCULOGENESIS AND EARLY HEMATOPOIESIS

VEGF/VEGFR: Lessons from the Knockouts

In lower and higher vertebrates, cells of the lateral mesoderm were found to express high levels of VEGF receptor mRNA just after gastrulation (Eichmann et al., 1993; Yamaguchi et al., 1993; Breier et al., 1995; Dumont et al., 1995; Flamme et al., 1995a; Fong et al., 1996). In Northern blot analysis of quail embryo, mRNA maximum expression of flk-1 was found at the definitive primitive streak to head-fold stages (Flamme et al., 1995a). By whole mount *in situ* hybridization, flk-1 mRNA is detected virtually in all cells of the posterolateral mesoderm (Yamaguchi et al., 1993; Eichmann et al., 1997; Liao et al., 1997). Similar results were obtained in zebrafish, mouse, and avian embryos and also for flt-1 and the avian homologue of flt-4. Since the posterolateral mesoderm is known to be the source of hemangioblastic cells of the yolk sac, the expression pattern is highly indicative of a role for VEGF receptors during differentiation of hemangioblastic cells. During differentiation of the vascular plexus, VEGF receptor expression becomes restricted to the endothelial lineage, while hematopoietic cells of the blood islands and perivascular cells (mesothelial cells and primitive smooth muscle cells) do not express the receptor. In cultures of individual flk-1 expressing cells sorted from mouse embryonic stem (ES) cell–derived embryoid bodies, it was shown that a portion of these cells that also express VE-cadherin can give rise to both blood and endothelial cells, while others differentiate into either blood or endothelial cells (Choi et al., 1998; Nishikawa et al., 1998). This is the first experimental evidence of the existence of cells with the potency of hemangioblasts. A subset of these cells, however, also gave rise to a third, undefined cell type (Hirashima et al., 1999). Whether flk-1–positive cells of the early embryo also compose a part of stem cells of the third basal compound of the vascular tree, the population of pericytes/vascular smooth muscle cells, remains to be examined (Fig. 2.2).

The pivotal role of the VEGF signal transduction system in vasculogenesis was proven by targeted deletion of its compounds in mice. Unexpectedly, even the loss of one allele of the VEGF gene was embryonically lethal (Carmeliet et al., 1996; Ferrara et al., 1996). While differentiation of angioblasts was not affected, the formation of intraembryonic blood vessels and remodeling of the yolk sac vasculature were severely impaired. The mutant yolk sacs revealed the honeycomb-like pattern of the primordial plexus and lacked large vitelline vessels connecting the yolk sac vasculature with intraembryonic blood vessels. In homozygous mutants that were generated by aggregation of VEGF–/– ES cells with tetraploid ES cells (the latter ones only give rise to the yolk sac endoderm, whereas the knockout cells establish the embryo proper), the formation of intraembryonic blood vessels failed even more severely than in the heterozygous embryos (Carmeliet et al., 1996). These data are indicative of the tight dose-dependent regulation of vasculogenesis by VEGF. Whether the importance of VEGF for vasculogenesis is dependent on its function as a survival factor or as a mitogen specific for endothelial cells is not yet clear.

The critical role of the VEGF dosage in vasculogenesis was also shown when in *Xenopus* embryos after overexpression of VEGF vascular hyperproliferation and fusion of vessels of the primordial vascular plexus was induced (Cleaver et al., 1997). Similar results were obtained by application of VEGF protein to early avian embryos (Drake and Little, 1995). These data show that endothelial VEGF receptors are not saturated by their ligand in the early embryo. In contrast, concomitant upregulation of flk-1 mRNA was observed after retroviral overexpression of VEGF in E4 chicken embryos when hypervascularization was induced (Flamme et al., 1995b). This upregulation of flk-1 in response to the ligand also occurs in capillaries growing into VEGF-producing malignant tumors and can be induced experimentally on brain capillaries (Plate et al., 1994; Kremer et al., 1997). The

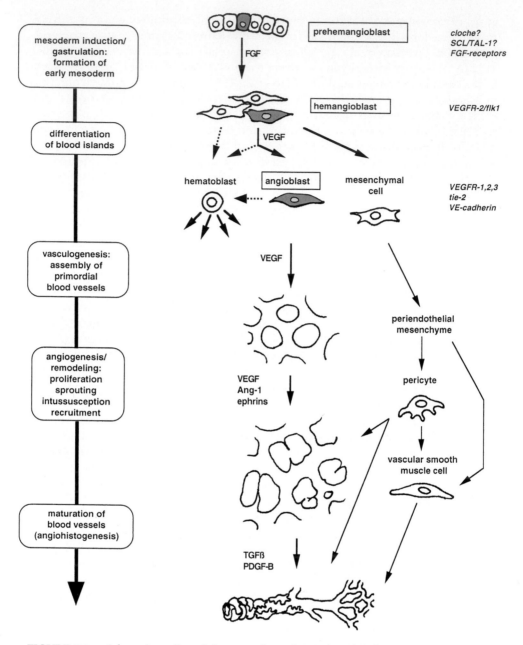

FIGURE 2.2.   Schematic outline of the steps of vasculogenesis and early angiogenesis. The pre-hemangioblastic and hemangioblastic cells are still hypothetical. Factors that are relevant for the individual steps of vascular development are indicated next to the arrows. Molecules characteristic of prehemangioblastic, hemangioblastic, and angioblastic cells are indicated in italics.

mechanisms by which VEGF regulates the expression level of its own receptor have not yet been elucidated.

Disruption of the genes of the cognate VEGF receptors caused different phe-notypes that were only partially complementary to the VEGF knockout phenotype. Mice homozygous for a deletion of the flk-1 gene died *in utero* before day 9.5 due to complete avascularity (Shalaby et al., 1995). This phenotype demonstrated that flk-1 is indispensable for vasculogenesis to occur. While mice heterozygous for the flk-1 deletion developed normally, in homozygous knockout embryos only a few blood cells but no differentiated endothelial or endocardial cells were formed. Hence, flk-1 is needed for the differentiation of angioblasts, whereas the effect on hematoblasts, as discussed later, may be indirect (Hidaka et al., 1999). The second message from these experiments is that the vascular system, in fact, is laid down before it is needed, because the mutant embryos, like the vascularized wild-type embryos, developed quite normally until day 8.5 without any vascular system. Conversely, targeted disruption of the VEGF receptor-1 (flt-1) gene resulted in disturbance of the endothelial lining and enlargement of primordial blood vessels (Fong et al., 1995). The knockout phenotype was characterized by irregular lacunary sinusoids in the yolk sac and endothelial cells inside the vessel lumen. The knockout of the VEGF receptor-3 (flt-4) was followed by failure of remodeling of the yolk sac vasculature and cardiac hemorrhage (Dumont et al., 1998). The data indicated that VEGF receptors are functionally not redundant. It was speculated that VEGF receptor-1 is important for assembly of endothelial cells in the primordial vasculature. This, however, was not the result of loss of signal transduction by flt-1, because transgenic mice in which both alleles of the flt-1 gene were replaced by an engineered gene that encodes a truncated form of flt-1 lacking the tyrosine kinase domain, develop normally without affecting the vascular system (Hiratsuka et al., 1998). This means that the extracellular domain of flt-1 exerts an important function during vascular development. Interestingly, the phenotype of vascular fusion that resulted from the application of VEGF protein in avian embryos closely resembled the flt-knockout phenotype (Drake and Little, 1995). Thus, it may be speculated that flt-1 is needed for balancing the effects of VEGF on early endothelial cells and that loss of this function leads to overshooting proliferation of the endothelium.

The Induction of Hemangioblast

While our knowledge of the role of VEGF signaling in the early steps of vascular development is continually increasing, only little is known about the determinants that lead epiblast cells to undergo differentiation into flk-1–expressing hemangioblastic cells of the mesoderm. Determination of hemangioblastic cells takes place during gastrulation, but, in contrast to constituents of the axial structures, is not dependent on gastrulation. This can be concluded from experiments in which either gastrulation was inhibited and blood islands still developed (Azar and Eyal-Giladi, 1979; Zagris, 1980), or pieces from premesodermal quail embryos were transplanted into limb buds of chick embryos and differentiation of quail endothelial cells was observed (Christ et al., 1991). However, in all these experiments epiblastic cells *in vivo* undergo epitheliomesenchymal transformation to give rise to hemangioblasts. Hemangioblastic cells are likely to be induced by the same molecular mechanisms that are also responsible for induction of ventral mesoderm in *Xenopus* mesoderm induction models (Slack et al., 1987). Data from experiments in pluripotent cells from E0 avian embryo have strongly suggested that fibroblast growth factors (FGFs) and their receptors constitute a pivotal signal transducing system for the induction of hemangioblastic cells (Flamme and Risau, 1992; Krah et al., 1994). Furthermore, protein kinase C appeared to be involved in downstream signaling since phorbolester mimicked the induction of blood islands by FGFs *in vitro* (Flamme et al., 1997a). During a search for FGF-inducible genes in this avian model of hemangioblast induction, the VEGF receptor-2/flk-1 was found among the immediate early genes induced by FGF. This induction underlines the specific role of FGF signaling in the initiation of vasculogenesis (Flamme et al., 1997a).

The fact that only a relatively small subset of epiblast cells undergoes differentiation into hemangioblastic cells after induction (our own unpublished observations) demonstrates that a pre-restriction must have taken place in the epiblast determining the number of cells that can differentiate into hemangioblasts. The term *prehemangioblasts* seems appropriate to denominate those fate restricted cells (Fig. 2.2). This hypothesis is further supported by the observation that the number of blood islands that can be induced from an individual avian E0 embryo increases concordantly with the cell number of the epiblast within the short period between laying and gastrulation (our own unpublished observations). The determinants of prehemangioblastic cells are still not known. Transcription factors that act upstream of flk-1 and are activated in response to FGF signaling are candidates. A zebrafish mutant called cloche (*clo*) has been identified that is not linked to the flk-1 locus and like the flk-1$^{-/-}$ mutant is lethal early and characterized by the complete lack of the hemangioblastic cell lineages including the endocardium (Liao et al., 1997). Consistently in the *clo* mutant flk-1 is not expressed orthotopically but by a small number of ill-defined cells. Interestingly, this mutant can be partially rescued by overexpression of the SCL/Tal-1 transcription factor (Liao et al., 1998), which in higher vertebrates was found to be a master regulator of hematopoiesis and also involved in early angiogenesis (Shivdasani et al., 1995; Porcher et al., 1996; Visvader et al., 1998). Moreover ectopic overexpression of SCL/Tal-1 in zebrafish embryos induced supernumerous hemangioblastic cells from the intermediate mesoderm, indicating a role for this transcription factor in cell lineage specification (Gering et al., 1998). However, SCL/tal-1 is not identical with the as yet unidentified *clo* gene product but acts downstream of it (Liao et al., 1998). Interestingly, in the endothelial lineage the *clo* mutation is cell autonomous, while for hematopoietic cells both cell-autonomous and nonautonomous reasons may be responsible for the reduction of blood cells (Parker and Stainier, 1999). This is very similar to the observations in cultures of flk-1$^{-/-}$ stem cells that are not able to differentiate into endothelial cells but retain their capability to generate all blood cells *in vitro* (although to a largely reduced extent) (Hidaka et al., 1999). These data point to an important role of the early endothelium for the maturation of blood cells and are suggestive of a role for endothelium in nursing cells in blood islands. However, the crosstalk between hematoblasts and angioblasts has not been deciphered as yet.

VEGF, Angioblasts, and the Early Vascular Pattern Formation

Angioblasts may be defined as cells with the characteristics of endothelial cells that have not yet formed a vascular lumen, or have not been integrated into a blood vessel, or have migrated outward from a preexisting capillary vessel. The contribution of migrating angioblasts to the formation of primordial blood vessels is considerable. In general, angioblasts have the shape of mesenchymal cells and are migratory, highly active, and highly invasive "prevascular" endothelial cells.

Angioblasts are involved in many angiogenetic processes in the embryo such as the vascularization of the head mesenchyme, which is of neuroectodermal origin and initially avascular (Noden, 1989), and the vascularization of the embryonic central nervous system, which in part is vascularized by the immigration of individual angioblasts detaching dorsally from the perineural vascular plexus (Kurz et al., 1996). In the extremities individual angioblasts migrate in the avascular subectodermal zone (Feinberg and Noden, 1991). The coronary plexus derives from angioblasts that migrate together with presumptive epicardial cells from the sinus venosus over the surface of the myocardium. By a process very similar to early vasculogenesis the coronary plexus is laid down and then becomes interconnected with the aortic outflow tract thus establishing the primordia of the main coronary arteries (Bogers et al., 1989; Mikawa and Fischman, 1992). It is notable that during the formation of the subepicardial plexus the highest amounts of VEGF protein are detectable in the epicardium and the peripheral myocardium (Tomanek et al., 1999).

Guidance of migrating angioblasts is an important aspect of vascular pattern formation during vasculogenesis. The pathways for angioblasts that are destined to form the principal vessels are determined during early development and may be used on opposite routes. Unilateral inhibition of the migration of angioblasts along the anterior intestinal portal in avian embryos is compensated by angioblasts from the other side that give rise to endocardium and connect with the dorsal aorta on the affected side (Coffin and Poole, 1991). In the head mesenchyme, which initially is avascular, angioblasts migrate over long distances to their final location. This is indicative of a high plasticity of early blood vessel primordia (Noden, 1989). Recent observations in zebrafish mutants support the concept that the development of the vascular macropattern is an epigenetic phenomenon. In lower vertebrates the formation of the dorsal aorta at the midline most likely depends on VEGF signals from the hypochord, an endoderm-derived structure closely associated with the notochord. In the zebrafish mutants floating head (flh) and no tail (ntl), which fail to form a notochord, no hypochord is present and no dorsal aorta is developed. However, formation of the aortic rudiment can be rescued in chimeras forming the wild-type notochord (Fouquet et al., 1997; Sumoy et al., 1997; Weinstein, 1999). Also in *Xenopus* hypochord-derived VEGF is a candidate factor for the attraction of angioblasts from the lateral plate mesoderm which form the aortic tube at the midline (Cleaver and Krieg, 1998). In higher vertebrates, in contrast, the notochord represents a barrier that in the trunk, by still unknown mechanisms, hinders angioblasts from crossing the sagittal body plane (Wilting et al., 1995). Thus, before fusion the dorsal aorta is laid down pairwise beneath the somites. Interestingly, the ventral parts of the somites express VEGF (Weinstein, 1999). Although whole-mount immunostainings of the nascent vasculature of quail embryos suggested that the dorsal aortae of higher vertebrates are formed by remodeling from a more or less uniform vascular network (Pardanaud et al., 1987; Coffin and Poole, 1988), electron microscopic analysis has revealed that as in lower vertebrates the dorsal aortae are formed as individual tubes by coalescence of individual angioblasts during the initial steps of vasculogenesis (Hirakow and Hiruma, 1983). However, gradients and differences in the local concentration of VEGF are unlikely to be the only cues that determine pattern formation of the early vasculature. Even massive overexpression of VEGF in the avian embryo wing bud did not interfere with general vascular pattern formation, although vascular density was remarkably increased (Flamme et al., 1995b). Likewise, the chronology of vascularization was not disturbed, not even in the cartilage blastema, which ectopically overexpressed VEGF. On the other hand, later during development VEGF is absolutely necessary for invasion of blood vessels during endochondral bone formation, as it was demonstrated by treatment of postnatal mice with soluble Flt-1–immunoglobulin G (IgG) chimeric protein (Gerber et al., 1999b). Thus other still unknown mechanisms, rather than simply the availability of an endothelial cell growth factor, must be important for details of vascular pattern formation and are supposed to regulate the onset of vascularization of an avascular tissue such as cartilage.

Angioblasts from early embryos are pluripotent for various vascular beds. Thus, angioblasts derived from the lateral mesoderm can contribute to endocardium as well as to extraembryonic vasculature (Coffin and Poole, 1991). In the adult, CD34-positive angioblastic cells have been isolated from the blood (Asahara et al., 1997). These angioblasts are derived from the bone marrow and are strongly stimulated by VEGF. It was shown that such angioblasts circulate in the blood, like their embryonic counterparts, and are also pluripotent, as they contribute to vascular growth in reproduction, wound healing, tumor growth, and repair of ischemic lesions (Asahara et al., 1999). Thus, postnatal vasculogenesis from "endothelial stem cells" may exist and may be complementary to vascular growth from preexisting vessels under physiologic and pathologic conditions.

Primitive Versus Definitive Hematoblasts and the Instructive Role of VEGF

While the extraembryonic yolk sac hematoblasts give rise to the primitive blood cells of the embryo, the definitive population of blood cells was demonstrated to be derived from hematoblasts of intraembryonic origin. The definitive hematopoietic cells bud from the floor of the aorta at early mid-gestation stages in embryos of higher vertebrates (Dieterlen et al., 1988; Olah et al., 1988; Medvinsky and Dzierzak, 1996; Tavian et al., 1996). It was hypothesized that their precursors immigrate to the aorta from the surrounding mesenchyme of the aorta/gonads/mesonephros (AGM) region. It was speculated that they are derivatives of coelomic epithelium (Munoz-Chapuli et al., 1999) or primordial germ cells (Rich, 1995). However, experimental data have suggested that the definitive hematopoietic cells are descendants of the endothelium that forms the floor of the aorta (Jaffredo et al., 1998). This is in line with the observation that the floor endothelium expresses the AML-1 transcription factor, which is indispensable for differentiation of definitive haematoblasts (Okada et al., 1998; North et al., 1999). Thus the concept of the hemangioblast was modified in that hematoblasts are formed by layering from angioblasts or endothelial cells (Fig. 2.2). Also data from *in vitro* differentiation of sorted flk-1 positive cells from ES cell–derived embryoid bodies are in support of this new concept. (98)

The "hemangioblastic" endothelium of the floor of the dorsal aorta is of splanchnopleuric origin while the endothelial cells that form the dorsolateral lining of the dorsal aorta are derived from the somatopleural mesoderm (Pardanaud et al., 1996). This distinction is of importance insofar as by elegant transplantation experiments it was shown that the splanchnic mesoderm contains precursors of both blood and endothelial cells, while somatopleural mesoderm contains only angioblasts. (The somatopleura is the mesodermal layer that underlies the dorsal ectoderm, while the splanchnopleura is that part of the mesoderm that resides on the endoderm that forms the resorptive epithelium of the yolk sac and gut.) It was speculated that the splanchnic mesoderm contains hemangioblasts or angioblasts with hemangioblastic potency, while the somatopleura contains pure angioblasts. Moreover, it was shown that contact of mesoderm with endoderm was instructive in that hematopoietic potency was acquired, and this effect could be mimicked by exposure to not only VEGF, but also basic FGF (bFGF) and TGF-β, while contact with ectoderm or treatment with epidermal growth factor (EGF)/TGF-α abrogated the hemangioblastic potency (Pardanaud and Dieterlen-Lievre, 1999).

The Endoderm, a Source of VEGF

The fact that hemangioblastic potency is restricted to that part of the mesoderm that is in immediate contact with the endoderm suggests a role of the endoderm in the induction of hemangioblastic cells. In embryoid bodies derived from ES cells deficient for the transcription factor GATA-4, that are defective for the differentiation of endoderm, no vascular plexus was formed (Bielinska et al., 1996). However, as shown by experiments in quail embryoid bodies and embryos, the endoderm is not sufficient for this induction but rather is cooperative by producing stimulating factors (Krah et al., 1994). The existence of those factors was postulated for the first time in 1964; after removal of the yolk sac endoderm from primitive streak stage chicken embryos, the number of blood islands was reduced and remaining blood islands lacked the typical envelope of endothelial cells (Wilt, 1964). The same author demonstrated by transfilter experiments that these effects were due to the withdrawal of soluble endoderm derived factor(s) (Miura and Wilt, 1969). VEGF-A is expressed by the endoderm and may represent (at least one of) the soluble endoderm derived factor(s) necessary for the differentiation of hemangioblastic cells (Breier et al., 1995; Flamme et al., 1995a). Consistently, treatment of flk-1 positive cells either from primitive streak quail embryos or from cultures of murine embryonic stem cells with VEGF-A or -C led to an expansion of the endothelial lineage at the expense of hematopoietic differentiation (Eichmann et al., 1997, 1998). Thus, differentiation of hematopoietic cells may take place either by default, when signaling by VEGF is below a certain threshold, or in the presence of other factors that bias the diversification of flk-1–positive hemangioblastic cells toward blood cells in the yolk sac. Interestingly, in the avian embryo yolk sac hematopoiesis is correlated with the contact of mesoderm to visceral (area opaca) endoderm (Mato et al., 1964; Kessel and Fabian, 1985). In contrast, the central parts of the yolk sac mesoderm that are in contact with embryonic (area pellucida) endoderm lack blood islands. However, both endoderms express VEGF-A, and the embryonic endoderm is able to instruct somatopleural endoderm to give rise also to hematopoietic cells (Pardanaud and Dieterlen-Lievre, 1999). Moreover the borderline between area opaca and pellucida endoderm exactly corresponds to a borderline between high and low levels of angiogenesis in the yolk sac mesoderm, respectively (Flamme, 1989). Hence, the role of endoderm in vasculogenesis, embryonic angiogenesis, and hematopoiesis is still puzzling.

THE ROLE OF VEGF SIGNALING IN EMBRYONIC ANGIOGENESIS

What Is an Angiogenesis Factor?

In the adult organism the endothelial cells lining the inner surface of all blood vessels and the heart chambers represent a population of relatively quiescent cells. In adult rats, for example, less than 0.3% of aortic endothelial cells were found to incorporate ^{3}H-thymidine, in contrast to 13% at birth (Schwartz and Benditt, 1977). There are only a few exceptions in the adult organism where endothelial cells constitute a highly proliferative population and thus adopt a more or less embryonic status, such as during corpus luteum formation and proliferation of the uterine mucosa, and in wound healing. However, there are also many destructive and proliferative diseases that absolutely depend on endothelial proliferation to progress, such as cancer, proliferative retinopathies, and rheumatoid arthritis (Folkman and Shing, 1992; Risau, 1997). Chronic ischemic diseases, in contrast, are characterized by a deficit of vascular growth, a fact that points to a dysregulation of the adaptive potency that normally enables the vasculature to grow in response to hypoxia and increased metabolic demands (Deindl et al., 2000). It is commonly assumed that the molecular mechanisms underlying vascular growth in the adult are basically the same mechanisms that stimulate and coordinate vascular growth in the embyro. Candidate angiogenesis factors are expected to induce proliferation and invasiveness of endothelial cells, to be chemoattractants and inducers of sprouting or nonsprouting vascular growth, and to protect blood vessels from regression.

Historically, fibroblast growth factors (FGFs) 1 and 2 were the first angiogenesis factors to be purified. They were found to strongly induce proliferation of endothelial cells and many other cell types *in vitro* and to induce angiogenesis response in certain *in vivo* assays (e.g., the rabbit cornea assay) (Burgess and Maciag, 1989; Friesel and Maciag, 1995). Therefore, FGFs and their receptors, which like VEGF receptors are receptor tyrosine kinases, have been implicated in embryonic angiogenesis. However, expression of neither of the FGF receptors was detected on embryonic endothelium, with the exception of the endothelial cells lining large vessels (Peters et al., 1992; Patstone et al., 1993). Consistently, retroviral and transgenic overexpression of secreted FGFs failed to induce embryonic angiogenesis but rather induced changes in pattern formation (Riley et al., 1993; Coffin et al., 1995; Stolen et al., 1997; Fulgham et al., 1999), which is in line with the role of distinct secreted FGF family members as important morphogens (Baird, 2000). In addition, the founding members FGF-1 and -2 lack a signal peptide and are not secreted under conditions of forced expression (Thompson and Slack, 1992). However, upon damage of the cell membrane FGF-1 and -2 stored within muscle cells may be released and execute their function in repair and wound healing (Clegg et al., 1987; Husmann et al., 1996).

In contrast, the temporal and spatial expression pattern of VEGF-A and its receptors flk-1 and flt-1 are highly indicative of a paracrine mode of signaling during embryonic angiogenesis. The receptors are highly and ubiquitously expressed by the embryonic endothelium, while the ligand is expressed by the organs being vascularized. This correlation is most impressive in the brain, where VEGF is expressed in the periependymal layers when blood vessels invade the neuroepithelium from the meningeal plexus (Breier et al., 1992, 1995). These observations were also suggestive of a role for VEGF as a chemoattractant and sprouting factor for vascular endothelial cells *in vivo*. In the developing retina of rats and cats, expression of VEGF by glial cells precedes the advancing front of immigrating blood vessels (Stone et al., 1995). In epithelial cells of developing lung and liver, high levels of VEGF mRNA were detected, but expression was most abundant in the epithelial cells of the choroid plexus and the podocytes of kidney glomeruli (Breier et al., 1992, 1995; Brown et al., 1992; Monacci et al., 1993). In these latter locations expression of VEGF is maintained constitutively throughout life and was correlated with the fenestrated phenotype of the capillary endothelium. There is evidence from different experiments *in vivo* and *in vitro* that VEGF can induce endothelial cell fenestrae under permissive conditions provided by an appropriate basal lamina (Roberts and Palade, 1995; Esser et al., 1998). It is still a matter of debate whether induction of fenestrae is related to the permeability inducing function of VEGF. *In vitro* and *in vivo* increased permeability of endothelial monolayers in response to VEGF is mainly executed via transcellular transport mechanisms involving vesiculovacuolar organelles (VVOs) and probably not the result of loosening of intercellular contacts (Feng et al., 1999; Michel and Neal, 1999). Analysis of transgenic mice bearing a knock-in of the *lac-Z* gene from *Escherichia coli* in the VEGF gene substantiated the findings from expression studies by *in situ* hybridization and essentially confirmed the concept of paracrine signaling by VEGF (Miquerol et al., 1999). However, a small subset of endothelial cells located in the outflow tract of the heart was also found to express the transgene. This finding suggests an autocrine role of VEGF at least for this site of expression.

While in overexpression experiments all the VEGF isoforms were found to be equally active as angiogenesis factors (Schmidt and Flamme, 1998), mice homozygous for a deletion of exons VI and VII show failure of cardiac vascularization, which was followed by fatal ischemic cardiomyopathy and postnatal lethality (Carmeliet et al., 1999b). Those mice, as expected, lacked the VEGF 164 and 188 isoforms, and although they expressed elevated levels of the small soluble 120 amino acid isoform, vascularization of the heart was markedly reduced. These data indicate that, for unknown reasons, VEGF 120 is insufficient for angiogenesis at least in the heart, while angiogenesis of all other organs apparently was not impaired.

Depletion of VEGF from adolescent and adult mice by administration of mFlt(1-3)-IgG, a soluble VEGF receptor chimeric protein, has reversibly stopped angiogenesis-dependent processes such as endochondral ossification and corpus luteum formation and the ovulation cycle (Ferrara et al., 1998; Gerber et al., 1999b). In conclusion, the experimental evidence that VEGF-A is indispensable for angiogenesis to occur in the embryo and in the adult in health and disease is indisputable, and the question is raised whether VEGF independent angiogenesis may exist at all.

Also other endothelial receptor tyrosine kinases and their soluble or membrane-bound ligands are apparently indispensable for angiogenesis, although they obviously are dispensable for normal vasculogenesis (Fig. 2.1). Deletion of the Tie-2 receptor or its ligand angiopoietin-1 (Ang-1) abrogated vascularization of the neural epithelium (Dumont et al., 1994; Sato et al., 1995; Suri et al., 1996). Instead, the vascular plexus remained in a more primitive stage of differentiation including reduced trabeculation of the heart where the ligand is highly expressed by the cardiomyocytes. These data gave rise to the hypothesis that the Ang-1/Tie-2 signal transduction system is a key regulator of vascular remodeling and maturation of blood vessels. Unlike the VEGF/VEGFR system, the activated Ang-1/Tie-2 signal transduction system does not induce proliferation of endothelial cells. In an *in vitro* angiogenesis assay that employs fibrin gels (Nehls and Drenckhahn, 1995), VEGF and Ang-1 synergistically induced endothelial cell sprouting (Koblizek et al., 1998). Transgenic overexpression of Ang-1 in the skin of mice resulted in increased vascularity (Suri et al., 1998). Coexpression of Ang-1 and VEGF had an additive effect on angiogenesis but in contrast to overexpression of VEGF only resulted in leakage-resistant vessels (Thurston et al., 1999). This synergism may be important for maturation of newly formed vessels and is likely to play an important role also during early angiogenesis, but direct evidence is still missing.

Interestingly, the function of Ang-1 may be antagonized and balanced by a homologue, Ang-2, which is expressed during vascular regression, e.g., in the corpus luteum, and which can inhibit embryonic angiogenesis when overexpressed as a transgene (Maisonpierre et al., 1997). In the embryo, Ang-2 is predominantly found in mesenchymal cells surrounding the aorta and may have its function during the maturation of large blood vessels. Ang-2 is induced by VEGF and hypoxia (Oh et al., 1999) and also newly formed blood vessels in tumors express Ang-2 (Stratmann et al., 1998), and therefore its role during angiogenesis is not yet clear. Likewise the function of other, recently identified members of the angiopoietin family has not been characterized yet (Conklin et al., 1999; Grosios et al., 1999; Valenzuela et al., 1999).

Recently a novel class of receptor tyrosine kinases and their ligands, which are membrane bound—the Eph-receptors and ephrins, respectively—have been discovered to play a fundamental role during early angiogenesis (Wang et al., 1998; Adams et al., 1999) (Fig. 2.1). The role of Eph/ephrin signaling was well established for axon guidance during maturation of the central nervous system (Holder and Klein, 1999). Targeted mutation of both ephrin-B2 and its receptor Eph-B4 revealed lethal vascular defects very similar to those obtained by knockout of the Ang-1/Tie-2 signal transduction system. Interestingly, ephrin-B2, like Ang-1 and VEGF, induces endothelial sprouting *in vitro* (Adams et al., 1999). It may be speculated that the different signaling systems cooperate during vascular remodeling and may depend on each other to execute angiogenesis. This is suggested by the striking similarity of the knockout phenotypes. Intriguingly, Eph-B4 is expressed by venous endothelium, whereas the ligand is expressed by arterial endothelium. The Eph-receptor/ephrin interaction is likely to result in a bidirectional signaling (Holland et al., 1996). Thus, a morphogenetic crosstalk at the arteriovenous border was postulated (Yancopoulos et al., 1998). However, the expression pattern does not sufficiently explain the general defects that result from the targeted mutations of the genes. In view of recent data from experiments in endothelial cell cultures, it may be speculated that additional related ligands and receptors may be involved in vascular remodeling and embryonic angiogenesis (Pandey et al., 1995; Stein et al., 1998).

Consistent with the central role of receptor tyrosine kinases in embryonic angiogenesis, disruption of the p120-rasGAP, a downstream signaling molecule and a negative regulator of Ras, led also to alterations of vascular remodeling in the yolk sac and disturbed maturation of the embryonic vascular plexus with aberrant ventral sprouts from the dorsal aorta (Henkemeyer et al., 1995).

VEGF, a Survival Factor for Immature Endothelium

While the conventional knockouts of VEGF and VEGF receptors result in early embryonic lethality during vasculogenesis, in newborn mice both the conditional knockout of the VEGF gene and depletion of VEGF by administration of mFlt(1-3)-IgG, a soluble VEGF receptor chimeric protein, resulted in severe growth retardation, failure of organ development (most prominently of the liver and kidney), and eventually death (Gerber et al., 1999a). The apoptosis index of endothelial cells was increased, which indicated that VEGF is not only a mitogen but also an important survival factor for endothelial cells. The endothelial survival signal by VEGF is executed via activation of PI_3 kinase-Akt/PKB and induction of Bcl-2 and is dependent on a functional complex of flk-1 with VE-cadherin/β-catenin, as was demonstrated by targeted disruption or cytosolic truncation of the VE-cadherin gene in mice (Carmeliet et al., 1999a). Interestingly, withdrawal of VEGF 4 weeks after birth did not impair endothelial survival, which is in line with the concept that VEGF is an important survival factor only for immature or growing blood vessels but not needed for the maintenance of mature vessels (Gerber et al., 1999a). In grafted tumors that expressed VEGF under the control of an inducible promoter, only those blood vessels that had already been invested by pericytes withstood a withdrawal of VEGF and did not regress (Benjamin and Keshet, 1997). Recruitment of pericytes is thought to depend on endothelial-derived PDGF-B and on signaling via PDGF receptors (Fig. 2.2). Mobilization of pericytes from capillaries of the retina by intraocular administration of exogenous PDGF results in renewed VEGF dependence of the endothelium (Benjamin et al., 1998). Whether compounds of the basal lamina or VEGF or other factors such as TGF-β are the crucial survival factors provided by pericytes and vascular wall is not known.

Factors Regulating VEGF and VEGF Receptor Gene Expression

Maturation of blood vessels is also correlated with downregulation of VEGF receptor expression in endothelial cells. In the rat and mouse brain expression of flk-1 mRNA and protein are downregulated postnatally and their expression is almost completely absent 3 weeks after birth when the brain has acquired its final volume (Kremer et al., 1997). This time course exactly correlates with the decrease in endothelial cell division (Robertson et al., 1985). Thus, it may be speculated that the expression level of flk-1 is a limiting factor for vascular growth in the brain. Recent studies in newborn rats revealed that postnatal downregulation of flk-1 expression is in parallel with downregulation of VEGF expression in neurons (Ogunshola et al., 2000). In slice cultures from adult mouse brain, flk-1 expression was reactivated after addition of its ligand VEGF (Kremer et al., 1997). The underlying molecular mechanisms have not been elucidated; they may be the same ones that led to upregulation of VEGF receptors on endothelial cells during tumor angiogenesis, e.g., in glioblastomas (Plate et al., 1993) since glioblastomas and many other malignant tumors express abundant amounts of VEGF and thus re-expose mature brain capillaries to VEGF. Likewise, overexpression of VEGF in a transgene approach led to upregulation of its signal transducing receptor (Flamme et al., 1995b) and, as already mentioned, under physiologic conditions, high constitutive expression of VEGF in the podocytes and the plexus epithelial cells is correlated with expression of flk-1 and flt-1 in the capillary endothelial cells of the kidney glomeruli and the choroid plexus throughout life.

Thus, VEGF apparently is an important (if not the most important) regulator of VEGF receptor expression during developmental and pathologic angiogenesis. According to the concept of paracrine signaling, VEGF positively enhances the susceptibility of endothelial cells for the angiogenic signal. Therefore, the question of how the expression of VEGF is regulated during development is of central importance. Oxygen tension and intracellular glucose concentrations reflecting metabolic demands are thought to be key regulators of angiogenesis in tissues and tumors (Adair et al., 1990). Only few years ago, challenged by the observation that VEGF is upregulated in the palisading cells of glioblastomas around necroses (Plate et al., 1992; Shweiki et al., 1992), it was demonstrated that the expression of VEGF mRNA is induced by hypoxia and low intracellular glucose concentrations (Shweiki et al., 1995) by a dual mechanism: transcriptional activation and stabilization of its mRNA (Carroll et al., 1994; Ikeda et al., 1995; Levy et al., 1995). While transcriptional upregulation of VEGF involves elements within the promoter, stabilization of its mRNA depends on elements within the 3′-untranslated region of the mRNA. Thus, for VEGF a long-sought molecular link seemed to be found between increased metabolical demands and increase of angiogenesis. In addition, VEGF receptors also appeared to be inducible by hypoxia *in vivo* (Tuder et al., 1995), and, at least for flt-1, a direct induction was demonstrated to depend on a hypoxia responsive element in the 5′-flanking region (Gerber et al., 1997).

The oxygen tension of a developing tissue balances its vascular density via regulation of VEGF gene expression. This was demonstrated in the retina of postnatal rats. In newborn rodents vascularization of the retina (a neural tissue) is not completed. An increase in oxygen tension was followed by regression of immature blood vessels due to decreasing VEGF release, while subsequent normalization of oxygen tension was followed by excessive proliferation of capillaries, most likely as the consequence of shifted oxygen sensitivity of VEGF regulation (Alon et al., 1995; Stone et al., 1995). Thus, an animal model of retinopathy of prematurity was created and hypoxia regulation of VEGF was recognized as the central mechanism of this syndrome. It is not yet clear to what extent the oxygen pressure regulates angiogenesis in the embryo, but as can be concluded from the data described above in late fetal and postnatal life the oxygen-dependent system of angiogenesis regulation is well established. It is not known when this system starts to work during ontogeny. Experiments in chicken embryos suggest that the vascularity can be influenced by the environmental oxygen pressure at early stages (Hoper and Jahn, 1995). Notably, besides the VEGF/VEGF receptor signaling system, no other endothelial growth factor signaling system has been identified to depend on tissue oxygenation and metabolism in a similar manner. However, in some locations where VEGF is expressed at high levels constitutively, such as in the choroid plexus epithelia and podocytes from kidney glomeruli, for unknown reasons its expression is regulated paradoxically in response to hypoxia (Marti and Risau, 1998).

A heterodimeric transcription factor, named HIF-1 (hypoxia inducible factor-1) consisting of the basic helix-loop-helix (bHLH) proteins HIF-1α and ARNT (aryl hydrocarbon receptor nuclear translocator), was found to mediate transcriptional activation of hypoxia inducible genes such as EPO, VEGF, and enzymes of glycolysis in response to hypoxia and hypoglycemia by binding to a hypoxia-responsive element in the regulatory sequences of those genes (Wang et al., 1995; Forsythe et al., 1996) (Fig. 2.3). Thus, HIF-1α became a candidate regulator of

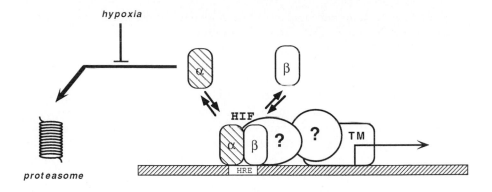

genes with functional HRE:	HRE sequence*	location:
erythropoietin	T_{11} A_{23} **CGTG** C_{18} G_{12} G_{14} C_9	5´FS
vascular endothelial growth factor	G_{10} G_2 G_4 T_7 T_3 T_7	3´FS
glucose transporter 1	C_4 A_3 C_5 C_5 G_7	5´UTR
phosphofructokinase L		intron
phosphoglycerate kinase 1		
enolase 1		
lactate dehydrogenase A		
heme oxygenase 1		
transferrin		

FIGURE 2.3.   Current model of gene activation by hypoxia inducible factors. The hypoxia signal prevents the α-subunit from being degraded by the proteasome. The cofactors (indicated by question marks) that link the HIF-complex to the basic transcriptional machinery (TM) are not yet defined. Target genes with functional hypoxia responsive elements (HREs) and sequences of HREs are listed. HREs are found in flanking sequences (FSs), untranslated regions (UTRs), or introns of hypoxia sensitive genes. (*From Wenger and Gassmann, 1997.)

VEGF expression *in vivo*. Consistently, targeted mutation of HIF-1α in mice led to a series of malformations of the embryo, including abnormal neural folding, cardiac and vascular malformations, and the complete lack of cephalic blood vessels (Iyer et al., 1998; Ryan et al., 1998). Unexpectedly, these defects could not be explained by a reduced expression of VEGF in locations where angiogenesis is needed to support tissue growth, because VEGF was even found upregulated in the HIF-1α null embryos (Kotch et al., 1999). In tissue culture cells induction of VEGF by low glucose concentration was retained. Thus, vascular regression was the result of cell death rather than VEGF deficiency. Conversely, results of the two published ARNT knockouts were inconsistent, revealing either early vascular defects in the yolk sac and loss of cellular responsiveness to hypoxia and hypoglycemia or defects in placental differentiation (Kozak et al., 1997; Maltepe et al., 1997). More recently a close homologue of HIF-1α was cloned by several groups and named EPAS-1/HRF/HLF/MOP-2 (Ema et al., 1997; Flamme et al., 1997b; Hogenesch et al., 1997; Tian et al., 1997) and renamed HIF-2α (Wenger and Gassmann, 1997). Unlike HIF-1α, which is expressed ubiquitously, HIF-2α is expressed most prominently in brain capillary endothelium and other blood vessels (Flamme et al., 1997b; Tian et al., 1997). High expression levels of HIF-2α in kidney glomeruli, choroid plexus, and lung—sites where VEGF is highly expressed—suggest a role of this factor in VEGF gene activation (Ema et al., 1997; Flamme et al., 1997b). Since HIF-2α activates the transcription of VEGF, Tie-2, and flk-1 (Ema et al., 1997; Tian et al., 1997; Kappel et al., 1999), a role of this transcription factor as a master regulator of vascular development was likely. However, targeted gene deletion of HIF-2α produced no vascular phenotype, and therefore, the role of HIF-2α in the transcriptional activation *in vivo* of these vascular target genes and its role in vasculogenesis and embryonic angiogenesis remain to be established (Tian et al., 1998). In HIF-1α null cells HIF-2α was upregulated, which could be the reason for increased VEGF expression (Ryan et al., 1998). Since HIF-1α is expressed ubiquitously but HIF-2α expression is very much restricted, it can be speculated that HIF-2α cannot fully compensate for the loss of HIF-1α, but, conversely, that HIF-1α can compensate for the loss of HIF-2α function. Both HIF-1 and -2 need to interact with the pleiotropic transcription factor complex forming protein CBP/p300 to transactivate the promoters of their target genes (Arany et al., 1996; Ebert and Bunn, 1998; Ema et al., 1999). The importance of this complex for vascular development is further underscored by the fact that homozygous mutant mouse embryos that express a truncated form of the CBP protein, like the HIF-1α mutants, exhibit defects in closure of the neural tube, vascular network formation, and hematopoiesis (Oike et al., 1999). As the hematopoietic and vasculogenetic defects could partially be rescued *in vitro* by addition of VEGF, a role for CBP in activating the transcription of VEGF appears conceivable.

The activity and availability of HIFs are mainly regulated on the protein level. Under normoxia the α-subunit is targeted to the proteasomal degradation, whereas under hypoxia targeting is inhibited (Huang et al., 1998; Kallio et al., 1999). The von Hippel–Lindau (VHL) tumor suppressor gene is likely to play an important role in targeting HIFs to the proteasome (Maxwell et al., 1999). While in the adult loss of function of the VHL gene product is associated with tumors overexpressing VEGF and HIF-2α (Wizigmann-Voos et al., 1995; Flamme et al., 1998), targeted deletion in mice resulted in embryonic lethality due to disturbed vascularization of the placenta, which, however, could not clearly be correlated with dysregulation of VEGF (Gnarra et al., 1997). Whether the VHL tumor suppressor gene plays a role in regulating embryonic expression of VEGF is not clear.

Interestingly, vertebrate HIFs are homologous to the *Drosophila* trachealess (*trh*) gene product that is a master regulator of tracheocyte differentiation (Isaac and Andrew, 1996; Wilk et al., 1996). The tracheal system of insects that serves as a pipeline for gas consists of a unique cell type, the tracheocyte, grows by sprouting, and its growth is enhanced in response to hypoxia and metabolic demands, thus resembling the vascular system of vertebrates in many respects (Manning and Krasnow, 1993). Moreover, human HIF-1α is functional in the *trh* complex after ectopic expression in the *Drosophila* embryo, which is indicative of a highly conserved function of these factors during formation of tubular systems (Zelzer et al., 1997).

CONCLUSION

According to its early destination to nourish the growing embryo, the cardiovascular system is the first functional organ system of the vertebrate organism. The integrity of vascular development is a requirement for normal embryogenesis and targeted mutations of molecules, which are essential for the formation, differentiation, and growth of blood vessels are early embryonically lethal. Gene targeting experiments have substantiated the concept that there is a fundamental difference between the establishment of the primordial vascular plexus from angioblasts and the formation of new blood vessels from preexisting vessels. The induction and differentiation of angioblasts during mesoderm formation, their coalescence into primordial blood vessels, and the remodeling and growth of these vessels are the distinct steps of early vascular development. Endothelial growth factors and their receptors have been identified as the key regulators of vascular development. While signaling by fibroblast growth factors is crucial for the induction of endothelial precursors, thus initiating vascular development, further steps of vascular growth and differentiation depend on signaling by VEGF, angiopoietins, and ephrins. VEGF initiates the morphogenesis of the early embryonic vasculature by stimulating angioblasts to proliferate, to sprout, and to migrate along preformed routes to their sites of destination, where, in concert with a multitude of different signals (angiopoietins, ephrins, PDGF, TGF-β, cues from the extracellular matrix, and cell-to-cell contacts), tube formation and blood vessel maturation are accomplished. This includes recruitment of pericytes/smooth muscle cells and differentiation of organotypic endothelial phenotypes. While VEGF, an endothelial specific mitogen and survival factor, is indispensable for the expansion and maintenance of the embryonic vasculature, its role as a vascular permeability factor during development is controversial. VEGF, the basic signal of angiogenesis, is tightly regulated by tissue oxygen and metabolical factors. Hypoxia-inducible transcription factors have been identified to mediate this regulation, thus always balancing vascular growth and metabolic demands of the growing tissues. Interaction, cooperation, and transcriptional regulation of the different signals during vascular development are still far from being understood and will be future issues in the developmental biology of blood vessels.

ACKNOWLEDGMENTS

This work was supported by the Zentrum für Molekulare Medizin, Universität zu Köln, the Max Planck Gesellschaft, and in part by grants from the Deutsche Forschungsgemeinschaft (Fl 223/2-1 and Fl223/3-1).

REFERENCES

Aase, K., Lymboussaki, A., Kaipainen, A., Olofsson, B., Alitalo, K., Eriksson, U. (1999). Localization of VEGF-B in the mouse embryo suggests a paracrine role of the growth factor in the developing vasculature. Dev Dyn 215:12–25.

Achen, M., Gad, J., Stacker, S., Wilks, A. (1997). Placenta growth factor and vascular endothelial growth factor are co-expressed during early embryonic development. Growth Factors 15:69–80.

Achen, M., Jeltsch, M., Kukk, E., et al. (1998). Vascular endothelial growth factor D (VEGF-D) is a ligand for the tyrosine kinases VEGF receptor 2 (Flk1) and VEGF receptor 3 (Flt4). Proc Natl Acad Sci USA 95:548–553.

Adair, T.H., Gay, W.J., Montani, J.P. (1990). Growth regulation of the vascular system: evidence for a metabolic hypothesis. Am J Physiol 259:R393–R404.

Adams, R.H., Wilkinson, G.A., Weiss, C., et al. (1999). Roles of ephrinB ligands and EphB receptors in cardiovascular development: demarcation of arterial/venous domains, vascular morphogenesis, and sprouting angiogenesis. Genes Dev 13:295–306.

Aiello, L.P., Pierce, E.A., Foley, E.D., et al. (1995). Suppression of retinal neovascularization *in vivo* by inhibition of vascular endothelial growth factor (VEGF) using soluble VEGF-receptor chimeric proteins. Proc Natl Acad Sci USA 92:10457–10461.

Alon, T., Hemo, I., Itin, A., Pe'er, J., Stone, J., Keshet, E. (1995). Vascular endothelial growth factor acts as a survival factor for newly formed retinal vessels and has implications for retinopathy of prematurity. Nature Med 1:1024–1028.

Arany, Z., Huang, L.E., Eckner, R., et al. (1996). An essential role for p300/CBP in the cellular response to hypoxia. Proc Natl Acad Sci USA 93:12969–12973.

Asahara, T., Masuda, H., Takahashi, T., et al. (1999). Bone marrow origin of endothelial progenitor cells responsible for postnatal vasculogenesis in physiological and pathological neovascularization. Circ Res 85:221–228.

Asahara, T., Murohara, T., Sullivan, A., et al. (1997). Isolation of putative progenitor endothelial cells for angiogenesis. Science 275:964–967.

Azar, Y., Eyal-Giladi, H. (1979). Marginal zone cells—the primitive streak-inducing component of the primary hypoblast in the chick. J Embryol Exp Morphol 52:79–88.

Baird, A. (2000). Fibroblast growth factors and their receptors. In: Rubanyi, G.M., ed. *Angiogenesis in health and disease*, pp. 75–88. New York: Marcell Dekker.

Barleon, B., Sozzani, S., Zhou, D., Weich, H.A., Mantovani, A., Marme, D. (1996). Migration of human monocytes in response to vascular endothelial growth factor (VEGF) is mediated via the VEGF receptor Flt-1. Blood 87:3336–3343.

Baumhueter, S., Dybdal, N., Kyle, C., Lasky, L. (1994). Global vascular expression of murine CD34, a sialomucine-like endothelial ligand for L-selectin. Blood 84:2554–2565.

Benjamin, L., Hemo, I., Keshet, E. (1998). A plasticity window for blood vessel remodelling is defined by pericyte coverage of the preformed endothelial network and is regulated by PDGF-B and VEGF. Development 125:1591–1598.

Benjamin, L.E., Keshet, E. (1997). Conditional switching of vascular endothelial growth factor (VEGF) expression in tumors: induction of endothelial cell shedding and regression of hemangioblastoma-like vessels by VEGF withdrawal. Proc Natl Acad Sci USA 94:8761–8766.

Bielinska, M., Narita, N., Heikinheimo, M., Porter, S.B., Wilson, D.B. (1996). Erythropoiesis and vasculogenesis in embryoid bodies lacking visceral yolk sac endoderm. Blood 88:3720–3730.

Bogers, A.J.J.C., Gittenberger-de Groot, A.C., Poelmann, R.E., Péault, B.M., Huysmans, H.A. (1989). Development of the origin of the coronary arteries, a matter of ingrowth or outgrowth? Anat Embryol 180:437–441.

Breier, G., Albrecht, U., Sterrer, S., Risau, W. (1992). Expression of vascular endothelial growth factor during embryonic angiogenesis and endothelial cell differentiation. Development 114:521–532.

Breier, G., Clauss, M., Risau, W. (1995). Coordinate expression of vascular endothelial growth factor receptor-1 (Flt-1) and its ligand suggests a paracrine regulation of murine vascular development. Dev Dyn 204:228–239.

Brown, L.F., Berse, B., Tognazzi, K., et al. (1992). Vascular permeability factor mRNA and protein expression in human kidney. Kidney Int 42:1457–1461.

Burgess, W.H., Maciag, T. (1989). The heparin-binding (fibroblast) growth factor family of proteins. Annu Rev Biochem 58:575–606.

Cao, Y., Chen, H., Zhou, L., et al. (1996). Heterodimers of placenta growth factor/vascular endothelial growth factor. Endothelial activity, tumor cell expression, and high affinity binding to Flk-1/KDR. J Biol Chem 271:3154–3162.

Carmeliet, P., Collen, D. (1999). Role of vascular endothelial growth factor and vascular endothelial growth factor receptors in vascular development. Curr Top Microbiol Immunol 237:133–158.

Carmeliet, P., Ferreira, V., Breier, G., et al. (1996). Abnormal blood vessel development and lethality in embryos lacking a single VEGF allele. Nature 380:435–439.

Carmeliet, P., Lampugnani, M.G., Moons, L., et al. (1999a). Targeted deficiency or cytosolic truncation of the VE-cadherin gene in mice impairs VEGF-mediated endothelial survival and angiogenesis. Cell 98:147–157.

Carmeliet, P., Ng, Y.S., Nuyens, D., et al. (1999b). Impaired myocardial angiogenesis and ischemic cardiomyopathy in mice lacking the vascular endothelial growth factor isoforms VEGF164 and VEGF188. Nat Med 5:495–502.

Carroll, S.M., White, F.C., Bloor, C.M. (1994). Hypoxia regulates vascular endothelial growth-factor (VEGF) messenger-RNA levels by a posttranscriptional mechanism. Circulation 90:521.

Chapman, W. (1918). The effect of the heart-beat upon the development of the vascular system in the chick. Am J Anat 23:175–203.

Choi, K., Kennedy, M., Kazarov, A., Papadimitriou, J., Keller, G. (1998). A common precursor for hematopoietic and endothelial cells. Development 125:725–732.

Christ, B., Grim, M., Wilting, J., von Kirschhofer, K., Wachtler, F. (1991). Differentiation of endothelial cells in avian embryos does not depend on gastrulation. Acta Histochem 91:193–199.

Clark, D.E., Smith, S.K., Licence, D., Evans, A.L., Charnock-Jones, D.S. (1998). Comparison of expression patterns for placenta growth factor, vascular endothelial growth factor (VEGF), VEGF-B and VEGF-C in the human placenta throughout gestation. J Endocrinol 159:459–467.

Clauss, M., Gerlach, M., Gerlach, H., et al. (1990). Vascular-permeability factor—a tumor-derived polypeptide that induces endothelial-cell and monocyte procoagulant activity, and promotes monocyte migration. J Exp Med 172:1535–1545.

Clauss, M., Grell, M., Fangmann, C., Fiers, W., Scheurich, P., Risau, W. (1996a). Synergistic induction of endothelial tissue factor by tumor necrosis factor and vascular endothelial growth factor: functional analysis of the tumor necrosis factor receptors. FEBS Lett 390:334–338.

Clauss, M., Weich, H., Breier, G., et al. (1996b). The vascular endothelial growth factor receptor Flt-1 mediates biological activities. Implications for a functional role of placenta growth factor in monocyte activation and chemotaxis. J Biol Chem 271:17629–17634.

Cleaver, O., Krieg, P. (1998). VEGF mediates angioblast migration during development of the dorsal aorta in Xenopus. Development 125:3905–3914.

Cleaver, O., Tonissen, K., Saha, M., Krieg, P. (1997). Neovascularization of the Xenopus embryo. Dev Dyn 210:66–77.

Clegg, C.H., Linkhart, T.A., Olwin, B.B., Hauschka, S.D. (1987). Growth factor control of skeletal muscle differentiation: commitment to terminal differentiation occurs in G1 phase and is repressed by fibroblast growth factor. J Cell Biol 105:949–956.

Coffin, J.D., Florkiewicz, R.Z., Neumann, J., et al. (1995). Abnormal bone growth and selective translational regulation in basic fibroblast growth factor (FGF-2) transgenic mice. Mol Biol Cell 6:1861–1873.

Coffin, J.D., Poole, T.J. (1988). Embryonic vascular development: immunohistochemical identification of the origin and subsequent morphogenesis of the major vessel primordia in quail embryos. Development 102:735–748.

Coffin, J.D., Poole, T.J. (1991). Endothelial cell origin and migration in embryonic heart and cranial blood vessel development. Anat Rec 231:383–395.

Conklin, D., Gilbertson, D., Taft, D.W., et al. (1999). Identification of a mammalian angiopoietin-related protein expressed specifically in liver. Genomics 62:477–482.

Connolly, D.T., Heuvelman, D.M., Nelson, R., et al. (1989). Tumor vascular permeability factor stimulates endothelial cell growth and angiogenesis. J Clin Invest 84:1470–1478.

Deindl, E., Fernández, B., Höfer, I.E., van Royen, N., Scholz, D., Schaper, W. (2000). Arteriogenesis, collateral blood vessels, and their development. In: Rubanyi, G.M., ed. Angiogenesis in health and disease, pp. 31–46. New York: Marcell Dekker.

De Lisser, H.M., Newman, P.J., Albelda, S.M. (1994). Molecular and functional aspects of PECAM-1/CD31. Immunology Today 15:490–495.

De Ruiter, M.C., Hogers, B., Poelmann, R.E., Vanlperen, L., Gittenberger, D.G.A. (1991). The development of the vascular system in quail embryos: a combination of micro-vascular corrosion casts and immunohistochemical identification. Scanning Microsc 5: 1081–1089.

De Ruiter, M.C., Poelmann, R.E., Mentink, M.M., Vaniperen, L., Gittenberger, D.G.A. (1993). Early formation of the vascular system in quail embryos. Anat Rec 235:261–274.

Detmar, M., Brown, L., Schon, M., et al. (1998). Increased microvascular density and enhanced leukocyte rolling and adhesion in the skin of VEGF transgenic mice. J Invest Dermatol 111:1–6.

De Vries, C., Escobedo, J.A., Ueno, H., Houck, K., Ferrara, N., Williams, L.T. (1992). The fms-like tyrosine kinase, a receptor for vasculare endothelial growth factor. Science 255: 989–991.

Dickson, M., Martin, J., Cousins, F., Kulkarni, A., Karlsson, S., Akhurst, R. (1995). Defec-tive haematopoiesis and vasculogenesis in transforming growth factor-beta 1 knock out mice. Development 121:1845–1854.

Dieterlen, L.F., Pardanaud, L., Yassine, F., Cormier, F. (1988). Early haemopoietic stem cells in the avian embryo. J Cell Sci Suppl 10:29–44.

Drake, C.J., Little, C.D. (1995). Exogenous vascular endothelial growth factor induces mal-formed and hyperfused vessels during embryonic neovascularization. Proc Natl Acad Sci USA 92:7657–7661.

Dumont, D.J., Fong, G.H., Puri, M.C., Gradwohl, G., Alitalo, K., Breitman, M.L. (1995). Vascularization of the mouse embryo—a study of Flk-1, Tek, Tie, and vascular endothe-lial growth factor expression during development. Dev Dyn 203:80–92.

Dumont, D., Gradwohl, G., Fong, G., et al. (1994). Dominant-negative and targeted null mutations in the endothelial receptor tyrosine kinase, tek, reveal a critical role in vascu-logenesis of the embryo. Genes Dev 8:1897–1909.

Dumont, D.J., Jussila, L., Taipale, J., et al. (1998). Cardiovascular failure in mouse embryos deficient in VEGF receptor-3. Science 282:946–949.

Ebert, B., Bunn, H. (1998). Regulation of transcription by hypoxia requires a multiprotein complex that includes hypoxia-inducible factor 1, an adjacent transcription factor, and p300/CREB binding protein. Mol Cell Biol 18:4089–4096.

Eichmann, A., Corbel, C., Jaffredo, T., et al. (1998). Avian VEGF-C: cloning, embryonic expression pattern and stimulation of the differentiation of VEGFR2-expressing endothelial cell precursors. Development 125:743–752.

Eichmann, A., Corbel, C., Nataf, V., Vaigot, P., Breant, C., Le Douarin, N. (1997). Ligand-dependent development of the endothelial and hemopoietic lineages from embryonic mesodermal cells expressing vascular endothelial growth factor receptor 2. Proc Natl Acad Sci USA 94:5141–5146.

Eichmann, A., Marcelle, C., Breant, C., Le Douarin, N.M. (1993). Two molecules related to the VEGF receptor are expressed in early endothelial cells during avian embryonic development. Mech Dev 42:33–48.

Ema, M., Hirota, K., Mimura, J., et al. (1999). Molecular mechanisms of transcription acti-vation by HLF and HIF-1alpha in response to hypoxia: their stabilization and redox signal-induced interaction with CBP/p300. EMBO J 18:1905–1914.

Ema, M., Taya, S., Yokotani, N., Sogawa, K., Matsuda, Y., Fujii-Kuriyama, Y. (1997). A novel bHLH-PAS factor with close sequence similarity to hypoxia-inducible factor-1 alpha regulates the VEGF expression and is potentially involved in lung and vascular development. Proc Natl Acad Sci USA 94:4273–4278.

Esser, S., Wolburg, K., Wolburg, H., Breier, G., Kurzchalia, T., Risau, W. (1998). Vascular endothelial growth factor induces endothelial fenestrations *in vitro*. J Cell Biol 140: 947–959.

Evans, H.M. (1909). On the development of the aorta, cardinal and umbilical veins and the other blood vessels of vertebrate embryos from capillaries. Anat Rec 3:498–519.

Farnebo, F., Piehl, F., Lagercrantz, J. (1999). Restricted expression pattern of vegf-d in the adult and fetal mouse: high expression in the embryonic lung. Biochem Biophys Res Commun 257:891–894.

Feinberg, R.N., Noden, D.M. (1991). Experimental analysis of blood vessel development in the avian wing bud. Anat Rec 231:136–144.

Feng, D., Nagy, J.A., Pyne, K., Hammel, I., Dvorak, H.F., Dvorak, A.M. (1999). Pathways of macromolecular extravasation across microvascular endothelium in response to VPF/VEGF and other vasoactive mediators. Microcirculation 6:23–44.

Ferrara, N. (2000). The role of vascular endothelial growth factor in angiogenesis. In: Rubanyi, G.M., ed. *Angiogenesis in health and disease*, pp. 47–74. New York: Marcell Dekker.

Ferrara, N., Carvermoore, K., Chen, H., et al. (1996). Heterozygous embryonic lethality induced by targeted inactivation of the VEGF Gene. Nature 380:439–442.

Ferrara, N., Chen, H., Davis-Smyth, T., et al. (1998). Vascular endothelial growth factor is essential for corpus luteum angiogenesis. Nat Med 4:336–340.

Flamme, I. (1989). Is extraembryonic angiogenesis in the chick embryo controlled by the endoderm? A morphology study. Anat Embryol Berl 180:259–272.

Flamme, I., Breier, G., Risau, W. (1995a). Vascular endothelial growth factor (VEGF) and VEGF receptor 2 (flk-1) are expressed during vasculogenesis and vascular differentiation in the quail embryo. Dev Biol 169:699–712.

Flamme, I., Fröhlich, T., Risau, W. (1997a). Molecular mechanisms of vasculogenesis and embryonic angiogenesis. J Cell Physiol 173:206–210.

Flamme, I., Fröhlich, T., von Reutern, M., Kappel, A., Damert, A., Risau, W. (1997b). HRF, a putative basic helix-loop-helix-PAS-domain transcription factor is closely related to Hypoxia-inducible factor-1α and developmentally expressed in blood vessels. Mech Dev 63:51–60.

Flamme, I., Krieg, M., Plate, K.H. (1998). Up-regulation of vascular endothelial growth factor in stromal cells of hemangioblastomas is correlated with up-regulation of the transcription factor HRF/HIF-2alpha. Am J Pathol 153:25–29.

Flamme, I., Risau, W. (1992). Induction of vasculogenesis and hematopoiesis *in vitro*. Development 116:435–439.

Flamme, I., von Reutern, M., Drexler, H.C.A., Syed-Ali, S., Risau, W. (1995b). Over-expression of vascular endothelial growth factor in the avian embryo induces hyper-vascularization and increased vascular permeability without alterations of embryonic pattern formation. Dev Biol 171:399–414.

Folkman, J., Shing, Y., (1992). Angiogenesis. J Biol Chem 267:10931–10934.

Fong, G.H., Klingensmith, J., Wood, C.R., Rossant, J., Breitman, M.L., (1996). Regulation of flt-1 expression during mouse embryogenesis suggests a role in the establishment of vascular endothelium. Dev Dyn 207:1–10.

Fong, G.H., Rossant, J., Gertsenstein, M., Breitman, M.L. (1995). Role of the flt-1 receptor tyrosine kinase in regulating the assembly of vascular endothelium. Nature 376:66–70.

Forsythe, J.A., Jiang, B.H., Iyer, N.V., et al. (1996). Activation of vascular endothelial growth factor gene transcription by hypoxia-inducible factor 1. Mol Cell Biol 16: 4604–4613.

Fouquet, B., Weinstein, B., Serluca, F., Fishman, M. (1997). Vessel patterning in the embryo of the zebrafish: guidance by notochord. Dev Biol 183:37–48.

Friesel, R.E., Maciag, T. (1995). Molecular mechanisms of angiogenesis: fibroblast growth factor signal transduction. FASEB J 9:919–925.

Fujio, Y., Walsh, K. (1999). Akt mediates cytoprotection of endothelial cells by vascular endothelial growth factor in an anchorage-dependent manner. J Biol Chem 274:16349–16354.

Fulgham, D.L., Widhalm, S.R., Martin, S., Coffin, J.D. (1999). FGF-2 dependent angio-genesis is a latent phenotype in basic fibroblast growth factor transgenic mice. Endothelium 6:185–195.

Gerber, H.P., Condorelli, F., Park, J., Ferrara, N. (1997). Differential transcriptional regulation of the two vascular endothelial growth factor receptor genes. Flt-1, but not Flk-1/KDR, is up-regulated by hypoxia. J Biol Chem 272:23659–23667.

Gerber, H.P., Dixit, V., Ferrara, N. (1998). Vascular endothelial growth factor induces expression of the antiapoptotic proteins Bcl-2 and A1 in vascular endothelial cells. J Biol Chem 273:13313–13316.

Gerber, H.P., Hillan, K.J., Ryan, A.M., et al. (1999a). VEGF is required for growth and survival in neonatal mice. Development 126:1149–1159.

Gerber, H.P., Vu, T.H., Ryan, A.M., Kowalski, J., Werb, Z., Ferrara, N. (1999b). VEGF couples hypertrophic cartilage remodeling, ossification and angiogenesis during endochondral bone formation. Nat Med 5:623–628.

Gering, M., Rodaway, A.R., Gottgens, B., Patient, R.K., Green, A.R. (1998). The SCL gene specifies haemangioblast development from early mesoderm. EMBO J 17:4029–4045.

Gnarra, J.R., Ward, J.M., Porter, F.D., et al. (1997). Defective placental vasculogenesis causes embryonic lethality in VHL-deficient mice. Proc Natl Acad Sci USA 94:9102–9107.

Gonzalez Crussi, F. (1971). Vasculogenesis in the chick embryo. An ultrastructural study. Am J Anat 130:441–460.

Grosios, K., Leek, J.P., Markham, A.F., Yancopoulos, G.D., Jones, P.F. (1999). Assignment of ANGPT4, ANGPT1, and ANGPT2 encoding angiopoietins 4, 1 and 2 to human chromosome bands 20p13, 8q22.3->q23 and 8p23.1, respectively, by *in situ* hybridization and radiation hybrid mapping. Cytogenet Cell Genet 84:118–120.

Haar, J.L., Ackerman, G. (1971). A phase and electron microscopic study of vasculogenesis and erythropoiesis in the yolk sac of the mouse. Anat Rec 170:199–224.

Henkemeyer, M., Rossi, D.J., Holmyard, D.P., et al. (1995). Vascular system defects and neuronal apoptosis in mice lacking ras GTPase-activating protein. Nature 377:695–701.

Hidaka, M., Stanford, W.L., Bernstein, A. (1999). Conditional requirement for the Flk-1 receptor in the *in vitro* generation of early hematopoietic cells. Proc Natl Acad Sci USA 96:7370–7375.

Hirakow, R., Hiruma, T. (1981). Scanning electron microscopic study on the development of primitive blood vessels in chick embryos at the early somite-stage. Anat Embryol Berl 163:299–306.

Hirakow, R., Hiruma, T. (1983). TEM-studies on development and canalization of the dorsal aorta in the chick embryo. Anat Embryol Berl 166:307–315.

Hirashima, M., Kataoka, H., Nishikawa, S., Matsuyoshi, N., Nishikawa, S. (1999). Maturation of embryonic stem cells into endothelial cells in an *in vitro* model of vasculogenesis. Blood 93:1253–1263.

Hiratsuka, S., Minowa, O., Kuno, J., Noda, T., Shibuya, M. (1998). Flt-1 lacking the tyrosine kinase domain is sufficient for normal development and angiogenesis in mice. Proc Natl Acad Sci USA 95:9349–9354.

His, W. (1900). Lecithoblast und Angioblast der Wirbelthiere. Abhandl K S Ges Wiss Math-Phys 22:171–328.

Hogenesch, J.B., Chan, W.K., Jackiw, V.H., et al. (1997). Characterization of a subset of the basic-helix-loop-helix-PAS superfamily that interacts with components of the dioxin signaling pathway. J Biol Chem 272:8581–8593.

Holder, N., Klein, R. (1999). Eph receptors and ephrins: effectors of morphogenesis. Development 126:2033–2044.

Holland, S., Gale, N., Mbamalu, G., Yancopoulos, G., Henkemeyer, M., Pawson, T. (1996). Bidirectional signalling through the EPH-family receptor Nuk and its transmembrane ligands. Nature 383:722–725.

Hoper, J., Jahn, H. (1995). Influence of environmental oxygen concentration on growth and vascular density of the area vasculosa in chick embryos. Int J Microcirc Clin Exp 15:186–192.

Houck, K.A., Ferrara, N., Winer, J., Cachianes, G., Li, B., Leung, D.W. (1991). The vascular endothelial growth factor family: identification of a fourth molecular species and characterization of alternative splicing of RNA. Mol Endocrinol 5:1806–1814.

Huang, L., Gu, J., Schau, M., Bunn, H. (1998). Regulation of hypoxia-inducible factor 1alpha is mediated by an O_2-dependent degradation domain via the ubiquitin-proteasome pathway. Proc Natl Acad Sci USA 95:7987–7992.

Hughes, A. (1935). Studies on the area vasculosa of the embryo chick. I. The first differentiation of the vitellina arteria. J Anat 70:76–122.

Husmann, I., Soulet, L., Gautron, J., Martelly, I., Barritault, D. (1996). Growth factors in skeletal muscle regeneration. Cytokine Growth Factor Rev 7:249–258.

Ikeda, E., Achen, M.G., Breier, G., Risau, W. (1995). Hypoxia-induced transcriptional activation and increased mRNA stability of vascular endothelial growth factor in C6 glioma cells. J Biol Chem 270:19761–19766.

Isaac, D.D., Andrew, D.J. (1996). Tubulogenesis in drosophila—a requirement for the trachealess gene product. Genes Dev 10:103–117.

Isner, J.M., Asahara, T. (2000). Therapeutic angiogenesis. In: Rubanyi, G.M., ed. *Angiogenesis in health and disease*, pp. 489–518. New York: Marcell Dekker.

Iyer, N.V., Kotch, L.E., Agani, F., et al. (1998). Cellular and developmental control of O_2 homeostasis by hypoxia-inducible factor 1 alpha. Genes Dev 12:149–162.

Jaffredo, T., Gautier, R., Eichmann, A., Dieterlen-Lievre, F. (1998). Intraaortic hemopoietic cells are derived from endothelial cells during ontogeny. Development 125:4575–4583.

Jakeman, L.B., Winer, J., Bennett, G.L., Altar, C.A., Ferrara, N. (1992). Binding sites for vascular endothelial growth factor are localized on endothelial cells in adult rat tissues. J Clin Invest 89:244–253.

Jeltsch, M., Kaipainen, A., Joukov, V., et al. (1997). Hyperplasia of lymphatic vessels in VEGF-C transgenic mice. Science 276:1423–1425.

Joukov, V., Pajusola, K., Kaipainen, A., et al. (1996). A novel vascular endothelial growth factor, VEGF-C, is a ligand for the Flt4 (VEGFR-3) and KDR (VEGFR-2) receptor tyrosine kinases. EMBO J 15:290–298.

Kaipainen, A., Korhonen, J., Mustonen, T., et al. (1995). Expression of the fms-like tyrosine kinase 4 gene becomes restricted to lymphatic endothelium during development. Proc Natl Acad Sci USA 92:3566–3570.

Kallio, P.J., Wilson, W.J., O'Brien, S., Makino, Y., Poellinger, L. (1999). Regulation of the hypoxia-inducible transcription factor 1alpha by the ubiquitin-proteasome pathway. J Biol Chem 274:6519–6525.

Kappel, A., Ronicke, V., Damert, A., Flamme, I., Risau, W., Breier, G. (1999). Identification of vascular endothelial growth factor (VEGF) receptor-2 (Flk-1) promoter/enhancer sequences sufficient for angioblast and endothelial cell-specific transcription in transgenic mice. Blood 93:4284–4292.

Kawasaki, T., Kitsukawa, T., Bekku, Y., et al. (1999). A requirement for neuropilin-1 in embryonic vessel formation. Development 126:4895–4902.

Kendall, R.L., Wang, G., Thomas, K.A. (1996). Identification of a natural soluble form of the vascular endothelial growth factor receptor, Flt-1, and its heterodimerization with KDR. Biochem Biophys Res Commun 226:324–328.

Kenyon, B.M., Voest, E.E., Chen, C.C., Flynn, E., Folkman, J., D'Amato, R.J. (1996). A model of angiogenesis in the mouse cornea. Invest Ophthalmol Vis Sci 37:1625–1632.

Kessel, J., Fabian, B. (1985). Graded morphogenetic patterns during the development of the extraembryonic blood system and coelom of the chick blastoderm: a scanning electron microscope and light microscope study. Am J Anat 173:99–112.

Kitsukawa, T., Shimono, A., Kawakami, A., Kondoh, H., Fujisawa, H. (1995). Over-expression of a membrane protein, neuropilin, in chimeric mice causes anomalies in the cardiovascular system, nervous system and limbs. Development 121:4309–4318.

Koblizek, T., Weiss, C., Yancopoulos, G., Deutsch, U., Risau, W. (1998). Angiopoietin-1 induces sprouting angiogenesis in vitro. Curr Biol 8:529–532.

Kotch, L.E., Iyer, N.V., Laughner, E., Semenza, G.L. (1999). Defective vascularization of HIF-1alpha-null embryos is not associated with VEGF deficiency but with mesenchymal cell death. Dev Biol 209:254–267.

Kozak, K.R., Abbott, B., Hankinson, O. (1997). ARNT-deficient mice and placental differentiation. Dev Biol 191:297–305.

Krah, K., Mironow, V., Risau, W., Flamme, I. (1994). Induction of vasculogenesis in quail blastodisc derived embryoid bodies. Dev Biol 164:123–132.

Kremer, C., Breier, G., Risau, W., Plate, K.H. (1997). Up-regulation of flk-1/vascular endothelial growth factor receptor 2 by its ligand in a cerebral slice culture system. Cancer Res 57:3852–3859.

Kukk, E., Lymboussaki, A., Taira, S., et al. (1996). VEGF-C receptor binding and pattern of expression with VEGFR-3 suggests a role in lymphatic vascular development. Development 122:3829–3837.

Kurz, H., Gartner, T., Eggli, P., Christ, B. (1996). First blood vessels in the avian neural tube are formed by a combination of dorsal angioblast immigration and ventral sprouting of endothelial cells. Dev Biol Vol 176:133–147.

Labastie, M.C., Poole, T.J., Peault, B.M., Le Douarin, N.M. (1986). MB1, a quail leukocyte-endothelium antigen: partial characterization of the cell surface and secreted forms in cultured endothelial cells. Proc Natl Acad Sci USA 83:9016–9020.

Lamoreaux, W.J., Fitzgerald, M.E., Reiner, A., Hasty, K.A., Charles, S.T. (1998). Vascular endothelial growth factor increases release of gelatinase A and decreases release of tissue inhibitor of metalloproteinases by microvascular endothelial cells in vitro. Microvasc Res 55:29–42.

Lanot, R. (1980). Formation of the early vascular network in chick embryo: microscopical aspects. Arch Biol Liege 91:423–438.

Larcher, F., Murillas, R., Bolontrade, M., Conti, C., Jorcano, J. (1998). VEGF/VPF over-expression in skin of transgenic mice induces angiogenesis, vascular hyperpermeability and accelerated tumor development. Oncogene 17:303–311.

Leung, D.W., Cachianes, G., Kuang, W.J., Goeddel, D.V., Ferrara, N. (1989). Vascular endothelial growth factor is a secreted angiogenic mitogen. Science 246:1306–1309.

Levy, A.P., Levy, N.S., Wegner, S., Goldberg, M.A. (1995). Transcriptional regulation of the rat vascular endothelial growth factor gene by hypoxia. J Biol Chem 270:13333–13340.

Liao, E., Paw, B., Oates, A., Pratt, S., Postlethwait, J., Zon, L. (1998). SCL/Tal-1 transcription factor acts downstream of cloche to specify hematopoietic and vascular progenitors in zebrafish. Genes Dev 12:621–626.

Liao, W., Bisgrove, B., Sawyer, H., et al. (1997). The zebrafish gene cloche acts upstream of a flk-1 homologue to regulate endothelial cell differentiation. Development 124:381–389.

Lindahl, P., Johansson, B., Leveen, P., Betsholtz, C. (1997). Pericyte loss and microaneurysm formation in PDGF-B-deficient mice. Science 277:242–245.

Lu, M., Perez, V.L., Ma, N., et al. (1999). VEGF increases retinal vascular ICAM-1 expression in vivo. Invest Ophthalmol Vis Sci 40:1808–1812.

Maglione, D., Guerriero, V., Viglietto, G., Delli Bovi, P., Persico, M.G. (1991). Isolation of a human placenta cDNA coding for a protein related to the vascular permeability factor. Proc Natl Acad Sci USA 88:9267–9271.

Maisonpierre, P., Suri, C., Jones, P., et al. (1997). Angiopoietin-2, a natural antagonist for Tie2 that disrupts in vivo angiogenesis. Science 277:55–60.

Makinen, T., Olofsson, B., Karpanen, T., et al. (1999). Differential binding of vascular endothelial growth factor B splice and proteolytic isoforms to neuropilin-1. J Biol Chem 274:21217–21222.

Maltepe, E., Schmidt, J.V., Baunoch, D., Bradfield, C.A., Simon, M.C. (1997). Abnormal angiogenesis and responses to glucose and oxygen deprivation in mice lacking the protein ARNT. Nature 386:403–407.

Manning, G., Krasnow, M. (1993). Development of the Drosophila tracheal system. In: *The Development of Drosophila melanogaster*, pp. 609–685. Cold Spring Harbor NY: Cold Spring Harbor Laboratory Press.

Marti, H.H., Risau, W. (1998). Systemic hypoxia changes the organ-specific distribution of vascular endothelial growth factor and its receptors. Proc Natl Acad Sci USA 95: 15809–15814.

Mato, M., Aikawa, E., Kishi, K. (1964). Some observations on interstice between mesoderm and endoderm in the area vasculosa of chick blastoderm. Exp Cell Res 35:426–428.

Maxwell, P.H., Wiesener, M.S., Chang, G.W., et al. (1999). The tumour suppressor protein VHL targets hypoxia-inducible factors for oxygen-dependent proteolysis. Nature 399:271–275.

McGann, J., Silver, L., Liesveld, J., Palis, J. (1997). Erythropoietin-receptor expression and function during the initiation of murine yolk sac erythropoiesis. Exp Hematol 25:1149–1157.

Medvinsky, A., Dzierzak, E. (1996). Definitive hematopoiesis is autonomously initiated by the AGM region. Cell 86:897–906.

Meyer, M., Clauss, M., Lepple-Wienhues, A., et al. (1999). A novel vascular endothelial growth factor encoded by Orf virus, VEGF-E, mediates angiogenesis via signalling through VEGFR-2 (KDR) but not VEGFR-1 (Flt-1) receptor tyrosine kinases. EMBO J 18:363–374.

Miao, H.Q., Soker, S., Feiner, L., Alonso, J.L., Raper, J.A., Klagsbrun, M. (1999). Neuropilin-1 mediates collapsin-1/semaphorin III inhibition of endothelial cell motility: functional competition of collapsin-1 and vascular endothelial growth factor-165. J Cell Biol 146:233–242.

Michel, C.C., Neal, C.R. (1999). Openings through endothelial cells associated with increased microvascular permeability. Microcirculation 6:45–54.

Migdal, M., Huppertz, B., Tessler, S., et al. (1998). Neuropilin-1 is a placenta growth factor-2 receptor. J Biol Chem 273:22272–22278.

Mikawa, T., Fischman, D.A. (1992). Retroviral analysis of cardiac morphogenesis: discontinuous formation of coronary vessels. Proc Natl Acad Sci USA 89:9504–9508.

Millauer, B., Wizigmann Voos, S., Schnurch, H., et al. (1993). High affinity VEGF binding and developmental expression suggest Flk-1 as a major regulator of vasculogenesis and angiogenesis. Cell 72:835–846.

Miquerol, L., Gertsenstein, M., Harpal, K., Rossant, J., Nagy, A. (1999). Multiple developmental roles of VEGF suggested by a LacZ-tagged allele. Dev Biol 212:307–322.

Miura, Y., Wilt, F.H. (1969). Tissue interaction and the formation of the first erythroblasts of the chick embryo. Dev Biol 19:201–211.

Monacci, W.T., Merrill, M.J., Oldfield, E.H. (1993). Expression of vascular permeability factor/vascular endothelial growth factor in normal rat tissues. Am J Physiol 264: C995–1002.

Munoz-Chapuli, R., Perez-Pomares, J.M., Macias, D., Garcia-Garrido, L., Carmona, R., Gonzalez, M. (1999). Differentiation of hemangioblasts from embryonic mesothelial cells? A model on the origin of the vertebrate cardiovascular system. Differentiation 64: 133–141.

Murphy, M.E., Carlson, E.C. (1978). An ultrastructural study of developing extracellular matrix in vitelline blood vessels of the early chick embryo. Am J Anat 151:345–375.

Murray, P.D.F. (1932). The development *in vitro* of the blood of the early chick embryo. R Soc London (Ser B III).

Nehls, V., Drenckhahn, D. (1995). A novel, microcarrier-based *in vitro* assay for rapid and reliable quantification of three-dimensional cell migration and angiogenesis. Microvasc Res 50:311–322.

Nishikawa, S., Nishikawa, S., Hirashima, M., Matsuyoshi, N., Kodama, H. (1998). Progressive lineage analysis by cell sorting and culture identifies FLK1+VE-cadherin+ cells at a diverging point of endothelial and hemopoietic lineages. Development 125:1747–1757.

Noden, D.M. (1989). Embryonic origins and assembly of blood vessels. Am Rev Respir Dis 140:1097–1103.

Nor, J.E., Christensen, J., Mooney, D.J., Polverini, P.J. (1999). Vascular endothelial growth factor (VEGF)-mediated angiogenesis is associated with enhanced endothelial cell survival and induction of Bcl- 2 expression. Am J Pathol 154:375–384.

North, T., Gu, T.L., Stacy, T., et al. (1999). Cbfa2 is required for the formation of intra-aortic hematopoietic clusters. Development 126:2563–2575.

Ogawa, S., Oku, A., Sawano, A., Yamaguchi, S., Yazaki, Y., Shibuya, M. (1998). A novel type of vascular endothelial growth factor, VEGF-E (NZ-7 VEGF), preferentially utilizes KDR/Flk-1 receptor and carries a potent mitotic activity without heparin-binding domain. J Biol Chem 273:31273–31282.

Ogunshola, O.O., Stewart, W.B., Mihalcik, V., Solli, T., Madri, J.A., Ment, L.R. (2000). Neuronal VEGF expression correlates with angiogenesis in postnatal developing rat brain. Brain Res Dev Brain Res 119:139–153.

Oh, H., Takagi, H., Suzuma, K., Otani, A., Matsumura, M., Honda, Y. (1999). Hypoxia and vascular endothelial growth factor selectively up-regulate angiopoietin-2 in bovine microvascular endothelial cells. J Biol Chem 274:15732–15739.

Oike, Y., Takakura, N., Hata, A., et al. (1999). Mice homozygous for a truncated form of CREB-binding protein exhibit defects in hematopoiesis and vasculo-angiogenesis. Blood 93:2771–2779.

Okada, H., Watanabe, T., Niki, M., et al. (1998). AML1(–/–) embryos do not express certain hematopoiesis-related gene transcripts including those of the PU.1 gene. Oncogene 17: 2287–2293.

Okamoto, N., Tobe, T., Hackett, S.F., et al. (1997). Transgenic mice with increased expression of vascular endothelial growth factor in the retina: a new model of intraretinal and subretinal neovascularization. Am J Pathol 151:281–291.

Olah, I., Medgyes, J., Glick, B. (1988). Origin of aortic cell clusters in the chicken embryo. Anat Rec 222:60–68.

Olofsson, B., Korpelainen, E., Pepper, M.S., et al. (1998). Vascular endothelial growth factor B (VEGF-B) binds to VEGF receptor-1 and regulates plasminogen activator activity in endothelial cells. Proc Natl Acad Sci USA 95:11709–11714.

Pandey, A., Shao, H., Marks, R., Polverini, P., Dixit, V. (1995). Role of B61, the ligand for the Eck receptor tyrosine kinase, in TNF-alpha-induced angiogenesis. Science 268: 567–569.

Pardanaud, L., Altmann, C., Kitos, P., Dieterlen, L.F., Buck, C.A. (1987). Vasculogenesis in the early quail blastodisc as studied with a monoclonal antibody recognizing endothelial cells. Development 100:339–349.

Pardanaud, L., Dieterlen-Lievre, F. (1999). Manipulation of the angiopoietic/hemangiopoietic commitment in the avian embryo. Development 126:617–627.

Pardanaud, L., Luton, D., Prigent, M., Bourcheix, L., Catala, M., Dieterlen-Lièvre, F. (1996). Two distinct endothelial lineages in ontogeny, one of them related to hemopoiesis. Development 122:1363–1371.

Pardanaud, L., Yassine, F., Dieterlen, L.F. (1989). Relationship between vasculogenesis, angiogenesis and haemopoiesis during avian ontogeny. Development 105:473–485.

Parker, L., Stainier, D.Y. (1999). Cell-autonomous and non-autonomous requirements for the zebrafish gene cloche in hematopoiesis. Development 126:2643–2651.

Patstone, G., Pasquale, E.B., Maher, P. (1993). Different members of the fibroblast growth factor receptor family are specific to distinct cell types in the developing chicken embryo. Dev Biol 155:107–123.

Peault, B.M., Thiery, J.P., Le Douarin, N.M. (1983). Surface marker for hemopoietic and endothelial cell lineages in quail that is defined by a monoclonal antibody. Proc Natl Acad Sci USA 80:2976–2980.

Peters, K.G., Werner, S., Chen, G., Williams, L.T. (1992). Two FGF receptor genes are differentially expressed in epithelial and mesenchymal tissues during limb formation and organogenesis in the mouse. Development 114:233–243.

Petrova, T.V., Makinen, T., Alitalo, K. (1999). Signaling via vascular endothelial growth factor receptors. Exp Cell Res 253:117–130.

Phillips, G.D., Stone, A.M., Jones, B.D., Schultz, J.C., Whitehead, R.A., Knighton, D.R. (1994). Vascular endothelial growth factor (rhVEGF165) stimulates direct angiogenesis in the rabbit cornea. In Vivo 8:961–965.

Plate, K., Breier, G., Millauer, B., Ullrich, A., Risau, W. (1993). Up-regulation of vascular endothelial growth factor and its cognate receptors in a rat glioma model of tumor angiogenesis. Cancer Res 53:5822–5827.

Plate, K.H., Breier, G., Weich, H.A., Mennel, H.D., Risau, W. (1994). Vascular endothelial growth factor and glioma angiogenesis—coordinate induction of VEGF receptors, distribution of VEGF protein and possible in vivo regulatory mechanisms. Int J Cancer 59:520–529.

Plate, K.H., Breier, G., Weich, H.A., Risau, W. (1992). Vascular endothelial growth factor is a potential tumour angiogenesis factor in human gliomas in vivo. Nature 359:845–848.

Poole, T.J., Coffin, J.D. (1989). Vasculogenesis and angiogenesis: two distinct morphogenetic mechanisms establish embryonic vascular pattern. J Exp Zool 251:224–231.

Porcher, C., Swat, W., Rockwell, K., Fujiwara, Y., Alt, F.W., Orkin, S.H. (1996). The T cell leukemia oncoprotein SCL/tal-1 is essential for development of all hematopoietic lineages. Cell 86:47–57.

Quinn, T.P., Peters, K.G., De, V.C., Ferrara, N., Williams, L.T. (1993). Fetal liver kinase 1 is a receptor for vascular endothelial growth factor and is selectively expressed in vascular endothelium. Proc Natl Acad Sci USA 90:7533–7537.

Rich, I. (1995). Primordial germ cells are capable of producing cells of the hematopoietic system in vitro. Blood 15:463–472.

Riley, B.B., Savage, M.P., Simandl, B.K., Olwin, B.B., Fallon, J.F. (1993). Retroviral expression of FGF-2 (bFGF) affects patterning in chick limb bud. Development 118:95–104.

Risau, W. (1997). Mechanisms of angiogenesis. Nature 386:671–674.

Risau, W., Flamme, I. (1995). Vasculogenesis. Annu Rev Cell Dev Biol 11:73–91.

Roberts, W.G., Palade, G.E. (1995). Increased microvascular permeability and endothelial fenestration induced by vascular endothelial growth factor. J Cell Sci 108:2369–2379.

Robertson, P.L., Du Bois, M., Bowman, P.D., Goldstein, G.W. (1985). Angiogenesis in developing rat brain: an in vivo and in vitro study. Brain Res 355:219–223.

Ryan, H.E., Lo, J., Johnson, R.S. (1998). HIF-1 alpha is required for solid tumor formation and embryonic vascularization. Embo J 17:3005–3015.

Sabin, F.R. (1917). Origin and development of the primitive vessels of the chick and of the pig. Carnergie Contrib Embryol 6:61–124.

Sabin, F.R. (1920). Preliminary note on the differentiation of angioblasts and the method by which they produce blood-vessels, blood plasma and red blood-cells as seen in the living chick. Anat Rec 13:199–204.

Sato, A., Iwama, A., Takakura, N., Nishio, H., Yancopoulos, G., Suda, T. (1998). Characterization of TEK receptor tyrosine kinase and its ligands, angiopoietins, in human hematopoietic progenitor cells. Int Immunol 10:1217–1227.

Sato, T.N., Tozawa, Y., Deutsch, U., et al. (1995). Distinct roles of the receptor tyrosine kinases Tie-1 and Tie-2 in blood vessel formation. Nature 376:70–74.

Schmidt, M., Flamme, I. (1998). The *in vivo* activity of vascular endothelial growth factor isoforms in the avian embryo. Growth Factors 15:183–197.

Schwartz, S.M., Benditt, E.P. (1977). Aortic endothelial cell replication.I. Effects of age and hypertension in the rat. Circ Res 41:248–255.

Shalaby, F., Rossant, J., Yamaguchi, T.P., et al. (1995). Failure of blood-island formation and vasculogenesis in Flk-1-deficient mice. Nature 376:62–66.

Shivdasani, R., Mayer, E., Orkin, S. (1995). Absence of blood formation in mice lacking the T-cell leukaemia oncoprotein tal-1/SCL. Nature 373:432–434.

Shweiki, D., Itin, A., Soffer, D., Keshet, E. (1992). Vascular endothelial growth factor induced by hypoxia may mediate hypoxia-initiated angiogenesis. Nature 359:843–845.

Shweiki, D., Neeman, M., Itin, A., Keshet, E. (1995). Induction of vascular endothelial growth factor expression by hypoxia and by glucose deficiency in multicell spheroids— implications for tumor angiogenesis. Proc Natl Acad Sci USA 92:768–772.

Slack, J.M., Darlington, B.G., Heath, J.K., Godsave, S.F. (1987). Mesoderm induction in early Xenopus embryos by heparin-binding growth factors. Nature 326:197–200.

Soker, S., Takashima, S., Miao, H.Q., Neufeld, G., Klagsbrun, M. (1998). Neuropilin-1 is expressed by endothelial and tumor cells as an isoform-specific receptor for vascular endothelial growth factor. Cell 92:735–745.

Stein, E., Lane, A., Cerretti, D., et al. (1998). Eph receptors discriminate specific ligand oligomers to determine alternative signaling complexes, attachment, and assembly responses. Genes Dev 12:667–678.

Stolen, C.M., Jackson, M.W., Griep, A.E. (1997). Overexpression of FGF-2 modulates fiber cell differentiation and survival in the mouse lens. Development 124:4009–4017.

Stone, J., Itin, A., Alon, T., et al. (1995). Development of retinal vasculature is mediated by hypoxia-induced vascular endothelial growth factor (VEGF) expression by neuroglia. J Neurosci 15:4738–4747.

Stratmann, A., Risau, W., Plate, K.H. (1998). Cell type-specific expression of angiopoietin-1 and angiopoietin-2 suggests a role in glioblastoma angiogenesis. Am J Pathol 153:1459–1466.

Sumoy, L., Keasey, J., Dittman, T., Kimelman, D. (1997). A role for notochord in axial vascular development revealed by analysis of phenotype and the expression of VEGR-2 in zebrafish flh and ntl mutant embryos. Mech Dev 63:15–27.

Suri, C., Jones, P.F., Patan, S., et al. (1996). Requisite role of angiopoietin-1, a ligand for the tie2 receptor, during early embryonic angiogenesis. Cell 87:1171–1180.

Suri, C., McClain, J., Thurston, G., et al. (1998). Increased vascularization in mice overexpressing angiopoietin-1. Science 282:468–471.

Takeshita, S., Tsurumi, Y., Couffinahl, T., et al. (1996). Gene transfer of naked DNA encoding for three isoforms of vascular endothelial growth factor stimulates collateral development *in vivo*. Lab Invest 75:487–501.

Takeshita, S., Zheng, L.P., Brogi, E., et al. (1994). Therapeutic angiogenesis. A single intra-arterial bolus of vascular endothelial growth factor augments revascularization in a rabbit ischemic hind limb model. J Clin Invest 93:662–670.

Tavian, M., Coulombel, L., Luton, D., San Clemente, H., Dieterlen-Lièvre, F., Péault, B. (1996). Aorta-associated CD34+ hematopoietic cells in the early human embryo. Blood 87:67–72.

Terman, B.I., Dougher Vermazen, M., Carrion, M.E., et al. (1992). Identification of the KDR tyrosine kinase as a receptor for vascular endothelial cell growth factor. Biochem Biophys Res Commun 187:1579–1586.

Terman, B.I., Khandke, L., Doughervermazan, M., et al. (1994). VEGF receptor subtypes kdr and flt1 show different sensitivities to heparin and placenta growth factor. Growth Factors 11:187–195.

Thompson, J., Slack, J.M.W. (1992). Over-expression of fibroblast growth factors in Xenopus embryos. Mech Dev 38:175–182.

Thurston, G., Suri, C., Smith, K., et al. (1999). Leakage-resistant blood vessels in mice transgenically overexpressing angiopoietin-1. Science 286:2511–2514.

Tian, H., Hammer, R.E., Matsumoto, A.M., Russell, D.W., McKnight, S.L. (1998). The hypoxia-responsive transcription factor EPAS1 is essential for catecholamine homeostasis and protection against heart failure during embryonic development. Genes Dev 12:3320–3324.

Tian, H., McKnight, S.L., Russell, D.W. (1997). Endothelial PAS domain protein 1 (EPAS1), a transcription factor selectively expressed in endothelial cells. Genes Dev 11:72–82.

Tischer, E., Gospodarowicz, D., Mitchell, R., et al. (1989). Vascular endothelial growth factor: a new member of the platelet-derived growth factor gene family. Biochem Biophys Res Commun 165:1198–1206.

Tomanek, R.J., Ratajska, A., Kitten, G.T., Yue, X., Sandra, A. (1999). Vascular endothelial growth factor expression coincides with coronary vasculogenesis and angiogenesis. Dev Dyn 215:54–61.

Tran, J., Rak, J., Sheehan, C., et al. (1999). Marked induction of the IAP family antiapoptotic proteins survivin and XIAP by VEGF in vascular endothelial cells. Biochem Biophys Res Commun 264:781–788.

Tuder, R.M., Flook, B.E., Voelkel, N.F. (1995). Increased gene expression for VEGF and the VEGF receptors KDR/Flk and Flt in lungs exposed to acute or to chronic hypoxia. Modulation of gene expression by nitric oxide. J Clin Invest 95:1798–1807.

Unemori, E.N., Ferrara, N., Bauer, E.A., Amento, E.P. (1992). Vascular endothelial growth-factor induces interstitial collagenase expression in human endothelial-cells. J Cell Physiol 153:557–562.

Valenzuela, D.M., Griffiths, J.A., Rojas, J., et al. (1999). Angiopoietins 3 and 4: diverging gene counterparts in mice and humans. Proc Natl Acad Sci USA 96:1904–1909.

Visvader, J., Fujiwara, Y., Orkin, S. (1998). Unsuspected role for the T-cell leukemia protein SCL/tal-1 in vascular development. Genes Dev 12:473–479.

Waltenberger, J., Clacsson-Welsh, L., Siegbahn, A., Shibuya, M., Heldin, C.H. (1994). Different signal transduction properties of KDR and Flt1, two receptors for vascular endothelial growth factor. J Biol Chem 269:26988–26995.

Wang, G.L., Jiang, B.H., Rue, E.A., Semenza, G.L. (1995). Hypoxia inducible factor 1 is a basic helix loop helix PAS heterodimer regulated by cellular O_2 tension. Proc Natl Acad Sci USA 92:5510–5514.

Wang, H., Chen, Z., Anderson, D. (1998). Molecular distinction and angiogenic interaction between embryonic arteries and veins revealed by ephrin-B2 and its receptor Eph-B4. Cell 93:741–753.

Weinstein, B.M. (1999). What guides early embryonic blood vessel formation? Dev Dyn 215:2–11.

Wenger, R.H., Gassmann, M. (1997). Oxygen(es) and the hypoxia-inducible factor-1. Biol Chem 378:609–616.

Wilk, R., Weizman, I., Shilo, B.Z. (1996). Trachealess encodes a bHLH-PAS protein that is an inducer of tracheal cell fates in Drosophila. Genes Dev 10:93–102.

Wilt, F.H. (1964). Erythropoesis in the chick embryo: the role of endoderm. Science 147:1588–1590.

Wilting, J., Brand-Saberi, B., Huang, R., et al. (1995). Angiogenic potential of the avian somite. Dev Dyn 202:165–171.

Wilting, J., Christ, B., Weich, H.A. (1992). The effects of growth factors on the day 13 chorioallantoic membrane (CAM): a study of VEGF165 and PDGF-BB. Anat Embryol Berl 186:251–257.

Wizigmann-Voos, S., Breier, G., Risau, W., Plate, K.H. (1995). Up-regulation of vascular endothelial growth factor and its receptors in von Hippel-Lindau disease-associated and sporadic hemangioblastomas. Cancer Res 55:1358–1364.

Yamaguchi, T.P., Dumont, D.J., Conlon, R.A., Breitman, M.L., Rossant, J. (1993). flk-1, an flt-related receptor tyrosine kinase is an early marker for endothelial cell precursors. Development 118:489–498.

Yancopoulos, G., Klagsbrun, M., Folkman, J. (1998). Vasculogenesis, angiogenesis, and growth factors: ephrins enter the fray at the border. Cell 29:661–664.

Yang, X.J., Cepko, C.L. (1996). Flk-1, a receptor for vascular endothelial growth factor (VEGF), is expressed by retinal progenitor cells. J Neurosci 16:6089–6099.

Yano, M., Iwama, A., Nishio, H., Suda, J., Takada, G., Suda, T. (1997). Expression and function of murine receptor tyrosine kinases, TIE and TEK, in hematopoietic stem cells. Blood 89:4317–4326.

Yoshida, A., Anand-Apte, B., Zetter, B.R. (1996). Differential endothelial migration and proliferation to basic fibroblast growth factor and vascular endothelial growth factor. Growth Factors 13:57–64.

Zagris, N. (1980). Erythroid cell differentiation in unincubated chick blastoderm in culture. J Embryol Exp Morphol 58:209–216.

Zelzer, E., Wappner, P., Shilo, B.Z. (1997). The PAS domain confers target gene specificity of Drosophila bHLH/PAS proteins. Genes Dev 11:2079–2089.

Ziegler, B.L., Valtieri, M., Porada, G.A., et al. (1999). KDR receptor: a key marker defining hematopoietic stem cells. Science 285:1553–1558.

CHAPTER 3

The Ties That Bind: Emerging Concepts About the Structure and Function of Angiopoietins and Their Receptors in Angiogenesis

Chitra Suri and George D. Yancopoulos

The Tie [*t*yrosine kinase with *i*mmunoglobulin and epidermal growth factor (*E*GF) homology domains] receptors, Tie-1 and Tie-2, have been found to be localized primarily to the endothelial and hematopoietic cells in all organisms (human, mouse, rat, zebrafish) that have been examined (Wilks, 1989; Dumont et al., 1992; Iwama, et al., 1993; Maisonpierre et al., 1993; Sato et al., 1993; Lyons et al., 1998). The only other receptors that share this feature are those that bind and are activated by the (VEGF) family members (discussed in the previous chapter). Tie-1 and Tie-2 are large (~160 kDa) multidomain proteins that are highly homologous. The rat genes share 32% sequence identity in their extracellular regions and 79% sequence identity in their intracellular regions (Maisonpierre et al., 1997). The mouse genes are both on chromosome 4, separated by only 12.2 centimorgans (cM), while the human Tie-1 gene is on 1p33–34, which is a syntenic location (Korhonen et al., 1994). Human Tie-2 is on 9p21. The ligands for Tie-2, the angiopoietins, comprise a unique family of proteins, the first member of which was cloned only a few years ago (Davis et al., 1996; Maisonpierre et al., 1997; Valenzuela et al., 1999). All the angiopoietins are highly homologous to each other; angiopoietin-1 (Ang1) and angiopoietin-2 (Ang2) are ~60% identical in their amino acid sequence; angiopoietin-3 and -4 (Ang3 and 4) show ~54% identity to Ang1 (Maisonpierre et al., 1997; Valenzuela et al., 1999). They have diverged enough to serve different functions and to reside on different chromosomes; for Ang1, the human gene is on 8q22 and the mouse on chromosome 15 (syntenic); for Ang2, the human gene is on 8p21 and the mouse gene on chromosome 8 (syntenic), whereas human Ang4 is on 20p13 and mouse Ang3 is on chromosome 2 (which are also syntenic) (Valenzuela et al., 1999). So far, there are no definitive ligands for Tie-1.

All tissues express Tie-1 and Tie-2 in the vasculature, albeit at different levels (Suri et al., unpublished observations). Interestingly, both the expression patterns and the levels of the two receptors are very similar in the various tissues (Suri et al., unpublished observations). The ligands, on the other hand, are differentially expressed at all developmental stages with different tissues expressing different levels and combinations (Suri et al., unpublished observations). By now it is quite clear that these ligands and receptors have to be present for a functional blood vessel to be generated and maintained.

TIE RECEPTORS PLAY ESSENTIAL ROLES DURING ANGIOGENESIS

Genetic manipulations in mice have demonstrated that Tie-1, Tie-2, Ang1, and Ang2 are all required for the establishment of an intact vasculature. In the absence of Tie-1, mice die early—anywhere from embryonic day E14.5 to P1, depending on the strain (Puri et al., 1995; Sato et al., 1995; Partanen et al., 1996). The first symptom is the development of edema, after which there are multiple hemorrhages. Since there are no reported changes in the vascular patterning in the mutant mice, it has been suggested that Tie-1 plays an important role late in development (perhaps later than Tie-2) in establishing/maintaining the integrity of an existing vascular network. On the other hand, mice lacking Tie-2 die around E9.5 with defects in the heart and in the endothelial lining of blood vessels (Dumont et al., 1994). In the yolk sac and aorta, the number of endothelial cells appear to be reduced and the surviving cells look disorganized. There are blood cells in the yolk sac cavity as well as in the trunk of the embryo, suggesting that the vessels may have ruptured. The heart appears poorly developed with a sparsely populated endocardium and an immature myocardium. Subsequent detailed analysis has revealed that Tie-2 not only is essential for the survival and perhaps proliferation of the endothelial cells, but also may be required during distinct stages of vascular development (Sato et al., 1995). Upon careful examination, the vascular networks at different sites, such as in and over the heart, brain, and yolk sac appear abnormal and are composed of dilated vessels, suggesting that the sprouting and branching phenomena are compromised in the absence of a functional Tie-2.

ANG1 IS REQUIRED FOR ANGIOGENESIS DURING DEVELOPMENT

Mice lacking Ang1 have a phenotype that is very similar to that seen in the absence of Tie-2 (Suri et al., 1996). The mutant embryos die around E11.5 with cardiovascular abnormalities. Heart development is severely retarded such that by E11.5 the ventricles are very small, the atria are almost not visible, and the coronary artery bed is significantly sparse. Histologic analysis reveals the endocardial layer to have separated from the underlying myocardial layer as well as an impressive reduction in the myocardial trabecular formation. It would appear that Ang1, which is thought to be made by the muscle layer/myocardium, adversely affects the development of the endothelium/endocardium which expresses the cognate receptor. The immature endothelium, in turn, disrupts the maturation of the myocardium. Interaction between these two cellular compartments appears to be necessary for both to develop appropriately. Vascular beds at all other sites of Ang1 expression are also adversely affected. For instance, in the yolk sac and forebrain, the vasculature resembles an immature capillary plexus wherein the vessels are dilated and uniform sized without the hierachical division into larger and smaller vessels. At other sites, such as in the umbilical region and eye, there is a marked reduction in vascular density. The normal patterning of vessels is disrupted, which is most obviously demonstrated in the brain. The distinctive distribution of large and small vessels in this tissue is not seen in the mutant mice. In addition, when vessels in the intersegmental region are examined at E10.5 vs. E11.5, it appears that they might even be regressing. Interestingly, at the ultrastructural level, the extracellular matrix around the endothelial cells appears to be composed of a disorganized array of collagen fibers. Most, if not all, the above-mentioned morphologic changes can be explained if Ang1 is required for the later developmental processes—namely, branching and maturation.

ANG1 OVEREXPRESSION IN THE ADULT

An Enhanced Vasculature

The current hypothesis is that Ang1 is essential for the stabilization of the endothelial cell and for its interaction with the surrounding smooth muscle layer and the extracellular matrix (Fig. 3.1B). This hypothesis is supported by observations in transgenic mice in which Ang1 is overexpressed in a specific tissue (Suri et al., 1998). When Ang1 is made available at high levels in the skin, the resultant [K14-Ang1 tranogenic (tg)] mice are so highly vascularized that the skin appears red. The blood vessels in these mice are enlarged, their numbers are increased, and they are highly branched. Since Ang1 is thought to stabilize the vascular structure, the enhanced vascularization may very well be due to a lack of the normal pruning and regression that accompany angiogenesis.

Interestingly, VEGF overexpression in the skin also leads to an increase in the number of blood vessels, but these vessels resemble capillaries rather than venules, as is the case with Ang1 overexpression (Thurston et al., 1999). As expected, VEGF and Ang1 coexpression has an additive effect on the vascularization state of the tissue. This finding has obvious implications for any potential therapy with these reagents.

A Stable, Leak-Proof Vasculature

One of the most striking observations in the K14-Ang1 tg. mice is that their blood vessels are resistant to leakage (Thurston et al., 1999). This is in stark contrast to the vessels that are generated when there is excess VEGF. The vessels in that case are fragile and highly permeable (Detmar et al., 1998; Robinson and Aiello, 1998). It has now been demonstrated that their permeability increases even more in the presence of inflammatory agents, whereas the Ang1–induced vessels show no increase at all (Thurston et al., 1999). This is an important difference in the effects of the two growth factors—Ang1 and VEGF. Even though they both significantly increase the blood vessel density, the resultant vessels exhibit very different properties, which again, has implications for any successful therapy.

58   C. Suri and G.D. Yancopoulos

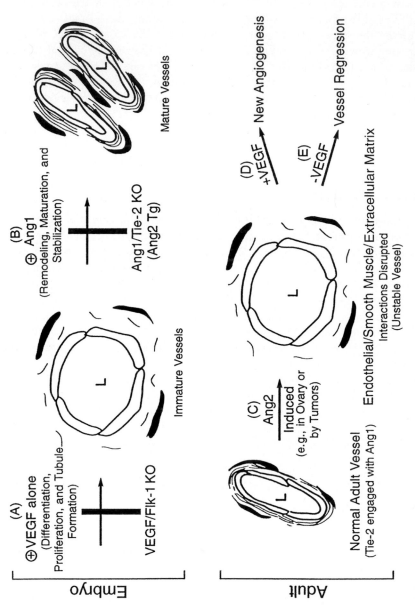

FIGURE 3.1.   Vascular architecture in the embryo as well as the adult, in normal and pathologic settings, is established and maintained by exquisite co-ordination between VEGF and angiopoietins.

ANG2 IS A NATURAL ANTAGONIST OF ANG1

Ang1 and Ang2 both bind to the Tie-2 receptor, but only the former activates the receptor as measured by receptor phosphorylation, the first event in the signaling cascade. Both *in vitro* and *in vivo*, Ang2 has the potential to antagonize the activation induced by Ang1 (Maisonpierre et al., 1997). *In vitro*, Ang2 suppresses the Tie-2 receptor phosphorylation induced by Ang1, and in transgenic mice, when Ang2 is overexpressed at sites of Ang1 expression, the mice die at ~E9.5 with a phenotype that is similar to that seen when the gene for Ang1 is removed (see above). The embryo shows massive vascular disruptions, which, as in the case when Ang1 is absent, appear to be caused by changes in both the endothelial and smooth muscle cell components. Interestingly, the Ang2 overexpressing mutant embryos have a more severe phenotype compared to the mutant embryos lacking Ang1. It is possible that in addition to acting as an antagonist at some sites, such as the heart, Ang2 serves as an agonist in certain other vascular beds. It is clear that *in vitro*, Ang2 can function either as an agonist or antagonist depending on the cellular context (Maisonpierre et al., 1997). The actual role of this complex molecule in angiogenesis may be better understood only once the mice that lack it have been analyzed in detail (Suri et al., in preparation). In any case, it is worth noting that Ang2 is the first example of a naturally occurring antagonist of a tyrosine kinase receptor. Examination of angiopoietin expression patterns *in vivo* suggest a very interesting role for Ang2 at sites of normal vascular remodeling in an otherwise stable adult vasculature (Maisonpierre et al., 1997).

AN APPARENT DESTABILIZING ROLE FOR ANG2 IN THE ADULT

While Ang1 is widely expressed in normal adult tissues (where Tie-2 is constitutively phosphorylated (Wong et al., 1997), consistent with it playing a continuously required stabilization role, Ang2 is typically expressed at sites of vascular remodeling in the adult, notably in the female reproductive tract (Maisonpierre et al., 1997). Detailed localization of Ang2 in the cycling ovary by *in situ* hybridization reveals that in regions of active vascular remodeling it is either expressed together with VEGF at sites of vessel sprouting and ingrowth (e.g., developing corpus lutei), or in the absence of VEGF at sites of frank vessel regression (e.g., atretic follicles and regressing corpus lutei). These expression patterns have led to the proposal of a model in which Ang2 plays a facilitative role at sites of vascular remodeling in the adult by blocking a constitutive stabilizing action of Ang1 (Fig. 3.1C). Further, it has been suggested that such destabilization by Ang2 in the presence of high VEGF levels primes the vessels to mount a robust angiogenic response reminiscent of that of early embryonic vessels prior to maturation (Fig. 3.1D). However, such destabilization by Ang2 in the absence of VEGF is instead proposed to lead to frank vessel regression (Fig. 3.1E).

STRUCTURAL CHARACTERISTICS OF THE TIE-2 RECEPTOR AND ITS LIGANDS

The complementary DNA (cDNA) for the Tie-2 receptor is 3.2 kilobase (kb), which encodes a protein of ~140 kDa. The size and stucture of the Tie-1 protein is very similar to that of Tie-2. In their extracellular domains, the receptors contain a unique combination of structural motifs; there is an immunoglobulin loop followed by three epidermal growth factor–like repeats, another immunoglobulin loop, and finally three fibronectin III–like repeats. The intracellular portions of the receptors, that are composed of two tyrosine kinase domains, are also very homologous to each other. The angiopoietins are fairly large proteins themselves (~70 kDa in monomeric form) that appear as higher-order aggregates within most cells. Their structure is unlike any other; there is a signal peptide at the N-terminus, followed by coiled-coil and fibrinogen-like domains (Davis et al., 1996). The former is probably responsible for generating the aggregate forms, whereas the latter, which bears homology to proteins such as fibrinogen, ficolin, and tenascin, appears to be required for binding to the receptor (Davis et al.; unpublished observations). The N-terminal contains sequences that distinguish an agonist Ang from an antagonist Ang.

SIGNALING EVENTS TRIGGERED BY TIE-2 RECEPTOR ACTIVATION

All the angiopoietins bind to the Tie-2 receptor [kDa ranging from 3 to 3.7 nanomolar (nM)] but only two of them (Ang1 and Ang4) acivate it. The first event is receptor autophosphorylation induced by receptor dimerization. Interestingly, the Tie-2 receptor has been found to be significantly phosphorylated in an E10.5 mouse embryo, consistent with a role for this receptor early in development (Koblizek et al., 1997). There are various intracellular proteins that bind to tyrosine phosphorylated receptors via specific [Src-homology 2 (SH2) and phospho tyrosine binding (PTB)] domains (Pawson, 1995). Two of the SH2-containing proteins, growth factor receptor bound protein 2 (GRB2) and Src-homology 2 containg protein tyrosine phosphatase 2 (SH-PTP2), appear to bind to the activated Tie-2 receptor *in vitro* (Huang et al., 1995). It has previously been shown that these proteins are in the signaling cascade leading to cell growth and differentiation (Neel, 1993; Feng and Pawson, 1994). A novel protein named Dok-R has now been shown to interact with the phosphorylated Tie-2 receptor via its PTB domain, *in vitro* and *in vivo* (Jones and Dumont, 1998). Upon binding, the Tie-2 receptor appears to phosphorylate Dok-R, which then binds to ras GTPase activating protein (rasGAP) and a small adaptor protein named NCK, which also contain SH2 or PTB domains. The role of NCK in mammalian signaling is not yet clear but its association with a variety of signaling molecules suggests that it is involved in processes ranging from cell proliferation to cell migration (Rockow et al., 1996). rasGAP is a negative regulator of ras activity, hastening the conversion of active guanosine triphosphate (GTP)-bound ras to the inactive guanosine diphosphate (GDP)-bound form (Trahey et al., 1988). Its role in angiogenesis has been well documented (Henkemeyer et al., 1995). Mice lacking this gene develop clear vascular deficits that are apparent by E9.5, and resemble some of the defects seen in the absence of either Tie-2 or Ang1.

Recently, the Janus tyrosine kinase (JAK)-signal transducer and activator of transcription (STAT) pathway has been implicated in Tie-2 receptor-mediated signaling (Korpelainen et al., 1999). In transfection studies, Tie-2 receptor activation causes a weak phosphorylation of STAT1 and robust phosphorylations of STAT3 and STAT5. Interestingly, a mutant form of the receptor that is thought to be constitutively active and causes venous malformations in humans (Vikkula et al., 1996) shows a much stronger signal for all the above-mentioned STATs. In addition, p21, a cell cycle protein previously shown to be regulated by STAT1 and STAT5 and implicated in cell growth inhibition and maturation in other systems, is upregulated in this system in a STAT5-dependent manner. The physiologic implications of STAT activation by the Tie-2 receptor remain unclear.

It should be noted that in all the above studies, the Tie-2 receptor was activated in a ligand-independent fashion, and it is yet to be determined if the angiopoietins initiate the same signaling events.

ROLE OF ANGIOPOIETINS IN TUMOR ANGIOGENESIS

It is now well accepted that the blood supply to the tumor is critical for the maintenance and progression of the tumor (Folkman, 1990). The apparently coordinated actions of VEGF and the angiopoietins during normal embryonic vascular development, coupled with the suggestion of a key destabilizing role for Ang2 during normal adult vascular remodeling, demanded that the potential actions of the angiopoietins be examined during pathologic adult angiogenesis, such as occurs in tumors. In the course of such examinations in tumors, substantial new insights have been obtained leading to an entirely new model of tumor angiogenesis, in which VEGF and Ang2 appear to play key roles (Holash et al., 1999; Zagzag et al., 1999).

Existing dogma had suggested that most tumors and metastases initially grow as avascular masses, up until they reach 1 to 2 mm in size, at which point they need to elicit angiogenic ingrowth in order to grow further (Fig. 3.2A). The new insights suggest that this existing dogma still applies to many tumor settings, particularly those in which tumor cells arise or are implanted into avascular areas or spaces. Such tumors include most commonly studied experimental tumors that are implanted into the avascular subcutaneous space. On the other hand, the new findings suggest that in many cases tumor cells appear to have immediate access to blood vessels, such as when they metastasize to or are implanted within a vascularized tissue (Holash et al., 1999; Zagzag et al., 1999). In such settings, tumor cells do not avoid vessels and grow avascularly, but instead seem to immediately co-opt adjacent existing vessels, often growing as cuffs around these existing vessels (Fig. 3.2B). Thus, even tiny tumors in these settings are well vascularized in the absence of any new angiogenesis. However, while its vessels are being co-opted by tumor, the host does not seem to simply sit idly by. Instead, a robust host defense mechanism is apparently activated, in which the co-opted vessels apparently sense an abnormality and initiate an apoptotic suicide cascade; endothelial cell apoptosis is initiated early in the process, followed by vessel regression (Fig. 3.2C). Regression of co-opted vessels then seems to take much of the dependent tumor with it, resulting in massive tumor death (Figure 3.2C). However, successful tumors seem to overcome host vessel regression by initiating angiogenic growth of new vessels to sustain the surviving tumor (Fig. 3.2C).

In this new view of tumor angiogenesis, the host is not a passive partner during tumor invasion, but rather a worthy adversary in a battle for the vasculature. But how does the host initiate endothelial cell apoptosis and vessel regression? A major clue has come from careful temporal and spatial examination of Ang2 expression during tumor angiogenesis (Stratmann et al., 1998; Holash et al., 1999; Zagzag et al., 1999). This examination has resulted in the striking observation that, while Ang2 is not detected in the endothelium of normal tissues, it is dramatically

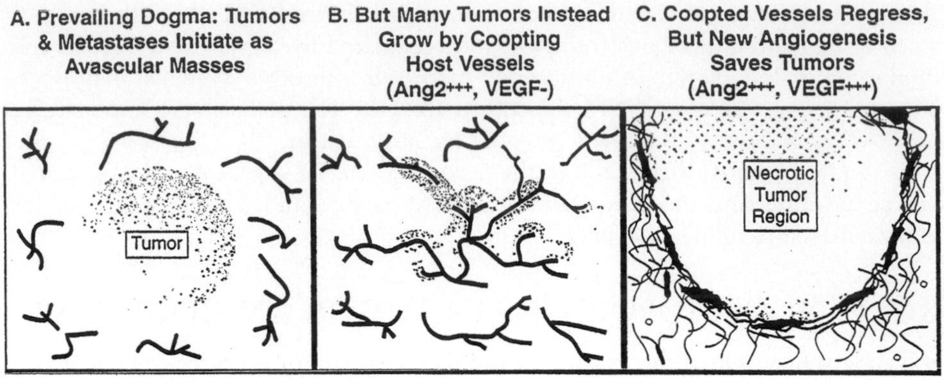

A. Prevailing Dogma: Tumors
& Metastases Initiate as
Avascular Masses

B. But Many Tumors Instead
Grow by Coopting
Host Vessels
(Ang2+++, VEGF-)

C. Coopted Vessels Regress,
But New Angiogenesis
Saves Tumors
(Ang2+++, VEGF+++)

Tumor

Necrotic
Tumor
Region

FIGURE 3.2.   Variable expression levels of Ang2 and VEGF, in and around tumors, determine whether vessels regress or grow.

induced in the endothelium of tumor–co-opted vessels shortly after vessel co-option, and just prior to endothelial apoptosis (Fig. 3.2B). Because Ang1 is involved in vessel stabilization *in vivo* and can support endothelial survival *in vitro*, autocrine induction of the Ang2 antagonist in co-opted vessels seems to be a likely initiator of apoptosis in the co-opted vessels (e.g., Fig. 3.1E). But if the induced Ang2 leads to vessel regression within the tumor, how does the tumor then subsequently mount an angiogenic response? In this case, VEGF expression patterns have provided the clue (Holash et al., 1999; Zagzag et al., 1999). Early on, the tumor is well vascularized by co-opted vessels (Fig. 3.2B). At this point, the tumor makes little if any detectable VEGF, while Ang2 levels are high in the co-opted vessels (Fig. 3.2B); thus, Ang2 in the absence of VEGF seems to drive vessel regression (Fig. 3.1E). However, shortly after vessel regression, the dying tumor dramatically upregulates its expression of VEGF, presumably because it is becoming hypoxic due to the loss of vascular support (Fig. 3.2C). The remaining tumor vessels, largely at the tumor edge, are now not only making Ang2 but are exposed to tumor-derived VEGF (Fig. 3.2C). As seen during normal vascular remodeling, the destabilizing signal provided by Ang2, which leads to vessel regression in the absence of VEGF (Fig. 3.1E), instead seems to potentiate the angiogenic response in combination with VEGF (Fig. 3.1D), allowing for robust new angiogenesis at the tumor rim (Fig. 3.2C). Thus, in tumors, Ang2 and VEGF apparently reprise the roles they play during normal vascular remodeling to regulate a previously underappreciated balance between vessel regression and vessel growth. Ang2 expression in the absence of VEGF expression apparently leads to tumor vessel regression (Fig. 3.1E), while Ang2 expression together with VEGF expression leads to robust tumor-associated angiogenesis (Fig. 3.1D).

ANGIOPOIETINS AS THERAPEUTIC AGENTS

The normal physiologic roles of the angiopoietins are just coming into focus. From studies with knockout mice, it seems clear that Ang1 works in a complementary, coordinated, and sequential manner with VEGF during embryonic vascular development. VEGF initiates endothelial differentiation and proliferation, as well as primitive tubule formation. But vessels formed with only VEGF, in embryos lacking Ang1, remain immature and leaky, and fail to undergo normal remodeling events. It appears as if Ang1 promotes endothelial interactions with surrounding support cells and stroma in a manner that may be necessary for normal vascular remodeling and maturation, as well as for the development of normal vascular integrity. Studies in transgenic mice confirm and extend the findings from the knockouts. While both VEGF and Ang1 can promote angiogenesis, vascular morphology is distinctly different in the presence of excess levels of the two factors. Furthermore, so is the quality of the vessel wall. VEGF-induced vessels are leaky, while Ang1 vessels are resistant to leak, consistent with the notion that Ang1 promotes vascular integrity. While the morphologic and angiogenic activities of VEGF and Ang1 appear additive, the "anti-permeability" effect of Ang1 is dominant over the "pro-permeability" effect of VEGF. Furthermore, Ang1 seems able to protect vessels against leak, whether it is induced by VEGF or inflammatory agents. These findings suggest that VEGF and Ang1 be considered in combination to promote therapeutic angiogenesis in ischemic settings, as they are used by the body during normal development, particularly so as to avoid the production of the leaky and immature vessels that seem to form in response to VEGF alone. Furthermore, these findings suggest that Ang1 be considered as an "anti-permeability" agent in a variety of clinical settings in which vascular leak is a problem, whether this leak is due to VEGF or to inflammatory mediators.

Emerging evidence indicates that VEGF and the angiopoietins recapitulate their embryonic roles during normal vascular remodeling in the adult, as well as in pathologic angiogenesis as occurs in tumors. Reexamination of tumor angiogenesis suggests a modification of the prevailing view that most malignancies and metastases originate as avascular masses that only belatedly induce angiogenic support. Instead, many tumors seem to rapidly co-opt existing host vessels to form an initially well-vascularized tumor mass. Perhaps as part of a host defense mechanism, there is widespread regression of these initially co-opted vessels, leading to a secondarily avascular tumor and massive tumor loss; unfortunately for the host, the remaining tumor is ultimately rescued by robust angiogenesis at its rim. The expression patterns of VEGF and the natural Tie-2 receptor antagonist, Ang2, strongly implicate them in the above processes. Ang2 is strikingly induced in co-opted vessels, prior to VEGF induction in the adjacent tumor cells, providing perhaps the earliest marker of tumor vasculature. The intense autocrine expression of Ang2 by endothelial cells in tumor-associated vessels appears to counter a paracrine stabilization/survival signal provided by low level constitutive expression of Ang1 in normal tissues. This apparently marks the co-opted vessels for regression by an apoptotic mechanism that seems to involve disrupted interactions between endothelial cells and the surrounding extracellular matrix and supporting cells. Subsequently, VEGF upregulation coincident with Ang2 expression at the tumor periphery is associated with robust angiogenesis. This late expression of tumor-derived VEGF may nullify the regression signal provided by Ang2; alternatively, the effect of this VEGF may actually be facilitated when vessels are destabilized by Ang2. Interestingly, newly formed tumor vessels are often tenuous and poorly differentiated, and undergo regressive changes even as blood vessel proliferation continues. The failure of many solid tumors to form a well-differentiated and stable vasculature may indeed be attributable to the fact that newly formed tumor vessels continue to overexpress Ang2. Thus, a persistent blockade of Tie-2 signaling may prevent vessel differentiation and maturation, contributing to the generally tenuous and leaky quality of tumor vessels.

Our findings clearly bolster the case for anti-VEGF therapies in cancer. Other studies have suggested that VEGF may play a role not only in inducing angiogenesis, but also in promoting the survival of fragile new vessels that have not been stabilized by mature interactions with extracellular matrix and supporting cells, such as smooth muscle. This has been most carefully documented during vascularization of the retina (Benjamin et al., 1998) and more recently in tumors (Benjamin et al., 1999). Thus, anti-VEGF therapies may be considered not only for blocking angiogenesis, but perhaps also for their ability to cause the regression of extant tumor vessels that are immature and tenuous in nature.

Ang2 appears to be the earliest marker of blood vessels that have been perturbed by invading tumor cells. As such, Ang2 may prove to be useful in the imaging of very small tumors and metastases, and may even be useful in schemes designed to specifically target chemotoxic therapy to tumor vasculature. Antiangiopoietin therapies employing soluble Tie-2 receptors also have been reported to be efficacious in animal models of cancer (Lin et al., 1997 1998). However, it is somewhat difficult to envision how these soluble receptors might work, as they have the potential to inhibit both Ang1 and Ang2. One possibility is that soluble Tie-2 receptor therapy potentiates the blockade of endogenous Tie-2 receptors by endogenous Ang2. In addition, immature tumor vessels may be particularly susceptible to blockade of an Ang1-mediated stabilization or survival signal, explaining the lack of toxicity of such agents to the vasculature in general. Such a mechanism may also be relevant to other antiangiogenic therapies. Thus, understanding the mechanisms underlying vessel regression is of great importance, not only in the context of anti-VEGF or antiangiopoietin therapies, but for other antiangiogenic therapies as well. It will be of particular interest to see whether such therapies differentially affect regression of new, as opposed to mature, existing vessels, and whether they do this by altering expression of either VEGF or the angiopoietins.

CONCLUSION

The Tie receptor system is required for different aspects of vascular development. In the embryo, Ang1 is essential to the developing cardiovascular system so that the embryo will die from an immature heart and disrupted vasculature if Ang1 is not provided. At the cellular level, both the endothelial and smooth muscle cell layers are affected, suggesting that the communication between these two layers is critical for the generation of an intact vascular wall. Genetic studies have shown that Ang1 is not required for the early stages of vascular formation, but is vital for the remodeling that occurs later and leads to a stable and mature vessel. Surprisingly, if Ang1 is overexpressed during development, it causes the tissue to become hypervascularized, perhaps because there is less vascular pruning and/or regression that are normally seen during remodeling. As might be expected if Ang1 is a stabilization factor, the new vessels generated by Ang1 are very "tight" and resistant to leakage. Ang2, a close relative of Ang1, is a more complex molecule. It can function as an agonist or an antagonist depending on the tissue, or rather, the cellular context. Tie-1, on the other hand, is an even more confounding molecule. It has no known ligand yet but its activation is clearly required since in its absence the vessels hemorrhage and there is tissue edema, causing the mice to die on the day of birth. Whatever the mechanism of action, the angiopoietins, via Tie-2 and perhaps Tie-1, initiate a cascade of events that bind the different layers of the blood vessels, generating a mature vasculature.

REFERENCES

Benjamin, L.E., Golojanin, D., et al. (1999). Selective ablation of immature blood vessels in established human tumors follows vascular endothelial growth factor withdrawal. J Clin Invest 103:159–165.

Benjamin, L.E., Hemo, I., et al. (1998). A plasticity window for blood vessel remodeling is defined by pericyte coverage of the preformed endothelial network and is regulated by PDGF-B and VEGF. Development 125:1591–1598.

Davis, S., Aldrich, T.H., et al. (1996). Isolation of angiopoietin-1, a ligand for the Tie2 receptor, by secretion-trap expression cloning [see comments]. Cell 87(7):1161–1169.

Detmar, M., Brown, L.F., et al. (1998). Increased microvasculature density and enhanced leukocyte rolling and adhesion in the skin of VEGF transgenic mice. J Invest Dermatol 111(1):1–6.

Dumont, D.J., Gradwohl, G., et al. (1994). Dominant-negative and targeted null mutations in the endothelial receptor tyrosine kinase, tek, reveal a critical role in vasculogenesis of the embryo. Genes Dev 8:1897–1909.

Dumont, D.J., Yamaguchi, T.P., et al. (1992). tek, a novel tyrosine kinase gene located on mouse chromosome 4, is expressed in endothelial cells and their presumptive precursors. Oncogene 7:1471–1480.

Feng, G.-S., Pawson, T. (1994). Phosphotyrosine phosphatases with SH2 domains: regulators of signal transduction. Trends Genet 10(2):54–58.

Folkman, J. (1990). Endothelial cells and angiogenic growth factors in cancer growth and metastasis. Introduction. Cancer Metastasis Rev 3:171–174.

Henkemeyer, M., Rossi, D.J., et al. (1995). Vascular system defects and neuronal apoptosis in mice lacking Ras GTPase-activating protein. Nature 377:695–701.

Holash, J., Maisonpierre, P.C., et al. (1999). Vessel cooption, regression, and growth in tumors mediated by angiopoietins and VEGF. Science 284(5422):1994–1998.

Huang, L., Turck, C.W., et al. (1995). GRB2 and SH-PTP2: potentially important endothelial signaling molecules downstream of the TEK/TIE2 receptor tyrosine kinase. Oncogene 11:2097–2103.

Iwama, A.I., Hamaguchi, I., et al. (1993). Molecular cloning and characterisation of mouse Tie and tek receptor tyrosine kinase genes and their expression in hematopoietic stem cells. Biochem Biophys Res Commun 195:301–309.

Jones, N., Dumont, D.J. (1998). The Tek/Tie2 receptor signals through a novel Dok-related docking protein, Dok-R. Oncogene 17:1097–1108.

Koblizek, T.I., Runting, A.S., et al. (1997). Tie2 receptor expression and phosphorylation in cultured cells and mouse tissues. Eur J Biochem 244:774–779.

Korhonen, J., Polvi, A., et al. (1994). The mouse Tie receptor tyrosine kinase gene: expression during embryonic angiogenesis. Oncogene 9:395–403.

Korpelainen, E.I., Karkkainen, M., et al. (1999). Endothelial receptor tyrosine kinases activate the STAT signaling pathway: mutant Tie-2 causing venous malformations signals a distinct STAT activation response. Oncogene 18(1):1–8.

Lin, P., Buxton, J.A., et al. (1998). Antiangiogenic gene therapy targeting the endothelium-specific receptor tyrosine kinase Tie2. Proc Nat Acad Sci USA 95:8829–8834.

Lin, P., Polverini, P., et al. (1997). Inhibition of tumor angiogenesis using a soluble receptor establishes a role for Tie2 in pathologic vascular growth. J Clin Invest 100:2072–2078.

Lyons, M.S., Bell, B., et al. (1998). Isolation of the zebrafish homologues for the Tie-1 and Tie-2 endothelium-specific receptor tyrosine kinases. Dev Dyn 212(1):133–140.

Maisonpierre, P.C., Goldfarb, M., et al. (1993). Distinct rat genes with related profiles of expression define a Tie receptor tyrosine kinase family. Oncogene 8:1631–1637.

Maisonpierre, P.C., Suri, C., et al. (1997). Angiopoietin-2, a natural antagonist for Tie2 that disrupts in vivo angiogenesis [see comments]. Science 277(5322):55–60.

Neel, B.G. (1993). Structure and function of SH2-domain containing tyrosine phosphatases. Semin Cell Biol 4:419–432.

Partanen, J., Puri, M.C., et al. (1996). Cell autonomous functions of the receptor tyrosine kinase TIE in a late phase of angiogenic capillary growth and cell survival during murine development. Development 122:3013–3021.

Pawson, T. (1995). Protein modules and signalling networks. Nature 373:573–580.

Puri, M.C., Rossant, J., et al. (1995). The receptor tyrosine kinase Tie is required for integrity and survival of vascular endothelial cells. EMBO J 14(23):5884–5891.

Robinson, G.S., Aiello, L.P. (1998). Angiogenic factors in diabetic ocular disease: mechanisms of today, therapies for tomorrow. Int Ophthalmol Clin 38(2):89–102.

Rockow, S., Tang, J., et al. (1996). Nck inhibits NGF and basic FGF induced PC12 cell differentiation via mitogen-activated protein kinase-independent pathway. Oncogene 12(11):2351–2359.

Sato, T.N., Qin, Y., et al. (1993). Tie-1 and Tie-2 define another class of putative receptor tyrosine kinase genes expressed in early embryonic vascular system. Proc Natl Acad Sci USA 90:9355–9358.

Sato, T.N., Tozawa, Y., et al. (1995). Distinct roles of the receptor tyrosine kinases Tie-1 and Tie-2 in blood vessel formation. Nature 376(6535):70–74.

Stratmann, A., Risau, W., et al. (1998). Cell type-specific expression of angiopoietin-1 and angiopoietin-2 suggests a role in glioblastoma angiogenesis [see comments]. Am J Pathol 153(5):1459–1466.

Suri, C., Jones, P.F., et al. (1996). Requisite role of angiopoietin-1, a ligand for the Tie2 receptor, during embryonic angiogenesis [see comments]. Cell 87(7):1171–1180.

Suri, C., McClain, J., et al. (1998). Increased vascularization in mice overexpressing angiopoietin-1. Science 282(5388):468–71.

Thurston, G., Suri, S., et al. (1999). Leakage-resistant blood vessels in mice transgenically overexpressing angiopoietin-1. Science 286(5449):2511–14.

Trahey, M., Wong, G., et al. (1988). Molecular cloning of two types of GAP complementary DNA from human placenta. Science 242(4886):1697–1700.

Valenzuela, D.M., Griffiths, J.A., et al. (1999). Angiopoietins 3 and 4: diverging gene counterparts in mice and humans. Proc Natl Acad Sci USA 96(5):1904–1909.

Vikkula, M., Boon, L.M., et al. (1996). Vascular dysmorphogenesis caused by an activating mutation in the receptor tyrosine kinase Tie2. Cell 87:1181–1190.

Wilks, A.F. (1989). Two putative protein-tyrosine kinases identified by application of the polynerase chain reaction. Proc Nat Acad Sci USA 86:1603–1607.

Wong, A.L., Haroon, Z.A., et al. (1997). Tie2 expression and phosphorylation in angiogenic and quiescent adult tissues. Circ Res 81:567–574.

Zagzag, D., Hooper, A., et al. (1999). In situ expression of angiopoietins in astrocytomas identifies angiopoietin-2 as an early marker of tumor angiogenesis [in process citation]. Exp Neurol 159(2):391–400.

CHAPTER 4

Extracellular Matrix in the Regulation of Angiogenesis

Jingsong Xu and Peter C. Brooks

The cardiovascular system plays critical roles in vertebrate development and homeostasis. A crucial process within this system involves the establishment of functional blood vessels. Over the last decade vascular development has gained a great deal of attention in part due to the realization of the importance neovascularization plays in both physiologic as well as pathologic processes. While it is known that the establishment of blood vessels is critical for proper development and maintenance of tissues and organs, its role in pathologic processes such as tumor growth began to be realized in the early 1970s following an intriguing hypothesis by Dr. Juda Folkman (1971) that the growth and expansion of solid tumors beyond a minimal size required the acquisition of new blood vessels. Over the last decade a great deal of experimental evidence supporting this hypothesis has been generated. While the central contention that growing tumors require blood flow to provide nutrients and remove metabolic wastes is well accepted, intriguing new studies suggest that the mechanisms by which tumors obtain this blood flow may differ. For example, recent studies by Holash and colleagues (1999) have suggested that small tumor lesions develop by cuffing and surrounding preexisting blood vessels. In other studies, investigations of uveal melanomas have revealed the existence of blood channels that appear to be composed of melanoma cells expressing a repertoire of genes commonly associated with endothelial cells (Maniotis et al., 1999). Thus, some tumors may have the unique capacity to establish blood flow through alternative mechanisms.

Historically vascular development was thought to occur by either of two similar processes called vasculogenesis and angiogenesis (D'Amore and Thompson, 1987; Flamme et al., 1997; Tyagi, 1997). Vasculogenesis can be defined as the process by which blood vessels arise from precursor cells called angioblasts (Flamme et al., 1997; Tyagi, 1997). In contrast, angiogenesis is the process by which new blood vessels develop from preexisting vessels (D'Amore and Thompson, 1987). This chapter focuses on angiogenesis.

Angiogenesis depends on the coordinated activity of a number of distinct families of molecules. Growth factors and their receptors, proteolytic enzymes, cell adhesion molecules, and extracellular matrix (ECM) components have been well studied (Ingber and Folkman, 1989; Carey, 1991; Leek, 1994; Brooks, 1996a;

Mignatti and Rifkin, 1996). Importantly, these families of molecules do not function in isolation but are rather connected in complex networks that function cooperatively to regulate new blood vessel development. In simple terms, angiogenesis can be organized into three general stages: initiation, proliferation/invasion, and maturation (Fig. 4.1). A variety of growth factors and cytokines have been shown to stimulate angiogenesis. Two well-characterized examples are basic fibroblast growth factor (bFGF) and vascular endothelial growth factor (VEGF) (Klagsbrun, 1992; Neufeld et al., 1999). These growth factors can be produced from a number of sources including inflammatory cells, stromal cells, and tumor cells. These molecules can bind to their respective endothelial tyrosine kinase receptors, thereby stimulating cellular proliferation, upregulation of cell adhesion molecules, and expression of proteolytic enzymes. During the invasive stage of angiogenesis, the activated endothelial cells utilize a number of matrix altering enzymes to remodel the local ECM. This proteolytic remodeling helps to create a microenvironment that is conducive to new blood vessel development. A crucial process following proteolytic remodeling is the transfer of biochemical information from the modified ECM to specialized cells that compose the blood vessels. This critical connection between the ECM and vascular cells is predominantly mediated by a class of cell surface receptors termed integrins (Hynes, 1992). Integrins are a family of cell surface heterodimers composed of α and β chains that mediate cellular interactions with both ECM components and other cells (Hynes, 1992). Studies have provided evidence that integrins play an important role in the regulation of angiogenesis (Brooks, 1996). During the maturation phase of angiogenesis, activated endothelial cells undergo morphogenesis and reorganize into capillary-like tubes. Eventually the endothelial cells secrete new ECM components, form associations with accessory cells such as pericytes, and differentiate into mature vessels (Meininger and Zetter, 1992; Hirschi et al., 1997; Lampugnani and Dejana, 1997).

As one considers the complex cellular, biochemical, and molecular events that contribute to angiogenesis, an interesting concept begins to emerge, which is the central importance of the ECM in the coordination and regulation of nearly all the events that contribute to angiogenesis (Fig. 4.2). For example, studies have indicated that the ECM can regulate processes such as cell adhesion, migration, differentiation, signal transduction, and gene expression (Zetter and Brightman, 1990; Adams and Watt, 1993; Juliano and Haskill, 1993). Moreover, the ECM can also provide a reservoir of proteolytic enzymes, protease inhibitors, and growth factors. The physical localization of these factors to ECM may regulate their accessibility and perhaps their functional activity.

Historically, the ECM was thought to provide mechanical and structural support to cells and tissues. However, following the development of new molecular, cellular, and biochemical techniques, this limited view of the ECM has changed dramatically. In fact, the ECM can be defined in broad terms as a complex interconnected network of fibrous proteins, proteoglycans, and structural glycoproteins that provide both mechanical and biochemical regulatory functions to cells and tissues. The regulatory information contained within the three-dimensional structure of the ECM must be recognized and transferred to recipient cells capable of forming new blood vessels. To this end, integrin-mediated ligation of ECM components has been shown to activate distinct signal transduc-

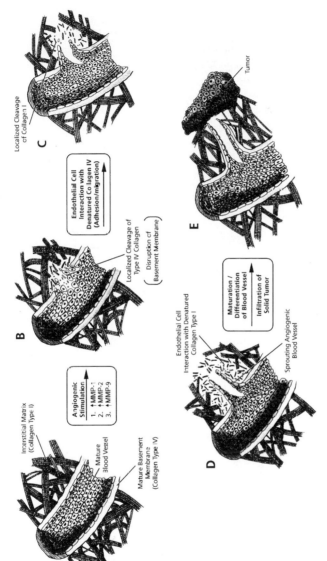

FIGURE 4.1.   The tumor-associated angiogenic cascade. Angiogenesis can be induced by a variety of growth factors that stimulate cell proliferation, elevated expression of matrix-altering enzymes, and cell adhesion receptors. Proteolytic remodeling of the basement membrane facilitates endothelial cell invasion and migration into the interstitial matrix. Endothelial cells release matrix-degrading enzymes to remodel the interstitial matrix components. Endothelial cells next interact with proteolytically modified interstitial matrix, undergo morphogenesis, and reorganize into capillary-like structures. Finally, the endothelial cells secrete new extracellular matrix (ECM) components, associate with accessory cells, and differentiate into functional vessels.

Cellular Processes Regulated by Extracellular Matrix

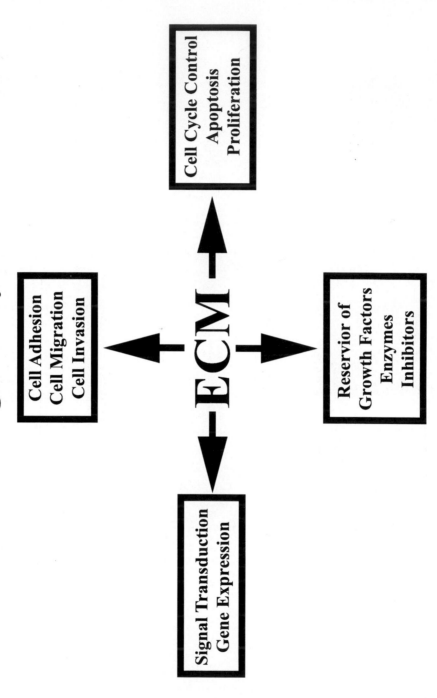

FIGURE 4.2.   Coordination of cellular processes by the extracellular matrix (ECM). Integrin receptors assist in the transfer of critical regulatory information from the ECM to vascular cells. Thus, the ECM can coordinate and regulate a number of cellular, biochemical, and molecular processes involved in vascular development.

TABLE 4.1. Major components of the extracellular matrix.[a]

Basement membrane
   Laminin
   Collagen-IV
   Enactin/nidogen
   SPARC
   Perlecan/proteoglycans

Interstitial ECM
   Fibronectin
   Vitronectin
   Thrombospondin
   Fiber-forming collagen
   Proteoglycans

[a] Partial list of the major components that help form the extracellular matrix.

tion pathways, which in turn may regulate neovascularization. Given the diversity of the molecules that compose the ECM, coupled with tissue-specific variations, an in-depth analysis of the various types of ECMs and their individual components is well beyond the scope of this chapter.

Therefore, for the purposes of our discussion, we will view the ECM in a vastly oversimplified way as being composed of two general compartments: the interstitial ECM and the basal lamina or basement membrane. While these two compartment can be studied as separate features, it is important to point out that they do not exist in isolation, but rather are interconnected by anchoring fibrils forming a continuum between the basal lamina and the interstitial matrix. The basement membrane is a specialized form of ECM that separates both epithelia and endothelia from their underlying mesenchyme (Timpl, 1989; Yurchenco and Schittny, 1990). The components of the vascular basement membrane can vary depending on the particular tissue microenvironments and whether the vessels are undergoing angiogenesis or are quiescent. The major components of the basement membrane include, laminin, collagen type IV, enactin/nidogen, secreted protein acidic and rich in cysteine (SPARC), perlecan, as well as other proteoglycans (Table 4.1). These components exhibit a complex pattern of molecular interconnections and supramolecular assemblies that are organized into a mesh-like network.

The mesh-like network of the basement membrane is connected to the underlying interstitial matrix by a series of anchoring fibers including collagen type VII and fibrilin (Timpl, 1989; Yurchenco and Schittny, 1990; Adachi et al., 1997). Some of the better characterized components include a variety of genetically distinct forms of collagen such as collagen types I, II, and III. Interestingly, at least 19 forms of collagen have been identified (Olsen, 1995; Prockop and Kivirkko, 1995). In addition, a number of noncollagenous glycoproteins also help compose the interstitial matrix including fibronectin, fibrinogen/fibrin, thrombospondin, and vitronectin (Olsen, 1995; Prockop and Kivirkko, 1995; Adachi et al., 1997). Finally, a number of proteoglycans also contribute to the complex architecture of the interstitial matrix. The network of proteins that make up the ECM, in conjunction with integrin receptors that connect the cells to the ECM, function cooperatively to regulate new blood vessel development. To this end, the primary focus of this chapter is on the critical roles of the intact supramolecular structures, and individual ECM molecules in neovascularization. While this chapter does not cover all the potential mechanisms that regulate angiogenesis, it reviews important concepts and insights concerning how the ECM may act to coordinate and integrate biochemical, cellular, and molecular mechanisms that ultimately lead to the development of functional blood vessels.

ORGANIZATION AND ASSEMBLY OF THE VASCULAR BASEMENT MEMBRANE

The basal lamina is composed or a variety of ECM components organized into a specialized mesh-like sheet or basement membrane. Individual molecules such as laminin, collagen IV, enactin/nidogen, and perlecan can be secreted by many cell types including endothelial, epithelial, and mesenchymal cells (Timpl, 1989; Yurchenco and Schittny, 1990; Adachi et al., 1997). However, for a functional basement membrane to form, these ECM components must be organized into an appropriate supramolecular structure. A number of elegant studies have begun to elucidate the complex mechanisms by which these components are organized (Yurchenco and O'Rear, 1994). We will not discuss in depth all the potential mechanisms, but rather briefly overview general aspects of molecular organization.

The basement membrane is composed of two distinct scaffolding networks of collagen IV and laminin (Timpl, 1989; Yurchenco and Schittny, 1990; Adachi et al., 1997). Individual triple helical collagen IV molecules are thought to self-assemble into higher-order supramolecular structures by at least three mechanisms. The N-terminal 7S domains can interact with each other forming tetramers. The C-terminal noncollagenous (NC1) domains are thought to promote dimerization (Tsilibary and Charonis, 1986; Timpl, 1989; Yurchenco and Schittny, 1990; Adachi et al., 1997). Finally, lateral associations along the central triple helical domain of collagen can also occur (Yurchenco and Ruben, 1987). Complex higher-order structures are stabilized by disulfide bonding and lysine aldehyde-derived cross-linking (Timpl, 1989; Yurchenco and Schittny, 1990; Adachi et al., 1997). These stabilized networks of collagen IV help to provide mechanical stability and binding sites for other molecules.

The second major scaffolding network within the basement membrane is composed of laminin (Yurchenco et al., 1992). Laminin is a complex multidomain molecule composed of three chains (Ekblom et al., 1998). Individual laminin molecules can self-assemble and oligermerize into high-ordered structures. Interestingly, studies *in vitro* have suggested that the polymerization of laminin is dependent on protein concentration, temperature, and calcium concentration (Ekblom et al., 1998; Mayer et al., 1998). Studies on laminin oligomerization have revealed that monomeric interactions between laminin molecules are low affinity and involve interactions between the N-terminal domains of the α, β, and γ arms (Yurcheno et al., 1992; Ekblom et al., 1998; Mayer et al., 1998).

While some studies have suggested that weak interactions occur between the collagen IV and laminin networks, other evidence indicates that enactin/nidogen molecules play an important role in connecting these two networks together into a complex suprastructure (Aumailley et al., 1993). In fact, studies indicate that the C-terminal globular domain of enactin/nidogen binds tightly to a high-affinity binding site in the short arm of laminin γ_1 chain. Moreover, enactin/nidogen has also been shown to interact with collagen IV, although with a lower affinity than that observed with laminin. The interaction between enactin/nidogen and collagen type IV involves the second globular domain (G2) near its N-terminus. Importantly, while enactin/nidogen appears to connect laminin and collagen IV, studies suggest that it can also bind a variety of other ECM components. In fact, enactin/nidogen has been shown to bind to fibronectin, fibrinogen, and perlecan (Chung and Durkin, 1990; Aumailley et al., 1993). Thus, enactin/nidogen may function as a critical organizing molecule for the formation of supramolecular structure of the basement membrane. Studies by Fleischmajer and colleagues (1997, 1998) suggest that integrin receptors may play a role in the structural organization and deposition of ECM molecules within the basement membrane. To this end, studies indicate that β_1 integrins specifically colocalized with collagen IV during the formation and organization of basement membrane (Fleischmajer et al., 1997, 1998). More importantly, function blocking antibodies directed to β_1 integrins disrupted the deposition and organization of collagen IV during the development of the basement membrane within culture systems (Fleischmajer et al., 1997, 1998).

ROLE OF BASEMENT MEMBRANE COMPONENTS IN ANGIOGENESIS

Laminin

Laminin is a large heterotrimer of approximately 900 kd composed of three distinct chains termed α, β, and γ (Yurchenco and Ruben, 1987; Yurchenco et al., 1992; Ekblom et al., 1998; Mayer et al., 1998). Multiple α, β, and γ chains have been identified that can give rise to an array of different laminin molecules. In fact, at least 11 different laminin molecules have been identified with distinct tissue distributions. These complex multidomain molecules not only contribute to the mechanical stability for cells and tissues, but also regulate cell behavior. For example, laminin has been shown to support cell adhesion, migration, regulate signal transduction, gene expression, and differentiation. Thus, it is likely that laminin plays an important role in new blood vessel development.

Laminin is known to support endothelial cell interactions by both β_1 integrins and nonintegrin receptors (Basson et al., 1990; Nomizu et al., 1998). Previous studies by Grant and colleagues (1989) demonstrated that different functional domains within laminin regulate distinct endothelial cell processes. In fact, it was shown that an RGD tripeptide containing domain within the α chain of laminin can promote endothelial cell adhesion. This adhesive promoting ability was shown to be dependent on ligation of β_1 integrins. Moreover, a second domain within the β_1 chain of laminin (YIGSR) was shown to induce cell-cell interactions, regulate morphogenesis, and reorganize endothelial cells into tube-like structures (Kubota et al., 1988; Grant et al., 1989). When added in soluble form, this peptide significantly inhibited angiogenesis. Interestingly, endothelial cells have been shown to bind to this YIGSR sequence via interactions with the nonintegrin laminin receptor (Basson et al., 1990). In further studies, a laminin-derived peptide from the α_1 chain (IKVAV) was shown to regulate protease activity and promote angiogenesis (Grant et al., 1992; Kibbey et al., 1992). Finally, other laminin peptides were also shown to promote endothelial cell adhesion and migration *in vitro* (Malinda et al., 1999). Taken together, several distinct sequences within laminin may function to regulate angiogenesis both positively and negatively.

While many of the angiogenesis regulatory sequences within laminin are readily available to cells, it is also possible that biologically relevant sequences or domains may be cryptic and require proteolytic remodeling for exposure. To this end, recent studies by Giannelli et al. (1997, 1999) demonstrated that matrix metalloproteinase-2 (MMP-2)-mediated cleavage of laminin-5 exposed a cryptic site within the γ_2 subunit that induced cellular motility. Thus, it is possible that cryptic sites within the three-dimensional structure of laminin may regulate angiogenesis as well.

Further support for a role for laminin in the regulation of angiogenesis comes from studies by Nicosia et al. (1994), who demonstrated that a stable complex of laminin and enactin could modulate growth of microvessels within a rat aorta model *in vitro*. Interestingly, the laminin-enactin complex could either stimulate or inhibit microvessel growth depending on the concentrations used. Taken together, these studies as well as many others suggest that endothelial cell interaction with laminin plays a crucial role in vascular development.

Collagen Type IV

Collagen type IV is the second major protein that forms a scaffolding network within the basement membrane. Collagen IV is composed of three chains organized in a triple helical fashion. At least six genetically distinct collagen IV chains have been identified and demonstrate specific patterns of tissue distribution (Hudson et al., 1993). These six chains of collagen IV can assemble in different combinations to provide distinct trimers. The most common form is composed of two α_1 (IV) chains and one α_2 (IV) chain. This form of collagen IV is thought to reside in virtually all basement membranes. Collagen IV can self-assemble into high-order suprastructures. These networks can be formed by associations of individual triple helical molecules via the 7S N-terminal domains resulting in tetrameric structures and via the C-terminal noncollagenous (NC1) domains forming hexameric structures (Tsilibary and Charonis, 1986; Yurchenco and Ruben, 1987). Studies have suggested that the 7S domain may function in collagen IV network assembly. The central triple helical region is known to mediate cellular interactions by providing a number of integrin and nonintegrin cellular binding sites (Vandenberg et al., 1991; Miles et al., 1995). Finally, the NC1 domains are thought to be involved with individual chain selection and network formation (Ries et al., 1995).

As was observed with laminin, collagen IV not only provides mechanical support and stability to cells and tissues, but also may actively regulate angiogenesis. To this end, endothelial cells are known to directly interact with collagen IV predominantly by β_1 integrin receptors. In fact, studies have identified a number of integrin binding sites within the triple helical region, including binding sites for integrins $\alpha_1\beta_1$, $\alpha_2\beta_1$, and $\alpha_3\beta_1$ sites (Vandenberg et al., 1991; Miles et al., 1995). Blocking ligation of β_1 integrins including $\alpha_1\beta_1$ and $\alpha_2\beta_1$ inhibits endothelial cord formation *in vitro* and angiogenesis *in vivo* (Senger et al., 1997). Early studies implicating collagen IV in angiogenesis came from work performed by Madri and Williams (1983). These studies indicated that endothelial cells cultured on crude

mixtures of interstitial collagen types I and III resulted in endothelial cell proliferation but showed little if any reorganization into cord-like structures. In
contrast, endothelial cells cultured on a mixture of basement membrane collagen
types IV and V resulted in rapid reorganization into cord-like structures. These
experiments suggested that different types of collagen may have distinct effects on
endothelial cell behavior. In further studies, Nicosia and Madri (1987) showed that
microvessels sprouting from rat aortic plasma clot cultures were associated with
collagen IV. Moreover, Ingber and Folkman (1988) showed that specific antagonists of collagen biosynthesis could potently inhibit angiogenesis *in vivo*. These
studies suggested that synthesis and presumably cellular interactions with collagen IV is required for new blood vessel development.

While many of the early studies implicated collagen IV in the control of angiogenesis, little information was available concerning whether specific domains
within collagen IV regulate endothelial cell behavior. To this end, studies by Herbst
and colleagues (1988) began to address this issue. These studies indicated that
pepsin-solublized triple-helical fragments of collagen IV could support endothelial cell adhesion and migration *in vitro*. Moreover, it was shown that the
noncollagenous (NC1) domain of collagen IV could also promote endothelial cell
interactions, but was much less potent than either intact or pepsin-solublized
triple-helical fragments (Chelberg et al., 1990; Tsilibary et al., 1990). Importantly,
our laboratory has recently identified unique α_v and β_1 integrin binding sites within
the NC1 domains of distinct forms of collagen IV (Petitclerc et al., 2000). Our
recent work has demonstrated that the NC1 domains from the α_2(IV), α_3(IV), and
α_6(IV) chains of collagen IV can support integrin-dependent endothelial cell adhesion and migration, while similar NC1 domains from α_1(IV), α_4(IV), and α_5(IV)
chains did not (Petitclerc et al., 2000). More importantly, we have demonstrated
that systemic administration of recombinant NC1 domains from α_2(IV), α_3(IV),
and α_6(IV) chains of human collagen IV potently inhibit angiogenesis and tumor
growth *in vivo*. These findings suggest that unique domains within collagen IV
may play crucial roles in angiogenesis.

While specific regions of collagen IV contribute to the regulation of angiogenesis, it is possible that cryptic regulatory sites may also exist. To this end, our laboratory has identified a cryptic site hidden within the three-dimensional structure
of collagen IV (Xu et al., 2000). This unique site can be exposed within the basement membrane of blood vessels by proteolytic enzymes such as MMP-2. More
importantly, a function-blocking monoclonal antibody (Mab) directed to this
cryptic site potently inhibits angiogenesis and tumor growth *in vivo*.

The mechanisms by which collagen IV regulate angiogenesis is not completely
understood. However, several laboratories have suggested that cellular interactions
with collagen may regulate cell survival and apoptosis (Mooney et al., 1999). In
fact, β_1 integrin–mediated interactions with collagen regulates survival of tumor
and mesanglial cells. Moreover, cellular interactions with collagen IV can also regulate cellular proliferation, adhesion, and migration, all of which could impact
angiogenesis. Finally, collagen IV can support endothelial cell spreading and reorganization of the actin cytoskeleton, which can activate distinct integrin-
dependent signal transduction pathways (Alessandro et al., 1998). Taken together,
these findings suggest multiple mechanisms by which collagen IV may contribute
to new blood vessel development.

Enactin/Nidogen

Enactin was identified as a sulfated ECM glycoprotein with an apparent molecular weight of 150 kd (Chung and Durkin, 1990). Further studies of the ECM by different laboratories identified a protein with similar characteristics that was termed nidogen (Chung and Durkin, 1990). Complementary DNA (cDNA) cloning of these proteins eventually revealed that they were identical. Thus, the term *enactin/nidogen* is often used for this ECM protein. Enactin/nidogen is a multidomain glycoprotein composed of three structural domains: a large N-terminal globular domain linked to a cysteine-rich central rod-like domain, and a C-terminal globular domain (Chung and Durkin, 1990; Mayer et al., 1998). Calcium-binding sites have been identified within the N-terminal domain, while the central rod-like domain consists of several epidermal growth factor (EGF)-homology repeats, one of which contains an RGD tripeptide motif known to support integrin-dependent cell adhesion. The C-terminal domain is thought to be involved in collagen and laminin binding and is similar to the thyroglobulin homology repeat. As mentioned above, enactin/nidogen is an important component of the basement membrane and is thought to help link the collagen IV scaffolding network to the laminin network. Laminin and collagen IV are thought to bind to enactin/nidogen through distinct domains. In addition to binding collagen IV and laminin, it can also bind to other ECM molecules including fibronectin, fibrinogen, and perlecan. Thus, an important function of enactin/nidogen is to help connect complex sets of proteins into the functional suprastructure of the basement membrane.

Previous studies have suggested that enactin/nidogen can mediate cellular adhesion (Chakaravarti et al., 1990). Dedhar and colleagues (1992) showed that integrin $\alpha_3\beta_1$ could bind directly to enactin. Function-blocking antibodies directed to $\alpha_3\beta_1$ blocked cell adhesion to immobilized enactin, while antibodies to $\alpha_2\beta_1$ had little if any effect. Enactin and fragments of enactin have been shown to regulate cellular motility (Chakaravarti et al., 1990; Dedhar et al., 1992). Enactin/nidogen is highly susceptible to proteolytic degradation (Chung and Durkin, 1990; Aumailley et al., 1993; Mayer et al., 1998). During angiogenesis, proteolytic remodeling of the ECM may result in the generation of enactin fragments, which in turn may regulate endothelial cell motility. Interestingly, in studies by Niquet and Represa (1996) it was shown that enactin/nidogen was expressed within the walls of blood vessels. Moreover, an increase in enactin staining was also detected within growing vessels of rat brains as compared to quiescent vessels. These data suggest that enactin may be preferentially upregulated during angiogenesis.

Additional evidence that enactin/nidogen may play a role in angiogenesis comes from studies performed by Nicosia and colleagues (1994). These studies indicated that either enactin-laminin complexes or purified enactin could modulate microvessel outgrowth from cultures of rat aorta. Enactin was also shown to stimulate an increase in the number and length of new microvessels. Enactin/nidogen stimulated angiogenesis at intermediate concentration yet showed little if any activity with either low or high concentrations. Further studies performed by these investigators indicated that enactin-laminin complexes modulated bFGF stimulated angiogenesis. Laminin and enactin may bind heparan sulfate proteoglycans. Since FGF activity and receptor binding can be modulated by heparan sulfate proteoglycans, it is possible that enactin may facilitate binding of heparin to bFGF, thereby enhancing angiogenesis. Taken together, these findings suggest that enactin/nidogen plays an important role in regulating neovascularization. While the specific mechanisms by which enactin/nidogen modulate angiogenesis is not completely understood, it may involve the regulation of a number of biochemical processes such as cellular adhesion, migration, and perhaps modulation of integrin-dependent signaling pathways.

SPARC

While not a structural protein of the ECM, numerous studies have identified SPARC (secreted protein acidic and rich in cysteine) as being transiently associated with the basement membrane (Sage and Bronstein, 1991). In fact, SPARC has been shown to bind to a number of ECM proteins in a calcium-dependent manner (Sage et al., 1989). Some of the ECM components that SPARC binds to include distinct types of collagen such as types I, III, IV, and V (Sage et al., 1989; Sage and Bronstein, 1991). Moreover, SPARC is also thought to bind to the cell surface. The capacity to interact with ECM molecules such as collagen likely contributes to its localization within the ECM. SPARC is a 43-kd calcium-binding protein that can be expressed by a number of cell types including endothelial cells. Previous studies have indicated that SPARC is identical to the previously identified proteins osteonectin and BM-40 (Sage et al., 1989; Sage and Bronstein 1991). Early investigations of SPARC have shown its expression patterns to be associated with cells and tissues undergoing proliferation, wound healing, and morphogenesis (Sage et al., 1989; Iruela-Arispe et al., 1991a). Moreover, SPARC may help regulate cellular proliferation, adhesion, migration, and cell cycle progression (Funk and Sage, 1991). SPARC is organized into at least four functional domains: an acidic cation-binding domain, a cysteine-rich domain containing glycosylation sites, a serine protease-sensitive neutral α-helical domain, and a calcium-binding EF-hand–like domain (Sage et al., 1989; Sage and Bronstein 1991).

Elevated expression of SPARC was observed with endothelial cells that were reorganizing into cord or tube-like structures *in vitro* (Iruela-Arispe et al., 1991, 1995). In other studies, SPARC was shown to be associated with angiogenic blood vessels *in vivo* (Lane et al., 1994). To study the possible roles of SPARC in angiogenesis, experiments were performed to assess the effects of SPARC on endothelial cell behavior. Addition of SPARC to cultures of endothelial cells caused reductions in focal contacts, prevented endothelial cell spreading, and induced endothelial cell rounding (Lane et al., 1994). Interestingly, in studies by Funk and Sage (1991), specific peptides derived from SPARC inhibited endothelial cell DNA synthesis *in vitro*. In fact, a 20 amino acid peptide (amino acids 54 to 73) appeared to block cell proliferation. The inhibition of DNA synthesis was thought to be due to disruption of the G_1 to S transition of the cell cycle. In contrast, a second region of SPARC containing amino acids 113 to 130 from the cationic region were shown to induce endothelial cord formation *in vitro* and angiogenesis *in vivo* (Lane et al., 1994). These findings suggested that specific regions within SPARC may modulate distinct cellular processes that regulate vascular development.

While several biologically active peptides have been identified from SPARC, they may be structurally masked since intact SPARC exhibited minimal stimulatory or inhibitory activity on angiogenesis. However, as we have discussed above, during angiogenesis a number of ECM-altering proteases are released that may modify the ECM. To this end, studies by Lane and colleagues have demonstrated that SPARC can be proteolytically cleaved by a number of serine proteases (Lane et al., 1994; Iruela-Arispe et al., 1995). Proteolytic cleavage of SPARC can release biologically active peptides including the stimulatory peptide containing amino acids 113 to 130. The active region of this peptide was shown to be the copper-binding peptide KGHK. Taken together, these studies show that SPARC may play an important role in the development of new blood vessels by actively regulating endothelial cell interactions with ECM proteins. Alterations of cell-ECM interactions may in turn lead to changes in cell shape, motility, and regulation of cell cycle progression.

Perlecan

Proteoglycans are another important class of molecules that compose the basement membrane. Several large proteoglycans have been found associated with the basement membrane, including bamacan, agrin, and the highly complex molecule perlecan (Timpl, 1993). Perlecan is a heparan sulfate proteoglycan composed of a 400-kd core protein linked to at least three heparan sulfate side chains (Iozzo et al., 1994; Hopf et al., 1999). The perlecan core protein is organized into five distinct domains. The N-terminal domain I is associated with a glycosaminoglycan (GAG)-binding site while domain II is composed of a number of low-density lipoprotein (LDL) receptor-like repeats. Domain III is composed of a number of laminin-A–like repeats, while domain IV is associated with repeating EGF-like structural motifs and immunoglobulin (Ig)-like repeats. Finally, domain V also has Ig-like repeats as well as laminin-G–like regions providing binding sites for GAGs. This complex proteoglycan provides distinct binding sites for a variety of ECM molecules including laminin, collagen IV, fibronectin, enactin/nidogen-1 and -2 and fibulin. Thus, perlecan likely represents an important organizational and interconnecting molecule contributing to the functional assembly of the basement membrane.

Perlecan can be expressed by a number of cell types including endothelial cells. In fact, studies have indicated that perlecan was expressed and deposited along newly forming blood vessels *in vivo* (Sharma et al., 1998). While these findings suggested that perlecan is associated with new blood vessels during angiogenesis, they do not provide direct evidence of a functional role. Imamura and coworkers (1991) provided evidence that cellular interactions with perlecan may enhance transformed endothelial cell growth while suppressing the growth of normal cells. In separate studies, Aviezer et al. (1994) and Sharma et al. (1998) provided evidence that perlecan may regulate angiogenesis *in vivo*. In fact, perlecan can enhance bFGF-induced angiogenesis in the rabbit ear chamber model (Aviezer et al., 1994b). This enhancement of angiogenesis may have been a result of perlecan binding to bFGF, thereby facilitating bFGF binding to its receptor. To this end, studies have indicated that heparin binding to bFGF facilitates bFGF receptor binding and mitogenicity (Yayon et al., 1991; Aviezer et al., 1994). In support of this mechanism, bFGF was shown to exhibit enhanced binding to perlecan as compared to syndecan and fibroglycan (Aviezer et al., 1994). In studies by Sharma and colleagues (1998), antisense targeting of perlecan significantly inhibited colon carcinoma tumor growth and angiogenesis *in vivo*, providing further support for a role for perlecan in the regulation of angiogenesis.

While the exact mechanisms by which perlecan modulates angiogenesis is not completely understood, these studies and others suggest several possibilities. First, Hayashi and coworkers (1992) have demonstrated that endothelial cells can interact with the perlecan core protein by both α_1 and β_3 integrin receptors. α_1 and β_3 integrins play important roles during the formation of blood vessels (Brooks, 1996b). Finally, since perlecan is a major proteoglycan mediating bFGF binding, it may modulate angiogenesis by regulating the accessibility, receptor binding, and mitogenicity of bFGF in the vascular microenvironment.

INTERSTITIAL ECM COMPONENTS AND ANGIOGENESIS

It can be seen from the discussion above that the basement membrane plays an important role in blood vessel development. However, if one considers the two compartment view of the ECM, the basement membrane is but one-half of the complex extracellular environment within which angiogenesis occurs (Fig. 4.3). Thus, our remaining discussion focuses on the potential roles that interstitial ECM components play during neovascularization. A variety of glycoproteins and proteoglycans are preferentially distributed in the interstitial ECM as compared to the basement membrane. While differences in protein localization and deposition do exist between these compartments, some proteins may be present to varying degrees in both, such as fibronectin (Hynes, 1990). Keeping these distribution patterns in mind, the interstitial ECM is composed of a mixture of proteoglycans and glycoproteins such as fibronectin, vitronectin, thrombospondin, and a variety of distinct types of collagen (Mosher, 1992; Adams and Watt, 1993). It is beyond the scope of this chapter to analyze in depth all the constituents of the interstitial matrix and their potential roles in angiogenesis. Thus, we have selected a few of the better characterized molecules as examples for our discussion.

Fibronectin

Fibronectin is a large multidomain adhesive glycoprotein dimer held together by disulfide bonds (Hynes, 1990). Each chain of fibronectin is approximately 220 kd. Fibronectin can be expressed by numerous stromal cells as well as endothelial cells (Hynes, 1990; Damsky and Werb, 1992). In general, fibronectin can be classified into three distinct types: cellular fibronectin, which is thought to form filaments associated with the cell surface; plasma fibronectin, which is a soluble form found in the circulation where it functions in blood clotting and wound healing (Schwarzbauer et al., 1983; Hynes, 1990; Damsky and Werb, 1992); and insoluble matrix fibronectin, which is primarily associated with ECM. We focus here on matrix fibronectin. As mentioned above, fibronectin is composed of a number of functional domains. The N-terminal region of fibronectin is thought to contribute to multimerization and fibronectin network assembly as well as providing binding sites for heparin (Schwarzbauer et al., 1983; Hynes, 1990; Damsky and Werb, 1992). In addition, studies have identified specific binding sites for collagen within fibronectin (Balian et al., 1980; Damsky and Werb, 1992). These sites are likely to be of critical importance for the functional assembly of the interstitial matrix. Fibronectin also possesses distinct sites that promote cell adhesion. Of particular importance is a region of the molecule containing the tripeptide sequence RGD.

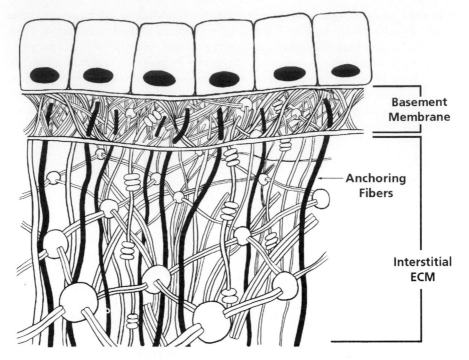

FIGURE 4.3.   Schematic diagram of the two-compartment model of the ECM. The ECM can be viewed as being composed of two general interconnected compartments including the basement membrane and interstitial ECM. The basement membrane is connected to the complex underlying interstitial matrix by a variety of anchoring fibers.

This site is known to support cell adhesion mediated by the $\alpha_5\beta_1$ integrin (D'Souza et al., 1991; Damsky and Werb, 1992). Interestingly, studies suggested that an accessory site in addition to the RGD sequence may be involved in stabilization of $\alpha_5\beta_1$-mediated cellular interactions with fibronectin (Leahy et al., 1996). In addition to the $\alpha_5\beta_1$ binding site, integrin $\alpha_4\beta_1$ binds to fibronectin within the IIICS region (Massia and Hubbell, 1992). Finally, fibronectin can also bind to a number of other molecules including CD44, heparan sulfate, and chondroitin sulfate proteoglycans (Moyano et al., 1999). Thus, as with many of the multidomain ECM molecules discussed above, fibronectin has been shown to regulate a wide variety of cellular and biochemical processes such as adhesion, migration, and signal transduction (D'Souza et al., 1991; Damsky and Werb, 1992).

Fibronectin that is incorporated into an insoluble fibrous network is thought to come from both cellular and plasma forms (Hynes, 1990; Damsky and Werb, 1992). In fact, studies suggest that soluble fibronectin may bind to the cell surface via interactions with integrin receptors and then be incorporated into the ECM by a mechanism that is not completely understood. Thus, the assembly and deposition of the insoluble fibronectin matrix likely involves complex interactions between cells and integrin receptors (Ruoslahti and Engvall, 1997). Soluble fibronectin is thought to be captured or bound to the surface of both fibroblasts and endothelial cells via interactions with the N-terminal 70-kd fragment rather than the RGD-containing cell-binding region (McKeown-Longo and Mosher, 1985). These findings are supported by the fact that a soluble 70-kd N-terminal fragment inhibits fibronectin matrix assembly (Fogerty et al., 1990; Nagai et al., 1991). Interestingly, the $\alpha_5\beta_1$ integrin may be involved since function blocking Mabs directed to either α_5 or β_1 could also block fibronectin matrix assembly. These intriguing results suggest that $\alpha_5\beta_1$ may bind to two distinct sites within fibronectin: a region within the N-terminus, and the central RGD cell-binding region.

In previous studies by Clark and colleagues (1982) it was shown that during wound healing, newly developing blood vessels were associated with elevated levels of fibronectin. Fibronectin was shown to accumulate within the walls of angiogenic vessels, while the levels of fibronectin rapidly dropped when endothelial cell proliferation and capillary growth decreased. In studies by Risau and Lemmon (1988) it was shown that newly growing chick brain capillaries were associated with elevated levels of fibronectin. Taken together, these results suggest that fibronectin may be an important component of the provisional ECM associated with newly developing capillaries. Further evidence to support this contention comes from studies by Elices and colleagues (1994), in which the alternatively spliced CS1 region of fibronectin was shown to be preferentially associated with rheumatoid arthritis microvasculature. Moreover, studies by Carnemolla and coworkers (1992, 1996) have shown that an alternatively spliced form of fibronectin containing the ED-B domain was specifically associated with angiogenic blood vessels. These studies suggested that ED-B fibronectin may be an important marker of angiogenesis *in vivo*.

While numerous studies have shown that specific forms of fibronectin are preferentially associated with the vascular provisional matrix, little mechanistic data was available concerning a direct functional role for fibronectin in angiogenesis. Important studies by Yang and coworkers (1993) provided compelling evidence that cellular interactions with fibronectin mediated by integrin $\alpha_5\beta_1$ play a crucial role in vascular development. For example, it was shown that transgenic mice deficient in the α_5 integrin subunit exhibited numerous mesodermal defects including disruption of fibronectin matrix assembly and defects in blood vessel formation. Moreover, Scheiner and colleagues (1993) demonstrated that CHO tumors that lacked the fibronectin receptor $\alpha_5\beta_1$ formed defective tumor associated blood vessels *in vivo*. Finally, in elegant studies by George et al. (1993), it was shown that transgenic mice lacking fibronectin had numerous blood vessel defects. Taken together, these and other findings provide evidence that fibronectin plays an important role in blood vessel formation.

The mechanisms by which fibronectin contributes to angiogenesis is not completely understood; however, studies have provided some clues. First, fibronectin is known to promote cell adhesion and motility. In fact, proteolytic fragments of fibronectin appear to regulate cellular adhesion, migration, and proliferation (Hu et al., 1997; Manabe et al., 1997; Penc et al., 1998). Moreover, in studies by Dike and Ingber (1996), endothelial cell ligation of fibronectin resulted in the induction of early growth response genes including c-*jun*, c-*fos*, and c-*myc*. In studies by Huhtala and colleagues (1995) it was shown that $\alpha_5\beta_1$-mediated ligation of the 120-kd cell-binding domain of fibronectin resulted in the induction of matrix degrading metalloproteinases (MMPs) such as MMP-1, MMP-9, and stromelysin. In contrast, $\alpha_4\beta_1$-mediated ligation of the CS1 region suppressed MMP production. In addition, studies by Stanton and coworkers (1998) showed that cellular ligation of fibronectin fragments from the cell-binding domain enhanced MMP-2 activation and processing of MT1-MMP. Importantly, these MMPs are thought to play critical roles in matrix remodeling and neovascularization (Liotta et al., 1991; Moses, 1997). Fibronectin has also been shown to play a role in the regulation of cell survival. In fact, ligation of fibronectin can upregulate the expression of Bcl-2, a protein known to suppress apoptosis and promote cell survival (Zhang et al., 1995). Finally, fibronectin has been shown to bind to a number or proteoglycans such as heparin as well as growth factors such as tumor necrosis factor-α (TNF-α) (Moyano et al., 1999). Taking into consideration the multiple functions and protein-protein interactions mediated by fibronectin, it is likely to contribute to angiogenesis by multiple mechanisms.

Vitronectin

Vitronectin is not a true structural ECM protein, but rather a major component of the plasma (Perissner, 1991; Perissner and Seiffert, 1998). Vitronectin is thought to make up approximately 0.2% to 0.5% of the total plasma proteins. However, like SPARC, it has been shown to be deposited into the ECM and bind to components of the interstitial matrix. Vitronectin is a 75-kd multidomain glycoprotein that is primarily synthesized by hepatocytes in the liver. It exists in the plasma in a monomeric form, but can form multimers. Interestingly, the formation of multimers can result in conformational changes in the molecule and expose specific binding sites for other proteins found in the ECM such as collagen type-I (Perissner and Seiffert, 1998). Vitronectin is composed of multiple structural domains including a somatomedin-B region in the N-terminus, and an RGD tripeptide cell-binding region (Perissner, 1991; Perissner and Seiffert, 1998). Other important domains of vitronectin include a series of seven hemopexin-like repeats as well as a glycosaminoglycan-binding site. Interestingly, vitronectin can bind to a diverse set of proteins, some of which depend on its multimeric state. Some of the molecules that bind to vitronectin include heparan sulfate proteogylcans, collagen types I and III, osteonectin (SPARC), and tenacin (Perissner, 1991; Perissner and Seiffert, 1998). Integrin receptors that mediate cellular interactions with the ECM also specifically bind vitronectin including the classical vitronectin receptor $\alpha_v\beta_3$ (Smith and Cheresh, 1990; Shattil, 1995). Other vitronectin-binding integrins include, $\alpha_{IIb}\beta_3$, $\alpha_v\beta_5$, and $\alpha_v\beta_1$. A number of other important molecules also interact with vitronectin such as the protease urokinase-type plasminogen activator (u-PA) and u-PA–u-PA receptor complexes. In addition, serine protease inhibitors such as plasminogen activator inhibitor-1 (PAI-1) can also bind to vitronectin (Shattil, 1995; Kanse et al., 1996; Chacakis et al., 1998; Perissner and Seiffert, 1998).

Studies have shown that vitronectin is associated with the ECM of newly developing blood vessels in a number of animal models (van Aken et al., 1997; Jang et al., 1998; Stefansson et al., 1999). Moreover, studies by Jang et al. (1998) showed that vitronectin deposition was highly elevated in association with the ECM surrounding proliferating human hemangiomas. Since the expression of vitronectin appears to be limited to only a few cell types such as hepatocytes, and little has been shown to be produced by endothelial cells, the vascular associated vitronectin is likely derived from the blood. To this end, numerous studies have shown that angiogenesis can be induced by VEGF, also known as vascular permeability factor (VPF) (Connolly, 1991; Dvorak et al., 1995). Moreover, angiogenic blood vessels are leaky, due in part to the activity of VEGF/VPF (Roberts and Palade, 1997). Thus, it is possible that increased permeability of these angiogenic blood vessels may allow extravasation of plasma proteins such as vitronectin, which in turn could be incorporated into the vascular ECM.

Studies on the roles of integrins and the ECM in angiogenesis have provided evidence that cellular interactions with vitronectin may play a role in the development of new blood vessels. For example, we and others have demonstrated that specific antibody and peptide antagonists that block endothelial cell interactions with vitronectin can potently inhibit endothelial cord formation *in vitro* and angiogenesis *in vivo* (Davis et al., 1993; Brooks et al., 1994a,b, 1995; Friedlander et al., 1995, 1996; Stromblad et al., 1996). Stefansson and colleagues (1999) showed that angiogenesis induced within the chorioallantoic membranes of chick embryos could be inhibited by antibodies directed to vitronectin.

Vitronectin may contribute to the regulation of angiogenesis by multiple mechanisms. First, endothelial cells can interact with vitronectin by ligation of integrins such as $\alpha_v\beta_3$ and $\alpha_v\beta_5$. We have shown that antagonists of integrin $\alpha_v\beta_3$ can induce apoptosis or programmed cell death in proliferating blood vessels (Brooks et al., 1994b, 1995; Stromblad et al., 1996). These results suggest that endothelial cell ligation of vitronectin may be associated with signaling events involved in the regulation of cell survival. In support of this hypothesis, we have previously shown that ligation of endothelial cell integrin $\alpha_v\beta_3$ can specifically modulate the expression of Bcl-2 and Bax, two proteins known to regulate cell survival (Park et al., 1996; Stromblad et al., 1996). Moreover, in studies by Isik and colleagues (1998) it was shown that endothelial cell interactions with immobilized vitronectin decreased apoptosis *in vitro*. Thus, an important role for vitronectin during angiogenesis may be associated with the regulation of endothelial cell survival. Studies have also indicated that vitronectin can promote endothelial cell motility (Stefansson et al., 1999). The deposition of vitronectin within the provisional matrix of newly forming capillary may potentiate endothelial cell migration. Stefansson and Lawrence (1996) demonstrated that PAI, which is known to bind to vitronectin, disrupted $\alpha_v\beta_3$-dependent cellular motility *in vitro*. Moreover, it was shown that the addition of vitronectin to matrigel enhanced endothelial cell migration (Stefansson et al., 1999).

While vitronectin may regulate endothelial cell survival and motility, other mechanisms could also be involved. For example, in studies by Chacakis et al. (1998) it was shown that vitronectin may localize u-PA and u-PA–u-PA receptor complexes to the vascular ECM. This localization may contribute to the regulation of proteolytic activity in the extracellular microenvironment of growing blood vessels. Thus, it is clear that vitronectin may contribute to the regulation of angiogenesis by multiple mechanisms.

Thrombospondin

Thrombospondin (TSP) was originally identified as a thrombin-sensitive protein (TSP) released from the α-granules of platelets (Lawler, 1986; Mosher, 1990). TSP is a large (450-kd) multidomain glycoprotein composed of three identical chains. While a major source of thrombospondin is platelets, a variety of cell types have been shown to express TSP, including fibroblasts, tumor cells, and endothelial cells. In general, TSP is organized into at least four distinct structural domains. These domains include a heparin-binding N-terminal globular domain, a region containing cysteine residues involved in cross-linking, a cysteine-rich segment exhibiting homology to the N-terminal propeptide of collagen, and a C-terminal calcium-sensitive globular domain. Distinct domains within TSP regulate a variety of functions that have been ascribed to TSP. For example, the C-terminal globular domain binds to heparin, while the central cysteine rich region binds to collagen types I, III, IV, and V, fibronectin, and fibrinogen (Canfield et al., 1990; Mosher, 1990). These interactions likely play an important role in the ability of TSP to be incorporated into the ECM. In addition to providing binding sites for a number of ECM associated proteins, TSP is thought to regulate other cellular and biochemical processes such as cell adhesion, migration, platelet aggregation, and regulation of protease production (Lawler, 1986; Mosher, 1990).

The exact role of TSP in the regulation of angiogenesis is unclear. In fact, a number of studies have provided conflicting data as to whether TSP stimulates or inhibits angiogenesis. Regardless of whether TSP is a stimulator or inhibitor, a wealth of information indicates that it can have impact on the angiogenic cascade. To this end, studies indicated that proliferating endothelial cells can express elevated levels of TSP as compared to quiescent cells (Canfield et al., 1990). Moreover, TSP was associated with newly growing blood vessels during wound healing and was also detected in tumor-associated blood vessels (Raugi et al., 1987; Miano et al., 1993).

In studies by Iruela-Arispe et al. (1993) it was shown that the levels of TSP decreased in association with endothelial cord formation. Moreover, antibodies directed to TSP resulted in enhancement of cord-like formation. In contrast, Qian and colleagues (1994) showed that TSP could stimulate endothelial cord development *in vitro*. Conflicting findings were also evident in animal models of angiogenesis. Tolsma and colleagues (1993) demonstrated that TSP or specific peptides derived from TSP could potently inhibit bFGF-induced angiogenesis. In studies by Dameron et al. (1994) it was shown that p53 can stimulate TSP-1 production in cultured fibroblasts. The increased expression of TSP-1 within these cells was shown to inhibit angiogenesis. These findings help define a mechanisms by which p53 and TSP-1 may regulate neovascularization. Finally, studies by Volpert et al. (1998) demonstrated that TSP-1 expression from HT1080 tumors could inhibit angiogenesis and tumor growth *in vivo*. A partial explanation for these conflicting findings may be associated with whether TSP-1 is in a soluble form or if it is matrix bound (Nicosia and Tuszynski, 1994). Soluble TSP-1 may bind to cells by interacting with integrin $\alpha_v\beta_3$. Thus, soluble TSP-1 may act as an antagonist, disrupting cellular interactions with the ECM. In contrast, matrix-localized TSP-1 may act as a local ECM substrate that supports cell adhesion, migration, and reorganization of endothelial cells into capillary sprouts. The cellular and biochemical mechanisms by which TSP-1 contributes to angiogenesis is still not clear.

TSP-1 can modulate cell adhesion, migration, and regulate protease production (Lawler, 1986; Mosher, 1990). Interestingly, studies by Gao and colleagues (1996) have suggested that integrin-associated protein (IAP) can bind to the C-terminal cell-binding domain of TSP-1. Moreover, IAP is known to bind to integrin $\alpha_v\beta_3$

and regulate its function. Thus, it is possible that TSP-1 may alter signal transduction pathways mediated by IAP-$\alpha_v\beta_3$ complex and thereby modulate angiogenesis. Importantly, signal transduction mediated by endothelial cell $\alpha_v\beta_3$ has been shown to regulate apoptosis (Stromblad et al., 1996). To this end, studies by Guo and coworkers (1997) demonstrated that TSP-1 and specific peptides from TSP-1 could induce apoptosis within endothelial cells. Given the multiple functions of TSP, it likely contributes to angiogenesis by several mechanisms. In conclusion, it is clear from the conflicting data presented that further investigation is needed to elucidate the mechanisms by which TSP regulates angiogenesis.

Collagen

As we have mentioned earlier in our discussion, collagen represents the most abundant ECM protein in the body. Collagen can exist in a vast array of forms. In fact, it has been estimated that approximately 30 different genes code for distinct α chains of collagen (Olsen, 1995; Prockop and Kivirkko, 1995). In general, collagen is composed of three individual chains folded into a triple-helical structure. The triple helix is dependent in part on the fact that the small amino acid glycine (G) is incorporated at every third position along the chain (G-X-Y) (Olsen, 1995; Prockop and Kivirkko, 1995). The amino acids at positions X and Y could be any residue but are often proline or hydroxyproline. These unique repeating units are a hallmark of collagen and allow for proper packing and folding. The quarter-staggered arrangement of the triple helix has been suggested to account in part for its ability to interact with a variety of the ECM molecules. To date at least 19 different forms of collagen have been identified with unique patterns of tissue distribution (Olsen, 1995; Prockop and Kivirkko, 1995). Collagen molecules can be grouped into at least three categories: fibrillar collagen, nonfibrillar collagens, and FACIT (fibril-associated collagens with interrupted triple helices) collagen.

Fibrillar collagens represent the major types within the interstitial ECM. Examples of fibrillar collagens include types I, II, and III. The nonfibrillar collagens, such as collagen IV are a major component of the basement membrane. Other members of this group include the transmembrane collagen XVII (Olsen, 1995; Prockop and Kivirkko, 1995). FACIT collagens such as IX, XII, and XIV associate with the fiber-forming collagens. As can been seen from the wide variety of forms of collagen, an in-depth analysis of their structural organization, distribution, and functions would be well beyond the scope of this chapter. Thus, we choose to focus on the most abundantly expressed form of collagen, which is collagen type I.

Collagen type I makes up approximately 25% of the total protein in the body and the vast majority of the protein within the interstitial ECM (Olsen, 1995; Prockop and Kivirkko, 1995). Collagen type I is primarily composed of two genetically distinct α chains organized into a triple helix composed of two α_1 (I) chains and one α_2 (I) chain. Collagen I is expressed by a variety of cell types including fibroblasts and endothelial cells. It is secreted as procollagen I and processed into its mature form by cleavage of the N- and C-terminal propeptides (Olsen, 1995; Prockop and Kivirkko, 1995; Kadler et al., 1996). This cleavage is mediated by specific N- and C-propeptidases. Importantly, the cleavage of the C- and N-terminal propeptides is thought to play a crucial role in the polymerization of collagen molecules into larger supramolecular structures and fibers (Fleischmajer et al.,

1987; Kadler et al., 1996). In addition, lateral associations between collagen fibers occur that contribute to the growing diameter of the fiber by interchain covalent cross-linking (Olsen, 1995; Prockop and Kivirkko, 1995). Finally, the growing collagen fibers can interact with other ECM molecules including fibronectin, thrombospondin, SPARC, proteoglycans, and other forms of collagen. These complex interconnections help in the assembly of the interstitial ECM network.

Since collagen I is ubiquitously expressed, it may play an important role in the development of new blood vessels. Early investigations of endothelial cells used collagen I–coated surfaces to propagate the cells in culture. In fact, when endothelial cells were cultured on two-dimensional collagen I, they form confluent monolayers (Madri and Williams, 1983). Interestingly, Montesano and colleagues (1983) demonstrated that if endothelial cells were embedded within a three-dimensional matrix of collagen I, they reorganize into capillary-like structures *in vitro*. Taken together, these studies provided evidence that collagen may regulate processes involved in the development of new blood vessels. Endothelial cell interactions with collagen I can be mediated by several β_1 integrins including $\alpha_1\beta_1$, $\alpha_2\beta_1$, and $\alpha_3\beta_1$ (Hynes, 1992). Moreover, function blocking antibodies directed to $\alpha_1\beta_1$ and $\alpha_2\beta_1$ have been shown to block endothelial cord formation *in vitro* and angiogenesis *in vivo* (Senger et al., 1997).

With the development of new molecular and genetic techniques, further evidence that collagen I is critical for the proper function and development of blood vessels has been provided. In elegant studies by Schnieke and coworkers (1987), retroviral insertion into the α_1 (I) collagen gene in mice caused embryonic lethality within the homozygous state. Importantly, these mice lacked functional triple-helical collagen I and demonstrate numerous vascular defects (Lohler et al., 1984). In fact, collagen I–deficient mice showed extensive rupture of their blood vessels (Lohler et al., 1984). In separate studies by Reed et al. (1998), aged mice exhibited significantly delayed angiogenesis. This delay in endothelial cell proliferation and angiogenesis was correlated with a decrease in expression of collagen type I. In still other studies, cellular interactions with collagen mediated by $\alpha_1\beta_1$ was shown to regulate proliferation (Pozzi et al., 1998).

As we have discussed with other ECM proteins, specific regions within a given molecule may help facilitate angiogenesis. Studies have shown that collagen I can bind to heparin, a known modulator of angiogenesis. Separate studies have identified both N-terminal as well as C-terminal regions of collagen I as binding sites for heparin (Sweeney et al., 1998). Interestingly, the N-terminal heparin binding site may play an important role in regulating endothelial tube formation *in vitro*.

While triple-helical collagen I may contribute to angiogenesis, interesting studies suggest that proteolytic remodeling of the collagenous microenvironment may also be of critical importance for neovascularization. To this end, specific inhibitors of collagen degrading enzymes called tissue inhibitors of metalloproteinases (TIMPs) can disrupt endothelial cord formation *in vitro* and angiogenesis *in vivo* (Liotta et al., 1991; Moses, 1997). Moreover, synthetic peptide and antibody antagonists of MMPs have been shown to inhibit angiogenesis and tumor growth. Thus, proteolytic remodeling of collagen may contribute to the development of new blood vessels. While the exact mechanisms by which proteolytic remodeling of collagen contributes to angiogenesis is not completely understood, studies are providing provocative clues. For example, proteolytic fragments of collagen can stimulate cellular proliferation, regulate cell cycle progression, and promote cellular motility, all of which could directly regulate angiogenesis (Koyama et al., 1996; Bhattacharyya-Pakrasi et al., 1998). Moreover, proteolytic cleavage of collagen can result in exposure of cryptic sites that are recognized by distinct integrin receptors (Davis, 1992; Montgomery et al., 1994; Petitclerc et al., 1999). While triple-helical collagen I is normally recognized by $\alpha_1\beta_1$ and $\alpha_2\beta_1$ integrins, proteolytically cleaved or denatured collagen was shown to be recognized by integrin $\alpha_v\beta_3$ (Davis, 1992; Montgomery et al., 1994; Petitclerc et al., 2000). Interestingly, antagonists of $\alpha_v\beta_3$ can potently inhibit angiogenesis *in vivo*. This inhibition of angiogenesis is thought in part to be due to the induction of endothelial cell apoptosis resulting from reduced expression of Bcl-2 and an enhanced expression of Bax (Stromblad et al., 1996). Thus, proteolytic remodeling of collagen I may modify certain β_1 integrin binding sites while exposing cryptic integrin binding sites within collagen. Altered integrin binding may function to activate distinct signaling pathways required for endothelial cell proliferation and motility, thereby facilitating angiogenesis. Thus, a wealth of experimental findings indicate that collagen I plays a critical role in the formation of new blood vessels, and furthermore that multiple mechanisms may contribute to the capacity of collagen I to regulate neovascularization.

CONCLUSION

As one can appreciate from our overview of the components, structural assembly, and basic functions of the ECM, this microenvironment represents a complex interconnected network that acts to coordinate and regulate nearly all aspects of blood vessel development. Moreover, it is also apparent that many of the ECM components have overlapping functions, which may provide crucial backup systems for specific cellular processes. While it was not our intention to analyze all the ECM molecules and their potential functions, we did try to provide the reader with some pertinent examples of how the ECM and its many components act to coordinate and regulate biochemical, molecular, and cellular events that contribute to angiogenesis. Considering the importance of angiogenesis in regulation of normal physiologic processes such as embryonic development and wound healing, combined with its critical role in pathologic processes such as tumor growth and metastasis, understanding the physiologic mechanisms regulating vascular development is of great importance. Given these facts, future studies of the ECM will likely provide us with new therapeutic approaches for the regulation of aberrant neovascularization.

ACKNOWLEDGMENTS

We would like to thank Kathryn Carner for her help in the preparation of this manuscript. We would also like to thank Dr. Eric Petitclerc, Loubna Hassanieh, and Dorothy Rodriguez for their helpful suggestions. P.C.B. was supported in part from grants CA74132-01 and CA086140 (NIH/NCI), the Baxter Foundation Award, and the V-Foundation Scholar's Award.

REFERENCES

Adachi, E., Hopkinson, I., Hayashi, T. (1997). Basement-membrane stromal relationships: interactions between collagen fibrils and lamina densa. Int Rev Cytol 173:73–156.

Adams, J.C., Watt, F.M. (1993). Regulation of development and differentiation by the extracellular matrix. Development 117:1183–1198.

Alessandro, R., Masiero, L., Lapidos, K., Spoonster, J., Kohn, E.C. (1998). Endothelial cell spreading on type IV collagen and spreading-induced FAK phosphorylation is regulated by Ca^{2+} influx. Biochem Biophys Res Commun 248:635–640.

Aumailley, M., Battaglia, C., Mayer, U., et al. (1993). Nidogen mediates the formation of ternary complexes of basement membrane components. Kidney Int 43:7–12.

Aviezer, D., Hecht, D., Safran, M., Eisinger, M., David, G., Yayon, A. (1994a). Perlecan, basal lamina proteoglycan, promotes basic fibroblast growth factor-receptor binding, mitogenesis, and angiogenesis. Cell 79:1005–1013.

Aviezer, D., Levy, E., Safran, M., et al. (1994b). Differential structural requirements of heparin and heparan sulfate proteoglycans that promote binding of bFGF to its receptor. J Biol Chem 269:114–121.

Balian, G., Click, E.M., Bornstein, P. (1980). Location of a collagen-binding domain in fibronectin. J Biol Chem 255:3234–3236.

Basson, C.T., Knowles, W.J., Bell, L., et al. (1990). Spatiotemporal segregation of endothelial cell integrin and non-integrin extracellular matrix-binding proteins during adhesion events. J Cell Biol 110:789–801.

Bhattacharyya-Pakrasi, M., Dickeson, S.K., Santoro, S.A. (1998). $\alpha 2\beta 1$ integrin recognition of the carboxyl-terminal propeptide of type I procollagen: integrin recognition and feedback regulation of matrix biosynthesis are mediated by distinct sequences. Matrix Biol 17:223–232.

Brooks, P.C. (1996a). Cell adhesion molecules in angiogenesis. Cancer Metastasis Rev 15: 187–194.

Brooks, P.C. (1996b). Role of integrins in angiogenesis. Eur J Cancer 32A:2423–2429.

Brooks, P.C., Clark, R.A.F., Cheresh, D.A. (1994a). Requirement of vascular integrin $\alpha v\beta 3$ for angiogenesis. Science 264:569–571.

Brooks, P.C., Montgomery, A.M.P., Rosenfeld, M., et al. (1994b). Integrin $\alpha v\beta 3$ antagonists promote tumor regression by inducing apoptosis of angiogenic blood vessels. Cell 79: 1157–1164.

Brooks, P.C., Stromblad, S., Klemke, R., Visscher, D., Sarkar, F.H., Cheresh, D.A. (1995). Antiintegrin $\alpha v\beta 3$ blocks human breast cancer growth and angiogenesis in human skin. J Clin Invest 96:1815–1822.

Canfield, A.E., Boot-Handford, R.P., Schor, A.M. (1990). Thrombospondin gene expression by endothelial cells in culture is modulated by cell proliferation, cell shape and the substratum. Biochem J 15:225–230.

Carnemolla, B., Leprini, A., Allemanni, G., Saginati, M., Zardi, L. (1992). The inclusion of the type III repeat ED-B in the fibronectin molecule generates conformational modifications that unmask a cryptic sequence. J Biol Chem 267:24689–24692.

Carnemolla, B., Neri, D., Castellani, P., et al. (1996). Phage antibodies with pan-species recognition of the oncofoetal angiogenesis marker fibronectin ED-B domain. Int J Cancer 68:397–405.

Carey, D.J. (1991). Control of growth and differentiation of vascular cells by extracellular matrix proteins. Annu Rev Physiol 53:161–177.

Castellani, P., Viale, G., Dorcaratto, A., et al. (1994). The fibronectin isoform containing the ED-B oncofetal domain: a marker of angiogenesis. Int J Cancer 59:612–618.

Chacakis, T., Kanse, S.M., Yutzy, B., Lijnen, H.R., Preisser, K.T. (1998). Vitronectin concentrates proteolytic activity on the cell surface and extracellular matrix by trapping soluble urokinase receptor-urokinase complexes. Blood 91:2305–2312.

Chakravarti, S., Tam, M.F., Chung, A.E. (1990). The basement membrane glycoprotein entactin promotes cell attachment and binds calcium ions. J Biol Chem 265:10597–10603.

Chelberg, M.K., McCarthy, J.B., Skubitz, A.P.N., Furcht, L.T., Tsilibary, E.C. (1990). Characterization of a synthetic peptide from type IV collagen that promotes melanoma cell adhesion, spreading, and motility. J Cell Biol 111:262–270.

Chung, A.E., Durkin, M.E. (1990). Entactin: structure and function. Am J Respir Cell Mol Biol 3:275–282.

Clark, R.A.F., Dellapelle, P., Manseau, E., Lanigan, J.M., Dvorak, H.F., Colvin, R.B. (1982). Blood vessel fibronectin increase in conjunction with endothelial cell proliferation and capillary in-growth during wound healing. J Invest Dermatol 79:269–276.

Connolly, D.T. (1991). Vascular permeability factor: a unique regulator of blood vessel function. J Cell Biochem 47:219–223.

D'Amore, P.A., Thompson, R.W. (1987). Mechanisms of angiogenesis. Annu Rev Physiol 49:453–464.

D'Souza, S.E., Ginsberg, M.H., Plow, E.F. (1991). Arginyl-glycyl-aspartic acid (RGD): a cell adhesion motif. Trends Biochem Sci 16:246–250.

Dameron, K.M., Volpert, O.V., Tainsky, M.A., Bouck, N. (1994). Control of angiogenesis in fibroblasts by p53 regulation of thrombospondin-1. Science 265:1582–1584.

Damsky, C.H., Werb, Z. (1992). Signal transduction by integrin receptors for extracellular matrix: cooperative processing of extracellular information. Curr Opin Cell Biol 4:772–781.

Davis, C.M., Danehower, S.C., Laurenza, A., Molony, J.L. (1993). Identification of a role of the vitronectin receptor and protein kinase C in the induction of endothelial cell vascular formation. J Cell Biochem 51:206–218.

Davis, G.E. (1992). Affinity of integrins for damaged extracellular matrix: $\alpha v \beta 3$ binds to denatured collagen type I through RGD sites. Biochem Biophys Res Commun 182:1025–1031.

Dedhar, S., Jewell, K., Rojiani, M., Gray, V. (1992). The receptor for the basement membrane glycoprotein entactin is the integrin $\alpha 3 \beta 1$. J Biol Chem 267:18908–18914.

Dike, L.E., Ingber, D.E. (1996). Integrin-dependent induction of early growth response genes in capillary endothelial cells. J Cell Sci 109:2855–2863.

Dvorak, H.F., Brown, L.F., Detmar, M., Dvorak, A.M. (1995). Vascular permeability factor/vascular endothelial growth factor, microvascular hyperpermeability, and angiogenesis. Am J Pathol 146:1029–1039.

Ekblom, M., Falk, M., Salmivirta, K., Durbeej, M., Ekblom, P. (1998). Laminin isoforms and epithelial development. Ann NY Acad Sci 857:194–211.

Elices, M.J., Tsai, V., Stahl, D., et al. (1994). Expression and functional significance of alternatively spliced CS1 fibronectin in rheumatoid arthritis microvasculature. J Clin Invest 93:405–416.

Flamme, I., Frolich, T., Risau, W. (1997). Molecular mechanisms of vasculogenesis and embryonic angiogenesis. J Cell Physiol 173:206–210.

Fleischmajer, R., Kuhn, K., Sato, Y., et al. (1997). There is temporal and spatial expression of $\alpha 1$ (IV), $\alpha 2$ (IV), $\alpha 5$ (IV), $\alpha 6$ (IV) collagen chains and $\beta 1$ integrins during the development of the basal lamina in an "in vitro" skin model. J Invest Dermatol 109:527–533.

Fleischmajer, R., Perlish, J.S., MacDonald, E.D., et al. (1998). There is binding of collagen IV to $\beta 1$ integrin during early skin basement membrane assembly. Ann NY Acad Sci 857:212–227.

Fleischmajer, R., Perlish, J.S., Olsen, B.R. (1987). The carboxylpropeptide of type I procollagen in skin fibrillogenesis. J Invest Dermatol 89:212–215.

Fogerty, F.J., Akiyama, S.K., Yamada, K.M., Mosher, D.F. (1990). Inhibition of binding of fibronectin to matrix assembly sites by anti-integrin ($\alpha 5 \beta 1$) antibodies. J Cell Biol 111:699–708.

Folkman, J. (1971). Tumor angiogenesis: therapeutic implications. N Engl J Med 285: 1182–1186.

Folkman, J. (1992). The role of angiogenesis in tumor growth. Cancer Biol Semin 3:65–71.

Friedlander, M., Brooks, P.C., Shaffer, R.W., Kincaid, C.M., Varner, J.A., Cheresh, D.A. (1995). Definition of two angiogenic pathways by distinct αv integrins. Science 270: 1500–1502.

Friedlander, M., Theesfeld, C.L., Sugita, M., et al. (1996). Involvement of integrin αvβ3 and αvβ5 in ocular neovascular diseases. Proc Natl Acad Sci USA 93:9764–9769.

Funk, S.E., Sage, E.H. (1991). The Ca^{2+}-binding glycoprotein SPARC modulates cell cycle progression in bovine aortic endothelial cells. Proc Natl Acad Sci USA 88:2648–2652.

Gao, A., Lindberg, F.P., Dimitry, J.M., Brown, E.J., Frazier, W.A. (1996). Thrombospondin modulates αvβ3 function through integrin-associated protein. J Cell Biol 135:533–544.

George, E.L., Georges-Labouesse, E.N., Patel-King, R.S., Rayburn, H., Hynes, R.O. (1993). Defects in mesoderm, neural tube and vascular development in mouse embryos lacking fibronectin. Development 119:1079–1091.

Giannelli, G., Falk-Marzillier, J., Schiraldi, O., Stetler-Stevenson, W.G., Quaranta, V. (1997). Induction of cell migration by matrix metalloprotease-2 cleavage of laminin-5. Science 277:225–228.

Giannelli, G., Pozzi, A., Stetler-Stevenson, W.G., Gardner, H.A., Quaranta, V. (1999). Expression of matrix metalloprotease-2-cleaved laminin-5 in breast remodeling stimulated by sex steroids. Am J Pathol 154:1193–1201.

Grant, D.S., Kinsella, J.L., Fridman, R., et al. (1992). Interaction of endothelial cells with a laminin A chain peptide (SIKVAV) in vitro and induction of angiogenesis in vivo. J Cell Physiol 53:614–625.

Grant, D.S., Tashiro, K., Segul-Real, B., Yamada, Y., Martin, G.R., Kleinman, H.K. (1989). Two different laminin domains mediate the differentiation of human endothelial cells into capillary-like structures in vitro. Cell 58:933–943.

Guo, N., Krutzsch, H.C., Inman, J.K., Roberts, D.D. (1997). Thrombospondin-1 and type I repeat peptides of thrombospondin 1 specifically induce apoptosis of endothelial cells. Cancer Res 57:1735–1742.

Hayashi, K., Madri, J.A., Yurchenco, P.D. (1992). Endothelial cells interact with the core protein of basement membrane perlecan through β1 and β3 integrins: an adhesion modulated by glycosaminoglycan. J Cell Biol 119:949–959.

Herbst, T.J., McCarthy, J.B., Tsilibary, E.C., Furcht, L.T. (1988). Differential effects of laminin, intact type IV collagen, and specific domains of type IV collagen on endothelial cell adhesion and migration. J Cell Biol 106:1365–1373.

Hirschi, K.K., Rohovsky, S.A., D'Amore, P.A.(1997). Cell-cell interactions in vessel assembly: a model for the fundamentals of vascular remodeling. Transplant Immunol 5:177–178.

Holash, J., Maisonpierre, P.C., Compton, D., et al. (1999). Vessel cooption, regression, and growth in tumors by angiopoietins and VEGF. Science 284:1994–1998.

Hopf, M., Gohring, W., Kohfeldt, E., Yamada, Y., Timpl, R. (1999). Recombinant domain IV of perlecan binds to nidogen, laminin-nidogen complex, fibronectin, fibulin-2 and heparin. Eur J Biochem 259:917–925.

Howell, S.J., Doane, K.J. (1998). Type VI collagen increases cell survival and prevents anti-β1 integrin-mediated apoptosis. Exp Cell Res 241:230–241.

Hu, M., Pollock, R.E., Nicolson, G.L. (1997). Purification and characterization of human lung fibroblast motility-stimulating factor for human soft tissue sarcoma cells: identification as an NH_2-terminal fragment of human fibronectin. Cancer Res 57:3577–3584.

Hudson, B.G., Reeders, S.T., Tryggvason, K. (1993). Type IV collagen: structure, gene organization, and role in human diseases. J Biol Chem 268:26033–26036.

Huhtala, P., Humphries, M.J., McCarthy, J.B., Tremble, P.M., Werb, Z., Damsky, C.H. (1995). Cooperative signaling by α5β1 and α4β1 integrins regulates metalloproteinases gene expression in fibroblasts adhering to fibronectin. J Cell Biol 129:867–879.

Hynes, R.O. (1990). Fibronectins, pp. 546–562. New York: Springer.

Hynes, R.O. (1992). Integrins: versatility, modulation, and signaling in cell adhesion. Cell 69:11–25.

Imamura, J., Tokita, Y., Mitsui, Y. (1991). Contact with the basement membrane heparin sulphate enhances the growth of transformed vascular endothelial cells, but suppresses normal cells. Cell Struct Funct 16:225–230.

Ingber, D., Folkman, J. (1988). Inhibition of angiogenesis through modulation of collagen metabolism. Lab Invest 59:44–51.

Ingber, D.E., Folkman, J. (1989). How does extracellular matrix control capillary morphogenesis? Cell 58:803–805.

Iozzo, R., Cohen, I.R., Grassel, S., Murdoch, A.D. (1994). The biology of perlecan: the multifaceted heparan sulphate proteoglycan of the basement membranes and pericellular matrices. Biochem J 302:625–639.

Iruela-Arispe, M.L., Bornstein, P., Sage, H. (1991a). Thrombospondin exerts an antiangiogenic effect on cord formation by endothelial cells in vitro. Proc Natl Acad Sci USA 88:5026–5030.

Iruela-Arispe, M.L., Hasselaar, P., Sage, H. (1991b). Differential expression of extracellular proteins is correlated with angiogenesis in vitro. Lab Invest 64:174–186.

Iruela-Arispe, M.L., Lane, T.F., Redmond, D., et al. (1995). Expression of SPARC during development of the chicken chorioallantoic membrane: evidence for regulated proteolysis in vitro. Mol Biol Cell 6:327–343.

Isik, F.F., Gibran, N.S., Jang, Y., Sandell, L., Schwartz, S.M. (1998). Vitronectin decrease microvascular endothelial cell apoptosis. J Cell Physiol 175:149–155.

Jang, Y., Arumugam, S., Ferguson, M., Gibran, N.S., Isik, F.F. (1998). Changes in matrix composition during the growth and regression of human hemangiomas. J Surg Res 80: 9–15.

Juliano, R.L., Haskill, S. (1993). Signal transduction from the extracellular matrix. J Cell Biol 120:577–585.

Kadler, K.E., Holmes, D.F., Trotter, J.A., Chapman, J.A. (1996). Collagen fibril formation. Biochem J 316:1–11.

Kanse, S.M., Kost, C., Wilhelm, O.G., Anderasen, P.A., Preissner, K.T. (1996). The urokinase receptor is a major vitronectin-binding protein on endothelial cells. Exp Cell Res 224:344–353.

Kibbey, M.C., Grant, D.S., Kleinman, H.K. (1992). Role of the SIKVAV site of laminin in promotion of angiogenesis and tumor growth: an in vivo matrigel model. J Natl Cancer Inst 84:1633–1638.

Klagsbrun, M. (1992). Mediators of angiogenesis: the biological significance of basic fibroblast growth factor (bFGF)-heparin and heparan sulfate interactions. Semin Cancer Biol 3:81–87.

Koyama, H., Raines, E.W., Bornfeldt, K.E., Roberts, J.M., Ross, R. (1996). Fibrillar collagen inhibits arterial smooth muscle proliferation through regulation of Cdk2 inhibitors. Cell 87:1069–1078.

Kubota, Y., Kleinman, H.K., Martin, G.R., Lawley, T.J. (1988). Role of laminin and basement membrane in the morphological differentiation of human endothelial cells into capillary-like structures. J Cell Biol 107:1589–1598.

Lampugnani, M.G., Dejana, E. (1997). Interendothelial junctions: structure, signaling and functional roles. Curr Opin Cell Biol 9:674–682.

Lawler, J. (1986). The structural and functional properties of thrombospondin. Blood 67: 1197–1209.

Lane, T.F., Iruela-Arispe, M.L., Johnson, R.S., Sage, E.H. (1994). SPARC is a source of copper-binding peptides that stimulate angiogenesis. J Cell Biol 125:929–943.

Leahy, D.J., Aukhil, I., Erickson, H.P. (1996). 2.0A crystal structure of a four-domain segment of human fibronectin encompassing the RGD loop and synergy region. Cell 84: 155–164.

Leek, R.D., Harris, A.L., Lewis, C.E. (1994). Cytokine networks in solid tumors; regulation of angiogenesis. J Leukoc Biol 56:423–435.

Liotta, L.A., Steeg, P.S., Stetler-Stevenson, W.G. (1991). Cancer metastasis and angiogenesis: an imbalance of positive and negative regulation. Cell 64:327–326.

Lohler, J., Timpl, R., Jaenisch, R. (1984). Embryonic lethal mutation in mouse collagen I gene causes rupture of blood vessels and is associated with erythropoietic and mesenchymal cell death. Cell 38:597–607.

Madri, J.A., Williams, S.K. (1983). Capillary endothelial cell cultures: phenotypic modulation by matrix components. J Cell Biol 97:153–165.

Malinda, K.M., Nomizu, M., Chung, M., et al. (1999). Identification of laminin α1 and β1 chain peptides active for endothelial cell adhesion, tube formation, and aortic sprouting. FASEB J 13:53–62.

Manabe, R., Oh-e, N., Maeda, T., Fukuda, R., Sekiguchi, K. (1997). Modulation of cell-adhesive activity of fibronectin by the alternatively spliced EDA segment. J Cell Biol 139:295–307.

Maniotis, A.J., Folberg, R., Hess, A., et al. (1999). Vascular channel formation by human melanoma cells in vivo and in vitro: vasculogenic mimicry. Am J Pathol 155:739–752.

Massia, S.P., Hubbell, J.A. (1992). Vascular endothelial cell adhesion and spreading promoted by the peptide REDV of the IIICS region of plasma fibronectin is mediated by integrin α4β1. J Biol Chem 267:14019–14026.

Mayer, U., Kohfeldt, E., Timpl, R. (1998). Structural and genetic analysis of laminin-nidogen interaction. Ann NY Acad Sci 857:130–142.

McKeown-Longo, P.J., Mosher, D.F. (1985). Interaction of the 70,000-mol-wt amino-terminal fragment of fibronectin with the matrix-assembly receptor of fibroblasts. J Cell Biol 100:364–374.

Meininger, C.J., Zetter, B.R. (1992). Mast cells and angiogenesis. Semin Cancer Biol 3:73–79.

Miano, J.M., Vlasic, N., Tota, R.R., Stemerman, M.B. (1993). Smooth muscle cell immediate-early gene and growth factor activation follows vascular injury: a putative in vivo mechanism for autocrine growth. Arterioscler Thromb 13:211–219.

Mignatti, P., Rifikin, D.B. (1996). Plasminogen activators and matrix metalloproteinases in angiogenesis. Enzyme Protein 49:117–137.

Miles, A.J., Knutson, J.R., Skubitz, A.P.N., Furcht, L.T., McCarthy, J.B., Fields, G.B. (1995). A peptide model of basement membrane collagen α1(IV) 531–543 binds the α3β1 integrin. J Biol Chem 270:29047–29050.

Montesano, R., Orci, L., Vassalli, P. (1983). In vitro rapid organization of endothelial cells into capillary-like networks is promoted by collagen matrices. J Cell Biol 97:1648–1652.

Montgomery, A.M.P., Reisfeld, R.A., Cheresh, D.A. (1994). Integrin αvβ3 rescues melanoma cells from apoptosis in three-dimensional dermal collagen. Proc Natl Acad Sci USA 91:8856–9960.

Mooney, A., Jackson, K., Bacon, R., et al. (1999). Type IV collagen and laminin regulate glomerular mesangial cell susceptibility to apoptosis via β1 integrin-mediated survival signals. Am J Pathol 155:599–606.

Moses, M.A. (1997). The regulation of neovascularization by matrix metalloproteinases and their inhibitors. Stem Cells 15:180–189.

Mosher, D.F. (1990). Physiology of thrombospondin. Annu Rev Med 41:85–97.

Moyano, J.V., Carnemolla, B., Albar, J.P., et al. (1999). Cooperative role for activated α4β1 integrin and chondroitin sulfate proteoglycans in cell adhesion to the heparin III domain of fibronectin. J Biol Chem 274:135–142.

Nagai, T., Yamakawa, N., Aota, S., et al. (1991). Monoclonal antibody characterization of two distant sites required for function of the central cell-binding domain of fibronectin in cell adhesion, cell migration, and matrix assembly. J Cell Biol 114:1295–1305.

Neufeld, G., Cohen, T., Gengrinovitch, S., Poltorak, Z. (1999). Vascular endothelial growth factor (VEGF) and its receptors. FASEB J 13:9–22.

Nicosia, R.F., Bonanno, E., Smith, M., Yurchenco, P. (1994). Modulation of angiogenesis in vitro by laminin-entactin complex. Dev Biol 164:197–206.

Nicosia, R.F., Madri, J.A. (1987). The microvascular extracellular matrix. Am J Pathol 128:78–90.

Nicosia, R.F., Tuszynski, G.P. (1994). Matrix-bound thrombospondin promotes angiogenesis is vitro. J Cell Biol 124:183–193.

Niquet, J., Represa, A. (1996). Entactin immunoreactivity in immature and adult rat brain. Brain Res 95:227–233.

Nomizu, M., Kuratomi, Y., Malinda, K.M., et al. (1998). Cell binding sequences in mouse laminin α1 chain. J Biol Chem 273:32491–32499.

Olsen, B.R. (1995). New insights into the function of collagens from genetic analysis. Curr Opin Cell Biol 7:720–727.

Park, D.S., Stefanis, L., Yan, C.Y.I., Farinelli, S.E., Greene, L.A. (1996). Ordering the cell death pathway. J Biol Chem 271:21898–21905.

Penc, S.F., Blumenstock, F.A., Kaplan, J.E. (1998). A 70-kDa amino-terminal fibronectin fragment supports gelatin binding to macrophages and decreases gelatinase activity. J Leukoc Biol 64:351–357.

Perissner, K.T. (1991). Structure and biological role of vitronectin. Annu Rev Cell Biol 7:275–310.

Perissner, K.T., Seiffert, D. (1998). Role of vitronectin and its receptors in heamostasis and vascular remodeling. Thromb Res 89:1–21.

Petitclerc, E., Boutaud, A., Prestayko, A., et al. (2000). New functions for NC1 domains of human collagen-IV: novel integrin ligands regulating angiogenesis and tumor growth in vivo. J Biol Chem 275:8051–8061.

Petitclerc, E., Stromblad, S., von Schalscha, T.L., et al. (1999). Integrin αvβ3 promotes M21 melanoma growth in human skin by regulating tumor cell survival. Cancer Res 59:2724–2730.

Pozzi, A., Wary, K.K., Giancotti, F.G., Gardner, H.A. (1998). Integrin α1β1 mediates a unique collagen-dependent proliferation pathway in vivo. J Cell Biol 142:587–594.

Prockop, D.J., Kivirikko, K.I. (1995). Collagens: molecular biology, diseases, and potentials for therapy. Annu Rev Biochem 64:403–434.

Qian, X., Nicosia, R.F., Bochenety, K.M., Rothman, V.L., Tuszyhski, G.P. (1994). The effects of thrombospondin on endothelial cell tube formation in vitro. Mol Biol Cell 2:179–186.

Raugi, G.J., Olerud, J.E., Gown, A.M. (1987). Thrombospondin in early human wound tissue. J Invest Dermatol 89:551–554.

Reed, M.J., Corsa, A., Pendergrass, W., Penn, P., Sage, E.H., Abrass, I.B. (1998). Neovascularization in aged mice: delayed angiogenesis is coincident with decreased levels of transforming growth factor β1 and type I collagen. Am J Pathol 152:113–123.

Ries, A., Engel, J., Lustig, A., Kuhn, K. (1995). The function of the NC1 domains in type IV collagen. J Biol Chem 270:23790–23794.

Risau, W., Lemmon, V. (1988). Changes in the vascular extracellular matrix during embryonic vasculogenesis and angiogenesis. Dev Biol 125:441–450.

Roberts, W.G., Palade, G.E. (1997). Neovasculature induced by vascular endothelial growth factor is fenestrated. Cancer Res 57:765–772.

Ruoslahti, E., Engvall, E. (1997). Cell adhesion in vascular biology: integrins and vascular extracellular matrix assembly. J Clin Invest 99:1149–1152.

Sage, E.H., Bronstein, P. (1991). Extracellular proteins that modulate cell-matrix interactions. J Biol Chem 266:14831–14834.

Sage, E.H., Vernon, R.B., Funk, S.E., Everitt, E.A., Angello, J. (1989). SPARC, a secreted protein associated with cellular proliferation, inhibits cell spreading in vitro and exhibits Ca^{2+}-dependent binding to the extracellular matrix. J Cell Biol 109:341–356.

Scheiner, C.L., Fisher, M., Bauer, J., Juliano, R.L. (1993). Defective vasculature in fibronectin-receptor-deficient CHO cell tumor in nude mice. Int J Cancer 55:436–441.

Schittny, J.C., Yurchenco, P.D. (1989). Basement membranes: molecular organization and function in development and disease. Curr Opin Cell Biol 1:983–988.

Schnieke, A., Dziadek, M., Bateman, J., et al. (1987). Introduction of the human pro α1(I) collagen gene into pro α1(I)-deficient Mov-13 cells leads to formation of functional mouse-human hybrid type I collagen. Proc Natl Acad Sci USA 84:764–768.

Senger, D.R., Claffey, K.P., Benes, J.E., Perruzzi, C.A., Sergiou, A.P., Detmar, M. (1997). Angiogenesis promoted by vascular endothelial growth factor: regulation through α1β1 and α2β1 integrins. Proc Natl Acad Sci USA 94:13612–13617.

Sharma, B, Handler, M., Eichstetter, I., Whitelock, J.M., Nugent, M.A., Iozzo, R.V. (1998). Antisense targeting of perlecan blocks tumor growth and angiogenesis in vivo. J Clin Invest 102:1599–1608.

Shattil, S.J. (1995). Function and regulation of the β3 integrins in hemostasis and vascular biology. Thromb Haemost 74:149–155.

Smith, J.W., Cheresh, D.A. (1990). Integrin (αvβ3)-ligand interaction. J Biol Chem 265: 2168–2172.

Stanton, H., Gavrilovic, J., Arkinson, S.J., et al. (1998). The activation of proMMP-2 (gelatinase A) by HT1080 fibrosarcoma cells is promoted by culture on a fibronectin substrate and is concomitant with an increase in processing of MT1-MMP (MMP-14) to a 45 kDa form. J Cell Sci 111:2789–2796.

Stefansson, S., Lawrence, D.A. (1996). The serpin PAI-1 inhibits cell migration by blocking integrin αvβ3 binding to vitronectin. Nature 383:441–443.

Stefansson, S., Wong, K.K., McMahon, G.A., et al. (1999). Plasminogen activator inhibitor-1 inhibits angiogenesis in vivo by both vitronectin and proteinase dependent pathways. (Submitted).

Stromblad, S., Becker, J.C., Yebra, M., Brooks, P.C., Cheresh, D.A. (1996). Suppression of p53 activity and p21[WAF1/CIP1] expression by vascular cell integrin αvβ3 during angiogenesis. J Clin Invest 98:1–8.

Sunderkotter, C., Steinbrink, K., Goebeler, M., Bhardwaj, R., Sorg, C. (1994). Macrophages and angiogenesis. J Leukoc Biol 55:410–422.

Sweeney, S.M., Guy, C.A., Fields, G.B., San Antonnio, J.D. (1998). Defining the domains of type I collagen involved in heparin-binding and endothelial tube formation. Proc Natl Acad Sci USA 95:7275–7280.

Timpl, R. (1989). Structure and biological activity of basement membrane proteins. Eur J Biochem 180:487–502.

Timpl, R. (1993). Proteoglycans of the basement membranes. Experientia 49:417–428.

Tolsma, S.S., Volpert, O.V., Good, D.J., Frazier, W.A., Polverini, P.J., Bouck, N. (1993). Peptides derived from two separate domains of the matrix protein thrombospondin-1 have anti-angiogenic activity. J Cell Biol 122:497–511.

Tsilibary, E.C., Charonis, A.S. (1986). The role of the main noncollagenous domain (NC1) in type IV collagen self-assembly. J Cell Biol 103:2467–2473.

Tsilibary, E.C., Reger, L.A., Vogel, A.M., et al. (1990). Identification of a multifunctional, cell-binding peptide sequence from the α1(NC1) of type IV collagen. J Cell Biol 111: 1583–1591.

Tyagi, S.C. (1997). Vasculogenesis and angiogenesis: extracellular matrix remodeling in coronary collateral arteries and the ischemic heart. J Cell Biochem 65:388–394.

van Aken, B.E., Seiffert, D., Thinnes, T., Loskutoff, D.L. (1997). Localization of vitronectin in the normal and atherosclerotic human vessel wall. Histochem Cell Biol 107:313–320.

Vandenberg, P., Kern, A., Ries, A., Luckenbill-Edds, L., Mann, K., Kuhn, K. (1991). Characterization of a type IV collagen major cell binding site with affinity to the α1β1 and the α2β1 integrins. J Cell Biol 113:1475–1483.

Volpert, O.V., Lawler, J., Bouck, N.P. (1998). A human fibrosarcoma inhibits systemic angiogenesis and the growth of experimental metastases via thrombospondin-1. Proc Natl Acad Sci USA 95:6343–6348.

Xu, J., Rodriguez, D., Petitclerc, E., et al. (2000). Proteolytic exposure of a cryptic domain within collagen-IV is required for angiogenesis and tumor growth in vivo. (Submitted).

Yang, J.T., Rayburn, H., Hynes, R.O. (1993). Embryonic mesodermal defects in α5 integrin-deficient mice. Development 119:1093–1105.

Yayon, A., Klagsbrun, M., Esho, J.D., Leder, P., Ornitz, D.M. (1991). Cell surface, heparin-like molecules are required for binding basic fibroblast growth factor to its high affinity receptor. Cell 64:841–848.

Yurchenco, P.D. (1990). Assembly of basement membranes. Ann NY Acad Sci 580:195–213.

Yurchenco, P.D., Cheng, Y., Colognato, H. (1992). Laminin forms an independent network in basement membranes. J Cell Biol 117:1119–1133.

Yurchenco, P.D., O'Rear, J.J. (1994). Basal lamina assembly. Curr Opin Cell Biol 6:674–681.

Yurchenco, P.D., Ruben, G.C. (1987). Basement membrane structure in situ: evidence for lateral associations in the type IV collagen network. J Cell Biol 105:2559–2568.

Zetter, B.R., Brightman, S.E. (1990). Cell motility and the extracellular matrix. Curr Opin Cell Biol 2:850–856.

Zhang, Z., Vuori, K., Reed, J.C., Ruoslahti, E. (1995). The α5β1 integrin supports survival of cells on fibronectin and up-regulates Bcl-2 expression. Proc Natl Acad Sci USA 92: 6161–6165.

CHAPTER **5**

The Role of Cell Adhesion Receptors in Vascular Development: An Overview

Paul Robson, Susan Pichla, Bin Zhou, and
H. Scott Baldwin

The establishment of the cardiovascular system represents an early, critical event essential for normal embryonic development. (Risau, 1997; Baldwin, 1998; Tallquist et al., 1999). Two different processes are thought to be involved in embryonic and extraembryonic blood vessel formation: *angiogenesis*, the budding and branching of vessels from preexisting vessels, and *vasculogenesis*, the *de novo* differentiation of endothelial cells from mesoderm and organization of endothelial progenitors into a primitive vascular plexus (Fig. 5.1). This plexus then expands through sprouting angiogenesis, intussusceptive growth, and intercalation of new endothelial cells. Finally, this vascular plexus is remodeled by pruning, fusion, and regression of preexisting vessels into an arborized vascular tree composed of arteries, capillaries, and veins (reviewed in Carmeliet and Collen, 1999). Formation of the yolk sac circulation, differentiation of the endocardium of the heart, and development of larger vascular networks occur by vasculogenesis, while other organs, such as the brain and kidney, appear to be vascularized primarily by angiogenesis.

Although angiogenesis can occur in both the embryo and adult (e.g., wound healing, tumor neovascularization, etc.), it was initially believed that vasculogenesis was restricted to embryonic development. However, recent identification of endothelial progenitor cells isolated from circulating blood (Asahara et al., 1997) and bone marrow (Shi et al., 1998; Takahashi et al., 1999) has raised speculation that vasculogenesis might also be involved in postnatal neovascularization (reviewed in Isner and Asahara, 1999; Tomanek and Schatteman, 2000). The organization of emerging endothelial cells into a vascular tree that will eventually mature into the adult cardiovascular system is the end result of a complex series of interactions between cells and the extracellular matrix (ECM) as well as cell-cell interactions. These interactions are mediated, at least in part, by specific cell adhesion receptors. Several different classes of cell adhesion molecules have been identified during vascular development (reviewed in Baldwin, 1996; Bazzoni et al., 1999) including members of the cadherin, immunoglobulin, integrin, and selectin families (Fig. 5.2). However, only a few of these receptors have been shown to be *endothelial*-specific cell adhesion molecules, and *in vivo* data documenting the role of these receptors in mediating either vasculogenic and/or angiogenic processes are

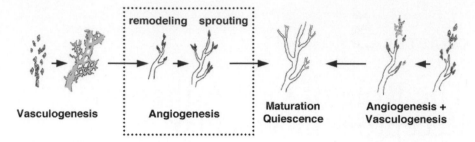

FIGURE 5.1.   Schematic depiction of the sequential development of mature vasculature.

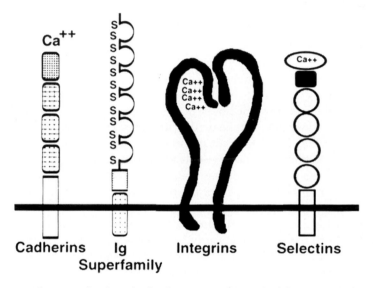

FIGURE 5.2.   The major families of cell adhesion receptors studied during vascular development. Each receptor is characterized by an extracellular region containing the ligand-binding domain, a short hydrophobic transmembrane region, and cytoplamsic tail. The cadherins are monomeric proteins involved in Ca^{2+}-dependent homophilic adhesion. They contain four homologous repeating units (large screens), and ligand specificity is determined primarily by the most distal repeat (small screen). The members of the immunogloboulin superfamily contain a characteristic, but variable, number of immunoglobulin-like domains formed by disulfide bonds, as well as a variable number of fibronectin type-III repeats (horizontal striped box). Integrins are heterodimers that participate in both cell-cell and cell-matrix adhesion. Selectins bind carbohydrate groups on adjacent cells and contain a Ca^{2+}-dependent lectin (oval) a single epidermal growth factor repeat (black square), and several complement-binding repeats (circles).

limited. This chapter reviews the mounting experimental data that are beginning to unravel the complicated cell adhesion–mediated interactions that are responsible for development of the vascular system.

VE-CADHERIN

Vascular endothelial (VE)-cadherin (CAD5) is a member of the cadherin family of transmembrane cell adhesion molecules that mediate homotypic cell-cell adhesion through their extracellular domain and are linked to the actin cytoskeleton via interactions of its cytoplasmic domain with α-catenin, β-catenin, and plakoglobin/γ-catenin (Lampugnani et al., 1995; reviewed in Yap et al., 1997). It is the primary cell adhesion molecule forming tight junctions and is exclusively produced by endothelial cells (Lampugnani et al., 1992). VE-cadherin is an early marker of endothelial differentiation as it is first detected among mesodermal derivatives destined to become endothelial cells during initial formation of the yolk sac blood islands (Breier et al., 1996). FACS analysis of embryonic stem cell differentiation (Nishikawa et al., 1998; Hirashima et al., 1999) and endothelial cells isolated from the early embryo proper (Nishikawa et al., 1998) document that expression of VE-cadherin correlates with hemangioblast commitment to the endothelial lineage. Similar conculsions have been reached by a detailed whole-mount immunonohistochemical analysis of VE-cadherin expression in embryonic day E6.5 to E9.5 embryos (Drake and Fleming, 2000). Following initial expression in the primitive vasculature, VE-cadherin is subsequently expressed by all endothelial cells in the adult (Breier et al., 1996).

In vitro experiments using cultured endothelial cells initially suggested that VE-cadherin might function in the formation of vascular-like structures in various extracellular matrices (Matsumura et al., 1997; Bach et al., 1998; Yang et al., 1999). *In vivo* data in the mature animal has documented an additional role for VE-cadherin in regulating leukocyte transmigration across the endothelial cell barrier as disruption of VE-cadherin–VE-cadherin interactions is essential for this process to occur (Del Maschio et al., 1996; Allport et al., 1997; Allport et al., 2000). In addition, it has been shown to regulate endothelial integrity and permeability as administration of a function blocking monoclonal antibody results in a time-dependent increase in vascular permeability in the heart and lungs with subsequent interstitial edema and inflammation (Corada et al., 1999). The first evidence that VE-cadherin might be important for vascular development came from initial experiments utilizing embryoid body formation from VE-cadherin −/− ES cells. These studies showed that while endothelial cells could differentiate from the mesoderm, they could not organize into a definitive vascular plexus, suggesting a defect in latter stages of vasculogenesis (Vittet et al., 1997). This observation was consistent with previous *in vitro* data. However, when mutant mice were generated from these same ES cells, they demonstrated delayed endothelial differentiation, but vasculogenesis did occur. The most striking defect was the absence of angiogenic remodeling of the primitive vascular plexus and the embryos died by E9.5 (Gory-Faure et al., 1999). The defects were most pronounced in the extraembryonic yolk sac circulation, and, surprisingly, VE-cadherin −/− embryos did not demonstrate defective cell-cell interactions, suggesting that VE-cadherin is dispensable for endothelial homophilic interaction but essential for some additional aspect of angiogenic remodeling.

The central role of VE-cadherin in vascular remodeling was further delineated by a second group of investigators evaluating both null mutant mice and those homozygous for the β-catenin–binding cytosolic domain truncation (Carmeliet et al., 1999). Again, endothelial differentiation and plexus formation occurred normally, but embryonic demise occurred by E9.5 with little angiogenic expansion of the vascular bed (Fig. 5.3). This group went on to demonstrate an increase in endothelial apoptosis in null mutant and truncation mutant embryos. *In vitro* studies with endothelial cells isolated from these embryos documented failure of

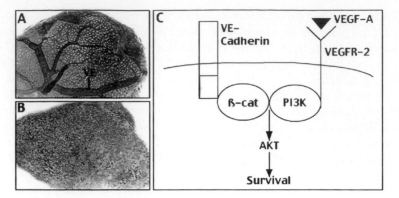

FIGURE 5.3.   Yolk sacs from embryonic day E9.25 wild-type embryos (A, a′) and embryos homozygous for a truncation mutation of the VE-cadherin cytoplasmic domain (B, b) stained with antibody to platelat endothelial cell adhesion molecule (PECAM). Significant endothelial differentiation has occurred but the primitive vascular plexus was unable to remodel to form an arborized vaculautre (VE, vitello-embryonic vessels). C: The proposed model of the role of VE-cadherin in vascular endothelial growth factor-A (VEGF-A)-mediated survival of endothelial cells via formation of a multicomponent junctional complex between VE-cadherin, β-catenin (β-cat), phosphetidylinositol (PI$_3$)-kinase (PI$_3$K), and VEGFR-2. VE-cadherin promotes VEGF-A–mediated activation of Akt and endothelial survival. (From Cameliet et al., 1999, with permission.)

these cells to transmit antiapoptotic signals of vascular endothelial growth factor-A (VEGF-A) through the Akt kinase and Bcl2 pathway. Disruption of this pathway was as a result of the inability of phosphatidylinositol (PI$_3$) kinase associated with activated VEGF receptor-1 (VEGFR-1) (flk-1) to complex with VE-cadherin–associated β-catenin (Fig. 5.3). Interestingly, one of the initial endothelial responses to apoptosis is cleavage of VE-cadherin and β-catenin (Brancolini et al., 1997; Herren et al., 1998), raising the possibility of accelerated apoptosis in VE-cadherin mutant embryos. Finally, there was no difference in response of these cells to basic fibroblast growth factor (bFGF). Thus, these experiments define the predominant role of VE-cadherin in mediating endothelial survival, not endothelial cell-cell recognition during vascular development, and further delineate different pathways for VEGF- and bFGF-mediated vascular ontogeny. This group also demonstrated extensive disorganization of the endocardium and myocardium, emphasizing the importance of myocardial-endocardial interactions during development.

N-CADHERIN

Neural (N)-cadherin is the only other major cadherin expressed by endothelial cells that apparently "competes" with VE-cadherin for localization to cell-cell borders. The exclusive presence of VE-cadherin in tight junctions is determined by unique recognition sequences in the cytoplasmic tail as defined by chimeric and deletion mutations. The presence of this cytoplasmic domain results in concentration of VE-cadherin to the tight junctions and diffuse distribution of N-cadherin over the surface of the cell (Navarro et al., 1998). Interestingly, mice lacking N-cadherin show a definitive defect in yolk-sac vascularization (Radice et al., 1997) with *in utero* death by E9.5. However, it is difficult to determine whether this is a primary phenotype, given that a major abnormality detected in the N-cadherin null mutant mice is dramatic alteration in the developing heart and myocardium. Thus, the yolk sac defects could be merely a reflection of altered cardiac performance resulting in decreased cardiac output. It has now been demonstrated that angiogenic remodeling is flow dependent (le Noble et al., 2000), supporting the yolk sac defects as a secondary phenomenon. Furthermore, replacement of N-cadherin in cardiac muscle only by transgenic insertion utilizing a myocardial specific promoter results in elimination of the yolk sac defect (Glen Radice, personal communication), emphasizing that the yolk sac defects are not the primary etiology of embryonic demise. Finally, no accentuation of N-cadherin expression is detected in VE-cadherin mutants, suggesting that the regulation of these two cadherins is not interdependent (Gory-Faure et al., 1999).

PECAM-1 (CD31)

Platelet endothelial cell adhesion molecule-1 (PECAM-1) is a member of the immunoglobulin (Ig) superfamily of cell adhesion receptors. It is expressed primarily at the cell-cell junctions of endothelial cells but is also detected at much lower levels on leukocytes, subsets of T lymphocytes, bone marrow stem cells, megakaryocytes, and platelets (reviewed in DeLisser et al., 1977). While PECAM-1 was initially thought to be primarily a cell adhesion molecule, evidence that tyrosine phosphorylation of the cytoplasmic domain results in binding of c-*src* (Lu et al., 1997) and the phosphatases SHP-1 and SHP-2 has raised considerable interest in the possible role of PECAM-1 in "outside-in" endothelial cell signaling (Newman, 1999). Furthermore, *in vitro* experiments suggesting that PECAM-1 can modulate levels of tyrosine phosphorylated β-catenin and can bind plakoglobin/γ-catenin have provided additional provocative evidence of the role of PECAM-1 in moderating endothelial cell phenotype (Ilan et al., 2000).

PECAM-1 is the first endothelial cell adhesion molecule to be expressed during endothelial differentiation (Baldwin et al., 1994; Vecchi et al., 1994; Vittet et al., 1996; Drake and Fleming, 2000) (Fig. 5.4) and it is expressed in multiple isoforms that result from alternative splicing of cytoplasmic domain exons during vascular development (Baldwin et al., 1994; Sheibani and Frazier, 1999), Interestingly, these isoforms are capable of either heterophilic or homophilic adhesion, dependent on the presence or absence of exon 14, in the cytoplasmic domain (Yan et al., 1995; Famiglietti et al., 1997). This exon contains one of the tyrosines thought to be critical in cytoplasmic signaling functions. *In vitro* and *in vivo* experiments using anti-PECAM antibodies have strongly suggested a role for PECAM in angiogenesis (DeLisser et al., 1997; Matsumura et al., 1997; Zhou et al., 1999). Furthermore, Madri and colleagues have documented that during normal vascular plexus development in the murine embryo, PECAM-1 undergoes tyrosine dephosphorylation of the cytoplasmic domain that correlates with the establishment of blood islands and definitive vessel formation (Pinter et al., 1997). This group has gone on to demonstrate that vascular changes in the yolk sac and embryo that are associated with hyperglycemia correlate with an inhibition of this tyrosine dephosphorylation (Pinter et al., 1999). Given the abundance of both *in vitro* and *in vivo* data to suggest a critical role for PECAM-1 in vascular development, the recent report of the PECAM null mutation is particularly surprising. Duncan and colleagues (1999)

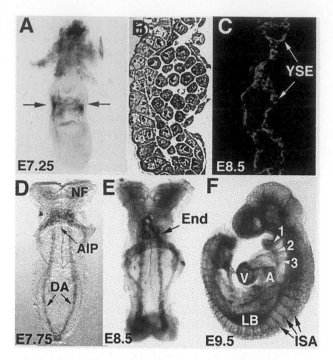

FIGURE 5.4.   Developmental distribution of PECAM. (A) *In situ* hybridization documenting initial expression of PECAM within the developing blood islands of the extraembryonic yolk sac in the E7.25 embryo. Photomicrograph (B) and immunohistochemical staining (C) of cross sections through the yolk sac of an E8.5 embryo showing that PECAM-1 is expressed by the endothelial cells lining the blood islands but is not detectable in the central hematopoietic precursors. D–F: Subsequent *in situ* expression of PECAM during early vasculogenesis of the endocardium (End) and dorsal aorta (DA) as well as during initial angiogenesis of the intersomitic arteries (ISA). NF, neural fold; AIP, anterior intestinal portal, or foregut; LB, limb bud; A, atria; V, ventricle; 1, 2, 3 = 1st, 2nd, and 3rd pharyngeal arches.

report no obvious defects in either vasculogenesis or angiogenesis in these mice. While they did observe subtle defects in leukocyte transmigration, and a subsequent report has identified bleeding abnormalities (Mahooti et al., 2000), this work suggests that PECAM-1 does not play an obligate role in vascular ontogeny. The possibility that VE-cadherin might be upregulated to compensate for the absence of PECAM has not been investigated.

Vascular cell adhesion molecule-1 (VCAM-1), another member of the Ig superfamily, thought to primarily function in white cell–endothelial cell interactions in response to inflammation, has been shown to be essential for normal vascular development in the embryo. Again, utilizing targeted disruption of the VCAM-1 gene, Kwee and colleagues (1995) and Gurtner et al. (1995) were able to demonstrate that a null mutation resulted in a failure of placental development. The allantois, which normally expresses VCAM-1, was unable to fuse with the chorion that expresses the α_4 integrin (see below), the only known ligand for VCAM-1. This resulted in death around embryonic day E9.5 to E10. Interestingly, a small number of VCAM-1 mutants did form a functional placenta and were able to survive until day E11.5 to E12.5, when they appeared to die from congestive heart failure. A close analysis of the hearts of these embryos revealed the loss of the epicardium and lack of subepicardial vascular development (Fig. 5.5). This suggests that the VCAM-1 (expressed by the subjacent myocardium) and α_4 (expressed by the epicardium) interactions are essential for normal epicardial and therefore coronary vascular development (Fig. 5.5) (reviewed in Mikawa, 1999). A reciprocal disruption of the α_4 gene gave a very similar phenotype (Yang et al., 1995). There was abnormal chorioallantoic development and some embryonic death at around day E10.5. However, the α_4 mutants that were able to develop functional placental circulation lived 1 to 2 days longer than the VCAM-1 mice *in utero* (E13.5), and clearly demonstrated loss of the epicardium and absence of coronary vascular development (Fig. 5.5). Thus, VCAM-1/α_4 interactions appear to be required for normal coronary vascular development as well as for normal placental formation.

INTEGRINS

The integrins are the best characterized group of cell-ECM adhesion molecules and are heterodimeric proteins that consist of an α subunit noncovalently associated with a β subunit. The combination of specific α and β subunits determines the particular ligands that are bound, and a given cell type usually expresses more than one integrin. Thus the ability of a cell to modulate adhesive characteristics is determined by both the types of integrins expressed as well as by the absolute number of any given integrin. The association of integrins with their ligands has been shown to affect cell proliferation, migration, differentiation, activation, and gene expression (reviewed in Hynes, 1992). Endothelial cells express several different integrins of the β_1 ($\alpha_1\beta_1$, $\alpha_2\beta_1$, $\alpha_3\beta_1$) and α_v families ($\alpha_v\beta_3$, $\alpha_v\beta_5$), as well as $\alpha_3\beta_6$, and $\alpha_6\beta_4$ (reviewed in Buck et al., 1993; Bazzoni et al., 1999; Hynes et al., 1999). Of these, the β_1 and α_v integrins have been the most heavily studied in the context of their role in vascular morphogenesis.

The *β_1 family of integrins* has received particular attention in the study of vascular development. Utilizing the monoclonal antibody CSAT (Cell Surface ATtachment), which interferes with ligand binding to all the β_1 integrins, Drake and colleagues (1992) were able to block formation of the lumen of the dorsal aorta in the early chick embryo. Interestingly, in these experiments the early stages of vasculogenesis were not obviously affected in that angioblast differentiation and spatial organization into chords appeared normal. Only the final stages of tube formation were inhibited, suggesting that while β_1 integrins are critical for final development of an endothelial tube, they are not involved in initial commitment of mesoderm to endothelial cell lineage or initial spatial organization. Elimination of all β_1 integrins by targeted deletion results in embryonic lethality prior to implantation and gastrulation; thus information on vascular specific deletions is not available. Conversely, null mutations of α_1 have no obvious phenotype (Gardner et al., 1996) while α_3 (Kreidberg et al., 1996) and α_6 (Georges-Labouesse et al., 1996) null mutants show perinatal lethality but no obvious vascular defects.

In a specific study of the α_5 integrins, Yang and coworkers (1993) deleted the α_5 gene by homologous recombination. The homozygous null mutations were embryonically lethal by E10 to E11, and while they developed a heart and vascular system, they showed defects in both the extraembryonic and embryonic vascular bed. The authors speculate that embryonic death was due to abnormal

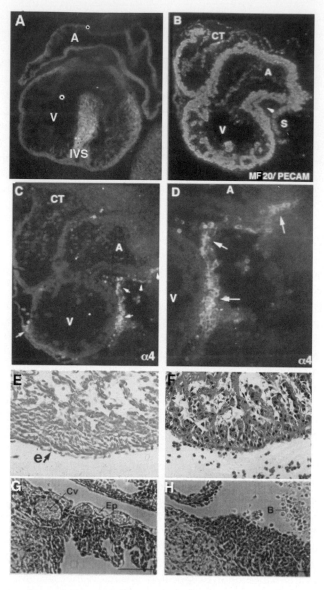

FIGURE 5.5.   Vascular cell adhesion molecule-1 (VCAM-1) and α_4 integrin expression during epicardial development. (A) Immunohistochemical localization of VCAM expression in a cross section through an E11.5 mouse heart. VCAM is expressed throughout the myocardium of the atrium (A), the compact layer of the ventricular (V) myocardium, and the interventricular septum (IVS). (B) Dual immunofluorescent photomicrograph of a sagittal section through an E10.5 heart defining the myocardium (red, antibody to sarcomeric myosin) and endocardium (green, antibody to PECAM). Low-power (C) and high-power (D) sections adjacent to that depicted in B, showing immunohistochemical localization of α_4-positive epicardial cells originating in the sinus venosus (S), contacting the heart at the atrioventricular sulcus, and migrating over the surface of the heart (arrows). CT, conotruncus. Cross sections through the heart of E12.5 wild-type (E) and VCAM −/− embryos (F) showing absence of epicardial formation and thinning of the compact myocardial layer. Cross sections through the heart of an E13.5 wild type (G) and α_4 −/− embryos (H) demonstrating inhibition of coronary vessel (Cv) formation and absence.

vascular integrity. Specifically, blood islands were able to form in the yolk sac but did not undergo remodeling by E9.5, which was observed in the wild-type embryos. This bears some resemblance to the processes affected by the injection of the CSAT antibody into the quail embryo discussed above. Furthermore, vessels that did form within the embryos appeared to be "leaky," and diffuse extravasation of blood cells was noted. Interestingly, the phenotype was much less severe than that seen in the fibronectin null embryos (George et al., 1993). This is not surprising, for while fibronectin is the only known ligand for $\alpha_5\beta_1$, several integrins can serve as fibronectin receptors as mentioned above. These additional fibronectin receptors may have been sufficient for many functions normally mediated by $\alpha_5\beta_1$ or perhaps even upregulated in the absence of $\alpha_5\beta_1$. This group went on to show that the dramatic effects of the fibronectin null mutation were strain dependent, with no heart and limited vessel formation on the 129/Sv background and formation of a definitive heart tube and endocardium on the C57/BL6 background (George et al., 1997). Thus, moderating factors other than just ligand recognition modulate the endothelial cell adhesion receptor interaction with the ECM.

The α_v *integrins* ($\alpha_v\beta_3$, $\alpha_v\beta_5$) have been extensively studied in relation to their role in angiogenesis. This is particularly true of $\alpha_v\beta_3$ (reviewed in Eliceiri and Cheresh, 1999, 2000). Because $\alpha_v\beta_3$ preferentially binds vitronectin, it is classified as one of the major vitronectin receptors. Yet it is one of the most promiscuous integrins and can bind several other ligands. Despite this ability to bind several different ligands, $\alpha_v\beta_3$ does not normally demonstrate a diffuse expression pattern in the quiescent vasculature (Hirsch et al., 1994), but it is prominently expressed by cytokine-stimulated endothelial and smooth muscle cells, and expression is particularly accentuated in newly formed vessels in the embryo (Drake et al., 1995). Brooks and colleagues (1994a,b) have demonstrated that $\alpha_v\beta_3$ antagonists (antibodies and specific cyclic peptides) block growth factor–induced *angiogenesis* in the chick chorioallantoic membrane (CAM) model (Brooks et al., 1994a), and subsequently showed that the $\alpha_v\beta_3$ antagonists blocked angiogenesis and induced tumor regression by selectively promoting apoptosis of angiogenic vascular endothelium (Brooks et al., 1994b). In addition, they have been able to show that VEGF and bFGF mediate distinct angiogenic mechanisms by utilizing either $\alpha_v\beta_3$ or $\alpha_v\beta_5$ signaling, respectively (Friedlander et al., 1995). *In vitro* evidence showing that $\alpha_v\beta_3$ can bind matrix metalloproteinase-2 (MMP-2) (Brooks et al., 1996), participate in the activation of the VEGF-A receptor, flk-1 (Soldi et al., 1999), and mediate EphB1-coupled activation of cell attachment (Huynh-Do et al., 1999) further support a role for $\alpha_v\beta_3$ in vascular development. Finally, reports that $\alpha_v\beta_3$ may be a ligand for PECAM (see discussion above) (Piali et al., 1995; Buckley et al., 1996; Testaz et al., 1999) and identification of Del1, a novel angiogenic ECM component expressed by endothelial cells during development that binds $\alpha_v\beta_3$ (Hidai et al., 1998; Penta et al., 1999), suggests the possibility of alternative roles for the α_v integrins in heterophilic interactions and autocrine regulation during vasculogenesis and angiogenesis.

An essential role for $\alpha_v\beta_3$ in vascular ontogeny appeared to be confirmed by the *in vivo* experiments of Drake and colleagues (1995), who demonstrated a dramatic disruption in formation of the dorsal aorta of chicken embryos by administration of the anti-$\alpha_v\beta_3$ antibody LM609. However, the central role of the α_v integrins in vascular development has been called into question by the evaluation of the α_v null mutant mice, which revealed extensive angiogenesis and vasculogenesis prior to embryonic demise (Hynes et al., 1999); 80% of the embryos died *in utero*, most likely as a result of placental insufficiency, and 20% were born alive but developed intestinal and intracerebral hemorrhages. However, the initial stages of embryonic and extraembryonic vascular development appeared normal through E9.5 Similarly, the β_3 null mutant mice showed normal vascular development with some placental insufficiency and thrombasthenia at birth (Hodivala-Dilke et al., 1999). Inactivation of the β_5 subunit has not been published. This dramatic difference between peptide and antibody blocking experiments and gene mutations was totally unexpected and initially quite difficult to explain. However, *in vitro* data from two groups have provided compelling evidence for a "transdominant inhibition" of integrin receptors (Diaz-Gonzalez et al., 1996; Hartwell et al., 1998). These experiments suggest that ligand binding of one integrin can inhibit the function of other target integrins. Thus, suppressive crosstalk between integrins is maintained by antibody blockade but removed when the integrin is genetically deleted. While this explanation has not been tested by *in vivo* experiments during development, it provides a more appealing explanation than mere "genetic redundancy."

SUMMARY

There has been an explosion of both *in vitro* and *in vivo* experimentation over the last 5 years that has focused on the role of cell adhesion in vascular ontogeny. But many questions remain unanswered. These studies have begun to unravel a complex set of interactions between distinct receptors and multiple ligands, receptor-receptor interactions, and interactions between cell adhesion receptors and vascular cytokines. In many cases, *in vitro* data appear to be contradicted by *in vivo* results, particularly when peptide and antibody experiments are compared to genetic deletion studies. However, evidence is beginning to surface to suggest that many of these discrepancies may be resolved as the interplay between the cell adhesion function and the cell signaling (outside in and inside out) function of these transmembrane proteins is deciphered. Furthermore, it is clear that interactions between different cell types in the vascular compartment (endothelial cells, neural crest cells, smooth muscle cells, perictyes), which will certainly be mediated by adhesion receptors, are critical in defining the patterning of the vascular tree (Waldo et al., 1996). Yet, little is known about the developmental regulation of interactions between these distinct cell populations. Tissue specific and temporally inducible mutations are certain to help in this arena. Finally, there is a host of new cell adhesion receptors yet to be clearly studied that will be essential for definition of the true regulation of vascular development mediated at the cell-cell and cell-ECM interfaces. While somewhat daunting, the importance of these future studies in providing the experimental rationale for novel therapeutic interventions is significant.

REFERENCES

Allport, J.R., Ding, H., et al. (1997). Endothelial-dependent mechanisms regulate leukocyte transmigration: a process involving the proteasome and disruption of the vascular endothelial-cadherin complex at endothelial cell-to-cell junctions. J Exp Med 186(4):517–527.

Allport, J.R., Muller, W.A., et al. (2000). Monocytes induce reversible focal changes in vascular endothelial cadherin complex during transendothelial migration under flow. J Cell Biol 148(1):203–216.

Asahara, T., Murohara, T., et al. (1997). Isolation of putative progenitor endothelial cells for angiogenesis. Science 275(5302):964–967.

Bach, T.L., Barsigian, C., et al. (1998). VE-cadherin mediates endothelial cell capillary tube formation in fibrin and collagen gels. Exp Cell Res 238(2):324–334.

Baldwin, H.S. (1996). Early embryonic vascular development. Cardiovasc Res 31(spec no): E34–45.

Baldwin, H.S. (1998). Molecular determinants of embryonic vascular development. In: Polin, R.A., Fox, W.W., eds. Fetal and neonatal physiology, vol. 1, pp. 801–813. Philadelphia: W.B. Saunders.

Baldwin, H.S., Shen, H.M., et al. (1994). Platelet endothelial cell adhesion molecule-1 (PECAM-1/CD31): alternatively spliced, functionally distinct isoforms expressed during mammalian cardiovascular development. Development 120(9):2539–2553.

Bazzoni, G., Dejana, E., et al. (1999). Endothelial adhesion molecules in the development of the vascular tree: the garden of forking paths. Curr Opin Cell Biol 11(5):573–581.

Brancolini, C., Lazarevic, D., et al. (1997). Dismantling cell-cell contacts during apoptosis is coupled to a caspase-dependent proteolytic cleavage of beta-catenin. J Cell Biol 139(3): 759–771.

Breier, G., Breviario, F., et al. (1996). Molecular cloning and expression of murine vascular endothelial-cadherin in early stage development of cardiovascular system. Blood 87(2):630–641.

Brooks, P.C., Clark, R.A., et al. (1994a). Requirement of vascular integrin alpha v beta 3 for angiogenesis. Science 264(5158):569–571.

Brooks, P.C., Montgomery, A.M., et al. (1994b). Integrin alpha v beta 3 antagonists promote tumor regression by inducing apoptosis of angiogenic blood vessels. Cell 79(7):1157–1164.

Brooks, P.C., Stromblad, S., et al. (1996). Localization of matrix metalloproteinase MMP-2 to the surface of invasive cells by interaction with integrin alpha v beta 3. Cell 85(5):683–693.

Buck, C.A., Baldwin, H.S., et al. (1993). Cell adhesion receptors and early mammalian heart development: an overview. C R Acad Sci III 316(9):838–859.

Buckley, C.D., Doyonnas, R., et al. (1996). Identification of alpha v beta 3 as a heterotypic ligand for CD31/PECAM-1. J Cell Sci 109(pt 2):437–445.

Carmeliet, P., Collen, D. (1999). Role of vascular endothelial growth factor and vascular endothelial growth factor receptors in vascular development. Curr Top Microbiol Immunol 237:133–158.

Carmeliet, P., Lampugnani, M.G., et al. (1999). Targeted deficiency or cytosolic truncation of the VE-cadherin gene in mice impairs VEGF-mediated endothelial survival and angiogenesis. Cell 98(2):147–157.

Corada, M., Mariotti, M., et al. (1999). Vascular endothelial-cadherin is an important determinant of microvascular integrity in vivo. Proc Natl Acad Sci USA 96(17):9815–9820.

Del Maschio, A., Zanetti, A., et al. (1996). Polymorphonuclear leukocyte adhesion triggers the disorganization of endothelial cell-to-cell adherens junctions. J Cell Biol 135(2):497–510.

DeLisser, H., Baldwin, H.S., et al. (1977). PECAM-1. Trends Cardiovasc Med 151:671–677.

DeLisser, H.M., Christofidou-Solomidou, M., et al. (1997). Involvement of endothelial PECAM-1/CD31 in angiogenesis. Am J Pathol 151(3):671–677.

Diaz-Gonzalez, F., Forsyth, J., et al. (1996). Trans-dominant inhibition of integrin function. Mol Biol Cell 7(12):1939–1951.

Drake, C.J., Cheresh, D.A., et al. (1995). An antagonist of integrin alpha v beta 3 prevents maturation of blood vessels during embryonic neovascularization. J Cell Sci 108(pt 7): 2655–2661.

Drake, C.J., Davis, L.A., et al. (1992). Antibodies to beta 1-integrins cause alterations of aortic vasculogenesis, in vivo. Dev Dyn 193(1):83–91.

Drake, C.J., Fleming, P.A. (2000). Vasculogenesis in the day 6.5–9.5 mouse embryo. Blood 95(5):1671–1679.

Duncan, G.S., Andrew, D.P., et al. (1999). Genetic evidence for functional redundancy of platelet/endothelial cell adhesion molecule-1 (PECAM-1): CD31-deficient mice reveal PECAM-1-dependent and PECAM-1-independent functions. J Immunol 162(5):3022–3030.

Eliceiri, B.P., Cheresh, D.A. (1999). The role of alpha-v integrins during angiogenesis: insights into potential mechanisms of action and clinical development. J Clin Invest 103(9):1227–1230.

Eliceiri, B.P., Cheresh, D.A. (2000). Role of alpha v integrins during angiogenesis [in process citation]. Cancer J Sci Am 6(suppl 3):S245–249.

Famiglietti, J., Sun, J., et al. (1997). Tyrosine residue in exon 14 of the cytoplasmic domain of platelet endothelial cell adhesion molecule-1 (PECAM-1/CD31) regulates ligand binding specificity. J Cell Biol 138(6):1425–1435.

Friedlander, M., Brooks, P.C., et al. (1995). Definition of two angiogenic pathways by distinct alpha v integrins. Science 270(5241):1500–1502.

Gardner, H., Kreidberg, J., et al. (1996). Deletion of integrin a_1 by homologous recombination permits normal murine development but gives rise to a specific deficit in cell adhesion. Dev Biol 175:301–313.

George, E.L., Baldwin, H.S., et al. (1997). Fibronectins are essential for heart and blood vessel morphogenesis but are dispensable for initial specification of precursor cells. Blood 90(8):3073–3081.

George, E.L., Georges-Labouesse, E.N., et al. (1993). Defects in mesoderm, neural tube and vascular development in mouse embryos lacking fibronectin. Development 119(4):1079–1091.

Georges-Labouesse, E., Messaddeq, N., et al. (1996). Absence of integrin alpha 6 leads to epidermolysis bullosa and neonatal death in mice. Nat Genet 13(3):370–373.

Gory-Faure, S., Prandini, M.H., et al. (1999). Role of vascular endothelial-cadherin in vascular morphogenesis. Development 126(10):2093–2102.

Gurtner, G.C., Davis, V., et al. (1995). Targeted disruption of the murine VCAM1 gene: essential role of VCAM-1 in chorioallantoic fusion and placentation. Genes Dev 9(1):1–14.

Hartwell, D.W., Butterfield, C.E., et al. (1998). Angiogenesis in P- and E-selectin-deficient mice. Microcirculation 5(2–3):173–178.

Herren, B., Levkau, B., et al. (1998). Cleavage of beta-catenin and plakoglobin and shedding of VE-cadherin during endothelial apoptosis: evidence for a role for caspases and metalloproteinases. Mol Biol Cell 9(6):1589–1601.

Hidai, C., Zupancic, T., et al. (1998). Cloning and characterization of developmental endothelial locus-1: an embryonic endothelial cell protein that binds the alpha-v-beta-3 integrin receptor. Genes Dev 12(1):21–33.

Hirashima, M., Kataoka, H., et al. (1999). Maturation of embryonic stem cells into endothelial cells in an in vitro model of vasculogenesis. Blood 93(4):1253–1263.

Hirsch, E., Gullberg, D., et al. (1994). A-v integrin subunit is predominantly located in nervous tissue and skeletal muscle during mouse development. Dev Dyn 210:108–120.

Hodivala-Dilke, K.M., McHugh, K.P., et al. (1999). Beta-3-integrin-deficient mice are a model for Glanzmann thrombasthenia showing placental defects and reduced survival. J Clin Invest 103(2):229–238.

Huynh-Do, U., Stein, E., et al. (1999). Surface densities of ephrin-B1 determine EphB1-coupled activation of cell attachment through alpha-v-beta-3 and alpha-5-beta-1 integrins. EMBO J 18(8):2165–2173.

Hynes, R.O. (1992). Integrins: versatility, modulation and signaling in cell adhesion. Cell 69:11–25.

Hynes, R.O., Bader, B.L., et al. (1999). Integrins in vascular development. Braz J Med Biol Res 32(5):501–510.

Ilan, N., Cheung, L., et al. (2000). Platelet-endothelial cell adhesion molecule-1 (CD31), a scaffolding molecule for selected catenin family members whose binding is mediated by different tyrosine and serine/threonine phosphorylation. J Biol Chem 275(28):21435–21443.

Isner, J.M., Asahara, T. (1999). Angiogenesis and vasculogenesis as therapeutic strategies for postnatal neovascularization. J Clin Invest 103(9):1231–1236.

Kreidberg, J.A., Donovan, M.J., et al. (1996). Alpha 3 beta 1 integrin has a crucial role in kidney and lung organogenesis. Development 122(11):3537–3547.

Kwee, L., Baldwin, H.S., et al. (1995). Defective development of the embryonic and extraembryonic circulatory systems in vascular cell adhesion molecule (VCAM-1) deficient mice. Development 121(2):489–503.

Lampugnani, M.G., Resnati, M., et al. (1992). A novel endothelial-specific membrane protein is a marker of cell-cell contacts. J Cell Biol 118:1511–1522.

Lampugnani, M.G., Corada, M., et al. (1995). The molecular organization of endothelial cell to cell junctions: differential association of plakoglobin, beta-catenin, and alpha-catenin with vascular endothelial cadherin (VE-cadherin). J Cell Biol 129(1):203–217.

le Noble, F., Frederid, P., et al. (2000). The transition from vasculogenesis to embryonic arteriogenesis is flow dependent. FASEB 14:A35.

Lu, T.T., Barreuther, M., et al. (1997). Platelet endothelial cell adhesion molecule-1 is phosphorylatable by c-Src, binds Src-Src homology 2 domain, and exhibits immunoreceptor tyrosine-based activation motif-like properties. J Biol Chem 272(22):14442–14446.

Mahooti, S., Graesser, D., et al. (2000). PECAM-1 (CD31) expression modulates bleeding time in vivo. Am J Pathol 157(1):75–81.

Matsumura, T., Wolff, K., et al. (1997). Endothelial cell tube formation depends on cadherin 5 and CD31 interactions with filamentous actin. J Immunol 158(7):3408–3416.

Mikawa, T. (1999). Cardiac lineages. In: Harvey, R.P., Rosenthal, N., eds. Heart development, pp. 19–33. San Diego: Academic Press.

Navarro, P., Ruco, L., et al. (1998). Differential localization of VE- and N-cadherins in human endothelial cells: VE-cadherin competes with N-cadherin for junctional localization. J Cell Biol 140(6):1475–1484.

Newman, P.J. (1999). Switched at birth: a new family for PECAM-1. J Clin Invest 103(1):5–9.

Nishikawa, S.I., Nishikawa, S., et al. (1998). Progressive lineage analysis by cell sorting and culture identifies FLK1 + VE-cadherin+ cells at a diverging point of endothelial and hemopoietic lineages. Development 125(9):1747–1757.

Penta, K., Varner, J.A., et al. (1999). Del1 induces integrin signaling and angiogenesis by ligation of alpha-V-beta-3. J Biol Chem 274(16):11101–11109.

Piali, L., Hammel, P., et al. (1995). CD31/PECAM-1 is a ligand for alpha v beta 3 integrin involved in adhesion of leukocytes to endothelium. J Cell Biol 130(2):451–460.

Pinter, E., Barreuther, M., et al. (1997). Platelet-endothelial cell adhesion molecule-1 (PECAM-1/CD31) tyrosine phosphorylation state changes during vasculogenesis in the murine conceptus. Am J Pathol 150(5):1523–1530.

Pinter, E., Mahooti, S., et al. (1999). Hyperglycemia-induced vasculopathy in the murine vitelline vasculature: correlation with PECAM-1/CD31 tyrosine phosphorylation state. Am J Pathol 154(5):1367–1379.

Radice, G.L., Rayburn, H., et al. (1997). Developmental defects in mouse embryos lacking N-cadherin. Dev Biol 181(1):64–78.

Risau, W. (1997). Mechanisms of angiogenesis. Nature 386:671–674.

Sheibani, N., Frazier, W.A. (1999). Thrombospondin-1, PECAM-1, and regulation of angiogenesis. Histol Histopathol 14(1):285–294.

Shi, Q., Rafii, S., et al. (1998). Evidence for circulating bone marrow-derived endothelial
cells. Blood 92(2):362–367.

Soldi, R., Mitola, S., et al. (1999). Role of alpha-v-beta-3 integrin in the activation of vascular endothelial growth factor receptor-2. EMBO J 18(4):882–892.

Takahashi, T., Kalka, C., et al. (1999). Ischemia- and cytokine-induced mobilization of bone
marrow-derived endothelial progenitor cells for neovascularization. Nat Med 5(4):434–
438.

Tallquist, M.D., Soriano, P., et al. (1999). Growth factor signaling pathways in vascular
development. Oncogene 18(55):7917–7932.

Testaz, S., Delannet, M., et al. (1999). Adhesion and migration of avian neural crest cells on
fibronectin require the cooperating activities of multiple integrins of the (beta)1 and
(beta)3 families. J Cell Sci 112(pt 24):4715–4728.

Tomanek, R.J., Schatteman, G.C. (2000). Angiogenesis: new insights and therapeutic potential. Anat Rec 261:126–135.

Vecchi, A., Garlanda, C., et al. (1994). Monoclonal antibodies specific for endothelial cells
of mouse blood vessels: their application in the identification of adult and embryonic
endothelium. Eur J Cell Biol 63:247–254.

Vittet, D., Buchou, T., et al. (1997). Targeted null-mutation in the vascular endothelial-
cadherin gene impairs the organization of vascular-like structures in embryoid bodies.
Proc Natl Acad Sci U S A 94(12):6273–6278.

Vittet, D., Prandini, M.H., et al. (1996). Embryonic stem cells differentiate in vitro to
endothelial cells through successive maturation steps. Blood 88(9):3424–3431.

Waldo, K.L., Kumiski, D., et al. (1996). Cardiac neural crest is essential for the persistence
rather than the formation of an arch artery. Dev Dyn 205(3):281–292.

Yan, H.C., Baldwin, H.S., et al. (1995). Alternative splicing of a specific cytoplasmic exon
alters the binding characteristics of murine platelet/endothelial cell adhesion molecule-1
(PECAM-1). J Biol Chem 270(40):23672–23680.

Yang, J.T., Rayburn, H., et al. (1993). Embryonic mesodermal defects in alpha 5 integrin-
deficient mice. Development 119(4):1093–1105.

Yang, J.T., Rayburn, H., et al. (1995). Cell adhesion events mediated by alpha 4 integrins
are essential in placental and cardiac development. Development 121(2):549–560.

Yang, S., Graham, J., et al. (1999). Functional roles for PECAM-1 (CD31) and VE-cadherin
(CD144) in tube assembly and lumen formation in three-dimensional collagen gels. Am
J Pathol 155(3):887–895.

Yap, A.S., Brieher, W.M., et al. (1997). Molecular and functional analysis of cadherin-based
adherens junctions. Annu Rev Cell Dev Biol 13:119–146.

Zhou, Z., Christofidou-Solomidou, M., et al. (1999). Antibody against PEcAM-1 inhibits
tumor angiogenesis in mice. Angiogenesis 3:181–188.

CHAPTER **6**

Development and Differentiation of Vascular Smooth Muscle

Mark W. Majesky, Xiu-Rong Dong, and Jun Lu

After a network of large and small endothelial channels suitable to conduct the nascent embryonic circulation has formed, the next step in vascular development is the assembly of a tunica media to provide mechanical support, prevent hemorrhage and rupture, and confer vasomotor and neurohumoral control of the circulation. Formation of the tunica media is a stepwise process involving recruitment and clustering of mesenchymal cells around endothelial vessels, activation of smooth muscle specific gene transcription, production of an elastin- and collagen-rich extracellular matrix, organization of smooth muscle cells (SMCs) into layers, and formation of an adventitia consisting of nerves, capillaries, fibroblasts, and connective tissue. Reciprocal signaling between endothelial cells and mesenchymal cells is critical for assembly of the tunica media, both to ensure endothelial cell survival and maturation and to stimulate mesenchymal cell differentiation and matrix production. The exchange of signals between endothelial cells and mesenchymal cells that is initiated during vascular development continues throughout life to ensure that changing target tissue demands for perfusion are coupled with corresponding adaptations in the structure and function of the tunica media.

ORIGINS OF VASCULAR SMOOTH MUSCLE

Multiple cell populations with different embryologic origins participate in the morphogenesis of the medial layer of vessel wall. Moreover, different vessels have selective and highly specialized functions that require structural and functional diversity of vascular SMCs. Lineage analysis studies of vascular development in the chick and mouse embryos have identified three general pathways by which vascular SMCs are recruited to form a tunica media during development: (1) selective recruitment from local mesenchyme surrounding nascent blood vessels, (2) downstream extension from preexisting SMC, and (3) transdifferentiation of endothelial cells into SMCs. The majority of vascular SMCs are thought to arise by local recruitment of mesenchymal progenitors surrounding embryonic blood vessels.[49,81] While blood flow is not required for angioblasts to form capillary-like vessels in the early embryo, development of a tunica media is typically associated with vessels through which blood flow can be demonstrated. As a result, it is assumed that critical genes required for tunica media formation will exhibit flow-dependent upregulation in endothelial cells of the embryonic vasculature. A second mechanism for tunica media formation is downstream extension of pre-existing SMCs along basement membranes of a newly forming vascular sprout or extension.[45,100] In this case, preexisting SMCs disengage from their cell-matrix and cell-cell connections and migrate along the outer surface of the nascent vessel and then reestablish a quiescent and differentiated phenotype at a new, downstream location. This process may or may not be accompanied by cell division. Finally, evidence in the early chick embryo suggests that endothelial cells in the dorsal aorta can undergo mesenchymal transformation and differentiate directly into SMCs.[29] It is unlikely that mesenchymal transformation of aortic endothelium generates the entire aortic SMC population since we know that the bulk of the proximal aorta is made up of cardiac neural crest–derived SMCs.[60] However, it is intriguing to consider that endothelial transdifferentiation may produce a specialized subpopulation of SMCs within the intima or inner media at selected sites in the vascular tree. It will also be important to determine if this process occurs during angiogenesis or microvascular remodeling in adult vessels.[70] More recently, evidence has been reported that pluripotent embryonic stem cells cultured on type IV collagen express a vasculogenic phenotype and can become either endothelial cells or SMCs depending upon the type of growth factor to which they are subsequently exposed.[118] The persistence of circulating vasculogenic stem cells of bone marrow origin in adults has also been suggested,[9] and may play important roles in vessel formation and repair in adult vasculature.

Recruitment of SMCs from Local Mesenchyme

It is generally thought that vascular SMC differentiation is initiated when diffusible factors released by endothelial cells act upon nearby mesenchymal cells to stimulate migration toward and clustering around developing blood vessels.[35] Recruitment of SMCs from local mesenchyme is probably how most large vessels acquire their smooth muscle coating in the early embyro. At a time when vascular SMC differentiation can first be recognized in development, a network of capillary-like vessels has already extended throughout the entire embryo. Upon initiation of contraction by the tubular heart, circulation of the blood begins. Shortly thereafter, these embryonic vessels are observed to acquire a tunica media to provide mechanical support for withstanding the increasing blood pressure and flow rates that accompany rapid growth of the embryo. Since SMCs are recruited locally and endothelial channels extend throughout the embryo, it follows that SMCs in different vessels are recruited from different sources of embryonic mesenchyme. Indeed, lineage mapping studies have provided evidence for at least four independent origins of vascular SMCs in development.[40,60] One lineage produces SMCs that make up the wall of the proximal aorta, pulmonary artery, common carotid artery, innominate and subclavian arteries, and ductus arteriosus. These SMCs differentiate from progenitors that originate in neural ectoderm-derived cardiac neural crest and migrate into the pharyngeal arches. These progenitor cells are called ectomesenchymal SMCs (Ect SMCs) to denote their origin from ectoderm-derived neural crest cells. Lateral and splanchnopleural mesoderm comprise two additional origins for vascular SMCs and both sources give rise to cells referred to as mesenchymal SMCs (Mes SMCs). In vessels where Ect and Mes SMCs coexist, the two SMC types are not equally distributed within the tunica media but exhibit sharp boundaries and transition zones.[114] Ect SMCs are the predominant or exclusive type of SMC in the proximal segments of the major outflow vessels, whereas Mes SMCs are the only SMC type in the distal segments of these vessels.[60,109]

A fourth independent origin for vascular SMCs in the embryo has been recently identified by studies employing retroviral markers, fluorescent tracers, or chick-quail chimeric embryos for lineage analysis. Each of these approaches showed that progenitors for coronary SMCs originate within extracardiac coelomic mesothelium and reside within a transient structure called the proepicardial organ (PEO).[39] The PEO contains a population of mesothelial cells that projects from the ventral body wall, makes contact with the looped heart tube, and covers the external surface of the heart to form the epicardial layer.[77] By a process of epithelial to mesenchymal transformation (EMT), epicardial cells give rise to a population of mesenchymal cells (epicardial-derived mesenchymal cells, EPDCs) that enter the myocardial layer and migrate widely within the developing heart to form the bulk of the nonmyocyte cell populations in the heart.[40] These epicardial-derived mesenchymal cells are progenitors for cardiac fibroblasts, coronary adventitial cells, coronary SMCs and a subset of valve mesenchymal cells. In addition, these cells are the source of instructive signals that direct competent myocardial cells to adopt a conduction tissue phenotype.[41,42,51] Thus, coronary SMCs are of a unique lineage that is distinct from that of any other type of smooth muscle in the vascular system. It should also be pointed out that since coronary adventitial fibroblasts and coronary SMCs originate from the same progenitor cells and share an extensive developmental history in common, it may not be surprising to find that after balloon angioplasty injury to the coronary wall in adult hearts, adventitial cells can migrate into the damaged media and reportedly differentiate into functional coronary SMCs.[92,94]

Recruitment of SMCs by Extension from Preexisting Medial Cells

A morphologic hallmark of capillary arterialization in adult tissues is the acquisition of a pericyte coating. Pericytes are SMC-like cells that encircle the walls of pre- and postcapillary arterioles and venules.[47,99] Endothelial cells and pericytes make characteristic tight junction contacts and communicate directly through connexin-containing gap junctions.[26,68] Arteriolar growth is an important event in vascular development and adaptation, but the mechanism of arteriolar wall formation has been unclear. Using a combination of antibodies directed against SMC-specific proteins, Price et al.[84] showed an age-dependent increase in the number of terminal arterioles that were coimmunostained with anti–smooth muscle myosin heavy chain (SM-MHC) and anti-SMαActin in developing rat anterior gracilis muscle. They also found that new terminal arterioles exhibited SMαActin and SM-MHC labeling that was always continuous with upstream arterioles.[84,100] These results implied that new arterioles were formed by pericyte extension from preexisting arteriolar medial cells.[100] The findings of Price et al. are reminiscent of those reported in the 1940s by Clark and Clark,[21] who directly viewed the formation and regression of arterioles in experimental chambers made in the skin of rabbit ears. They observed that upon delivery of a stimulus for increased blood flow through a particular capillary network, pericytes could be observed detaching from preexisting arterioles and migrating to more distal positions in close association with the microvascular wall. Upon reduction in blood flow, pericytes were seen detaching from the arteriole, migrating away from the microvessel and wandering into the interstitial matrix, where they were frequently seen to undergo cell death. Taken together, these studies provide support for the idea that developing microvessels acquire a medial layer by extension of a preexisting population of SMCs through pericyte detachment and chemotaxis along the capillary endothelial cell basement membrane. A similar process also appears to play an important role in formation of a smooth muscle coating around smaller vessels that branch off of the major arterial trunks in the developing embryo.[67]

What are the signals that induce pericytes to form close associations with, and migrate along, capillary endothelium? Gene deletion studies in mice have shown that pericytes depend on platelet-derived growth factor-B (PDGF-B) produced by capillary endothelial cells for their development or survival;[45,67] reviewed in ref. 66. PDGF-B expression by endothelial cells in angiogenic sprouts is much greater than in quiescent stable microvessels. PDGF-β receptor-positive pericytes are normally found clustered around PDGF-B–expressing endothelial cells and are induced to migrate along sprouting capillary basement membranes by chemotaxis. In PDGF-B–deficient mice, capillary vessels are more tortuous, pericytes are frequently missing, and capillary walls are distended and exhibit numerous microaneurysms.[67] PDGF-B null mice also have a deficiency in pericyte-like mesangial cells in renal glomeruli. Curiously, not all microvessels showed a requirement for PDGF-B signaling to support normal development. Vessels dependent on PDGF-B were found in brain, heart, lung, and skeletal muscle, whereas vessels in the liver, gastrointestinal tract, adrenal gland, and perineural plexus were not affected by the absence of PDGF-B. A detailed study of PDGF-B/PDGF-β receptor signaling in coronary development showed that while penetrating coronary arteries completely lack SMαActin-positive cells, subepicardial vessels have SMCs but they are much fewer in number than in wild-type hearts.[45] These findings are consistent with the idea that penetrating coronary arteries acquire their media by downstream extension of SMCs from proximal sites along the main subepicardial coronary vessels in a PDGF-B–dependent manner. A very similar mechanism may operate in brain vasculature where submeningeal vessels contribute SMCs to vessels that penetrate the brain itself by way of a PDGF-B–dependent downstream extension of preexisting SMCs. While the normal appearance of the aorta and other large artery walls in PDGF-B–deficient mice suggests that they are independent of PDGF-B signaling for normal development, the possibility that compensatory mechanisms in development rescue large vessel defects in PDGF-B (−/−) mice must not be ruled out. Indeed, a study of chimeric mice composed of a mixture of wild-type and PDGF-β receptor–deficient cells showed that SMCs in large vessel walls also respond to a PDGF-B signal for recruitment, proliferation, and/or survival during tunica media formation.[25] Taken together, these studies suggest that for some vessels, PDGF-B is required for initial recruitment of SMCs or pericytes from surrounding mesenchyme. For most vessels, it appears that other factors provide the initial stimulus for SMC recruitment, and PDGF-BB provides either a mitogenic/survival stimulus for expansion of the initial SMC pool or a chemotactic stimulus for downstream extension of preexisting SMCs.

Venous SMCs may also acquire a SMC coating by an endothelial-dependent process. A point mutation in the angiopoietin-1 receptor gene *Tie-2* is responsible for the human genetic deficiency venous malformation. This mutation leads to an amino acid substitution that constitutively activates the receptor in a ligand-independent manner.[113] Patients with venous malformation have thin-walled and extemely dilated veins in specific locations under the skin. Venous dilation is associated with severe reduction in the number of venous SMCs. Since the *Tie-2* receptor is not expressed in SMCs or their mesenchymal precursors, these data are consistent with the interpretation that endothelial-dependent signals that are under the control of angiopoietin-1/*Tie-2* signaling are required for recruitment of venous SMCs as well as arterial SMCs.

As with large arteries and veins, microvessels also require ongoing bidirectional signaling between endothelial cells and SMC/pericytes for formation and survival. One example of this interdependence is found in the postnatal development of the retinal vasculature. In this system, microvascular density is controlled primarily by availability of oxygen. Hypoxia promotes new vessel formation, while hyperoxia promotes vascular regression. Careful analysis of the timing of hyperoxia-induced "vascular pruning" showed that it corresponds to a transient period characterized by the appearance of a pericyte-free neovascular plexus.[12] Normally, retinal vessels acquire a pericyte coating by out-migration of preexisting pericytes from upstream arterioles. In the absence of pericytes, new capillary vessels that are formed are very labile and exhibit high rates of endothelial cell apoptosis. However, once the new vessels have acquired a pericyte cell layer, endothelial apoptosis rates drop sharply.[11] Vascular endothelial growth factor (VEGF) is known to be a survival factor for endothelial cells and is produced by pericytes and SMCs. Therefore, pericyte recruitement brings a source of VEGF in close contact with retinal capillary endothelium. Consistent with this idea, intraocular injection of VEGF accelerated pericyte coverage.[12] Likewise, heparin-binding epidermal growth factor (HB-EGF) was found to induce retinal neovascularization by stimulating VEGF production by SMCs/pericytes.[1] By contrast, intraocular injection of PDGF-BB caused vascular regression due to detachment of PDGF β-receptor–positive pericytes from newly coated vessels while having no effect on already established vessels. Therefore, endothelial cells and pericytes exchange critical factors to ensure coordinated retinal microvascular formation and survival during postnatal development.

A good example of how the study of vascular development helps us to understand how angiogenesis in the adult vasculature is regulated can be found in a study by Asahara et al.[9] Using a rabbit cornea micropocket assay for angiogenesis, these workers found that while VEGF alone induces neovascularization, the process is enhanced by exogenous addition of either angiopoietin-1 or angiopoietin-2. Detailed morphometric analysis of vessel length, diameter, and density showed that angiopoietin-1 enhances VEGF-induced angiogenesis by promoting maturation of newly formed vessels, whereas angiopoietin-2 works to initiate endothelial sprouting by downregulating the stabilizing influence of angiopoietin-1. Therefore, the coordinated production of VEGF, angiopoietin-1, and angiopoietin-2 by pericytes and SMCs closely controls the ability of endothelial cells to initiate angiogenesis and plays a major role in whether newly formed vessels either regress or survive.[1,12,18]

Transdifferentiation of Endothelial Cells into SMC

The notion that endothelial cells can directly transdifferentiate into vascular SMCs seems to run counter to the idea that formation of the tunica media occurs by recruitment of SMCs from surrounding mesenchymal cells. However, the possibility that endothelial cells can give rise to SMCs is not a new idea. In 1944, Altschul[5] reported that histologic analyses of human atherosclerosis revealed that endothelial cells could be observed to leave the surface layer of the aorta, undergo a mesenchymal change, and become embedded in the aortic subendothelial matrix. He remarked that the number of endothelial cell–derived mesenchymal cells are "frequently great enough to produce a distinct thickening of the subendothelial space." Subsequently, Arciniegas et al.[7] reported that >90% of bovine aortic endothelial cells treated with transforming growth factor-β_1 (TGF-β_1) *in vitro* convert to spindle-shaped cells that coordinately gain SMαActin and loose factor VIII–related antigen expression. This phenotypic change became "irreversible" after 20 days of incubation in TGF-β_1–containing medium. Moreover, this group observed direct conversion of von Willebrand factor–positive endothelial cells to SMαActin-positive mesenchymal cells in explants of chick embryo aorta.[6] Sarkisov et al.[91] reported observations of capillary endothelial cell conversion into SMC-like connective tissue cells (myofibroblasts) in granulation tissue during excisional wound repair. Moreover, Tuder et al.[111] observed that endothelial cells can convert to myofibroblasts and SMCs in intimal thickenings associated with pulmonary hypertension. Finally, it should be pointed out that a very similar process of endothelial to mesenchymal transformation in heart development is well established.[71] During formation of the cardiac valves, endocardial cells are known to respond to factors produced by competent myocardial cells, transform into mesenchymal cells, invade cushion tissue extracellular matrix, and differentiate into valve interstitial cells.[33] Indeed, defects in endocardial transformation underlie the phenotypes of the *Nuclear Factor of Activated T* cells (NF-AT),[28,87] Smad6,[37] and hyaluronic acid synthase-2 (HAS-2)[16] null mice. If endothelial to mesenchymal conversion occurs in the endocardium during heart development, the possibility that a similar process occurs in vascular endothelium warrants careful consideration.

Indeed, DeRuiter et al.[29] provided evidence for conversion of endothelial cells to SMCs during early stages of development of the dorsal aorta. They reported finding cells that coexpress markers for both endothelial cells (monoclonal antibody QH-1) and SMCs (SMαActin) directly beneath the endothelial layer in developing quail aorta. Furthermore, following injection of gold-labeled wheat germ agglutinin particles (Au-WGA) into the anterior vitelline vein, Au-WGA particles were detected by electron microscopy both in the surface lining endothelial cells and in cells located within the subendothelial layer of the tunica media. The authors conclude that the most likely interpretation of these data is that quail hybridoma-1–expressing cells endocytose Au-WGA particles at the luminal surface, undergo a mesenchymal transformation, invade the subendothelial matrix, and acquire an SMC phenotype. Thus the subendothelial intima and inner media are pictured to be built up of SMC-like cells that originate from the endothelial layer. If true, this could explain the characteristic organization of the first SMCs that appear immediately subjacent to the endothelial layer around large arteries in the early embryo.[50] While an origin of vascular SMCs from endothelial cells awaits confirmation by an independent approach, it is worth noting that endothelial cells from embryonic day E9.5 to E16 mouse embryo dorsal aorta have also been shown to produce satellite-like cells capable of entering the myogenic lineage, expressing MyoD, and participating in skeletal muscle fiber formation upon injection into adult muscle.[27,80] At first glance, the finding that endothelial cells could be a source of skeletal myoblasts seems remarkable. Yet it is important to note that the origins of somites and aortic endothelium are not so far removed from each other during early development. In fact, one source of endothelial cells in the embryo is from somites and the paired dorsal aortae are formed directly below, and in connection to, the bilateral pairs of somites that differentiate from paraxial mesoderm.[23,82] If confirmed, transdifferentiation of endothelial cells into SMCs during vascular development may portend a reevaluation of the way we think about formation of the tunica media in the embryo. In addition, the concept of endothelial conversion into SMCs has important implications for the origin of intimal SMCs in adult vessels and the role of vasa vasorum within developing atherosclerotic plaques as a possible source of plaque mesenchymal cells during progression of human atherosclerosis.[70]

ENDOTHELIAL-SMOOTH MUSCLE SIGNALING IN TUNICA MEDIA FORMATION

The process of vessel wall formation involves bidirectional signaling between mesenchymal cells and endothelial cells. Angioblast differentiation to primitive endothelial cells and expansion of the initial endothelial cell pool is critically dependent on VEGF produced by mesenchymal cells.[20,34] Angiopoietin-1 is produced by mesenchymal cells that are in close proximity to developing blood vessels and is required for normal endothelial cell development.[106] Angiopoietin-1 binds to and activates Tie-2/tek receptors on endothelial cells and promotes maturation and survival of developing endothelial cells. Angiopoietin-2 is a naturally occurring antagonist that blocks angiopoietin-1 binding to Tie-2/tek. Knockouts of angiopoietin-1 have severe vascular defects and embryonic lethality very similar to that exhibited by Tie-2 knockouts.[106] Most prominent are deficiencies in endothelial cell sprouting and vascular remodeling. Mesenchymal cells are also an important source of VEGF and, as described above, both VEGF and angiopoietin are required for optimal vascular development in the embryo and angiogenesis in the adult.

Genetic evidence that endothelial cells signal mesenchymal cells to form the tunica media is provided by the phenotypes of LKLF and endoglin knockout mice. LKLF (lung Kruppel-like factor) is a zinc finger transcription factor of the Drosophila Kruppel gene family that is selectively expressed in developing endothelial cells and to a lesser extent in lung buds and vertebral column.[58] LKLF null embryos die between E12.5 and E14.5 from vascular defects characterized by hemorrhage and rupture. Vasculogenesis and primary network formation are normal in LKLF (–/–) mice and SMαActin-positive cells are formed in close apposition to vascular channels. Therefore, LKLF is not required to specify either endothelial or smooth muscle lineages. However, the tunica media in LKLF null embryos is abnormally thin and disorganized, SMC differentiation is delayed, and extracellular matrix deposition is reduced.[58] Absence of LKLF does not alter the expression of PDGF-B, *Tie-2*, *Tie-1*, flk1, HB-EGF, or TGF-β_1 in endothelial cells nor disrupt red cell, platelet, or myelomonocytic development. Since LKLF is not normally expressed by SMCs or their mesenchymal precursors, the phenotype of LKLF (–/–) mice strongly suggests that LKLF activity is required for a pathway in endothelial cells that controls SMC recruitment, differentiation, or organization in the tunica media via paracrine mechanisms. Therefore, an important goal for future work will be to identify the critical gene products for vessel wall formation controlled by LKLF in endothelial cells.

Endoglin is a type III TGF-β receptor that is selectively expressed by endothelial cells.[72] Hypomorphic mutations in the endoglin locus cause human hereditary hemorrhagic telangectasia (HHT).[73] In mice, deletion of the endoglin gene produces an embryonic lethal phenotype with severe vascular defects[63] that resemble those in LKLF knockouts described above. SMC recruitment and differentiation are defective, causing a secondary failure of vascular development and early embryonic death. As in the LKLF knockout, vasculogensis and early capillary-like vascular network formation are normal in endoglin-null mice, endothelial differentiation markers are expressed normally, and blood cells are produced at levels similar to those in wild-type mice. Vascular defects begin to appear around E9.5 to E10.5 and correlate with failure of SMC recruitment. Since endoglin is expressed by endothelial cells and is absent in SMCs, the endoglin null phenotype suggests that critical SMC recruiting factors are produced by endothelial cells in response to a TGF-β–dependent signal. In this respect, it is noteworthy that about half of TGF-β_1–deficient embryos exhibit failure of vascular development at point when endothelial–mesechymal interactions are required.[30] Therefore, the phenotypes of the LKLF and endoglin-null mice described above, together with those of the PDGF-B–,[67] angiopoietin-1–,[106] PDGF receptor-β–,[45] and Smad-6–deficient[37] mice, all point to critical endothelial-derived signals that direct mesenchymal cell recruitment and organization of a functional tunica media. It can also be anticipated that in addition to secreted factors, endothelial cells will also be found to signal mesenchymal cells by short-range or cell surface molecules including notch and its ligands (serrate, delta),[56,95,98] ephs and their receptors,[4,115] and junctional proteins including connexins,[57] cadherins,[10] and PDZ (*postsynaptic density protein-95, disc large, zonula occludens-1*) domain–containing proteins.[38] Identification of those critical endothelial-derived signals for vessel wall formation is an important area for future work.

DIFFERENTIATION AND MATURATION OF THE TUNICA MEDIA

The final steps in vessel wall development are growth, differentiation, maturation, and innervation of medial SMCs to form a mature and functional tunica media. Extensive studies of SMC differentiation over the years have identified a large number of contractile and cytoskeletal proteins whose expression can be correlated with maturation of the tunica media *in vivo*. Despite this progress, critical regulatory genes and signal transduction pathways that control the transcriptional activation of these SMC specific structural genes in development remain unclear, and their identification is an important area of ongoing investigation. This section briefly summarizes SMC differentiation as it pertains to development of the tunica media. Emphasis is placed on identification of genes whose mutant phenotypes involve defects in formation of the tunica media. For more extensive coverage of the general topic of SMC differentiation, the reader is directed to several excellent reviews.[81,101,107]

The Role of Mechanical Forces and the Extracellular Matrix

A general rule in vessel wall formation is that the final wall thickness attained during development is proportional to wall stress. According to the law of Laplace, wall stress is the product of blood pressure times lumen diameter. Therefore, as lumen diameter increases, wall thickness correspondingly increases to normalize overall wall stress. Since blood flow is the primary determinant of lumen diameter, it follows that wall thickness is indirectly proportional to blood flow. In general, the mass of SMCs increases or decreases in tandem with that of extracellular matrix (ECM) during wall remodeling, presumably as a result of signaling pathways activated by SMC-ECM adhesion receptors. The importance of this tightly coupled interaction for control of overall wall dimensions was revealed in dramatic fashion by the phenotype of the tropoelastin knockout mouse.[61]

In large elastic arteries, tropoelastin is secreted into the extracellular space, where it binds to scaffold proteins called microfibril proteins, becomes processed and then cross-linked by lysyl oxidase into insoluble bands of elastin and microfibril proteins called elastic lamellae. Normally, inner and outer diameters of the aorta continue to increase throughout gestation and the perinatal period in response to increases in cardiac output and blood flow. This is accomplished by increases in the size of the fenestrations within elastic lamellae as the vessel circumference becomes progressively larger, rather than a continuous breakdown and resynthesis of the elastic fibers.[117] In mice unable to produce tropoelastin, pups are born but do not survive beyond postnatal day 4.5 due to progressive obstructive arterial malformations.[61] In elastin (−/−) mice, aortic diameters increase normally until around E17.5 at which time the outer and inner diameters of the ascending aorta start to become progressively smaller, despite continued growth of the fetus. This change in wall structure was accompanied by progressive subendothelial accumulation of SMCs that continued until eventually obliterating the aortic lumen.[61] No differences were found in endothelial cell injury, inflammation, or hemodynamics when aortas from E16.5 elastin (−/−) vs. wild-type embryos were compared. The process of subendothelial accumulation of SMCs was also found in systemic arteries of all sizes in elastin null mice, including distributing arteries and even arterioles.

It is interesting to note that the developmental age at which tropoelastin-null mice begin to show deviation from the wild type in artery wall formation (E17.5) is very close to the time that Majack's group[22] found aortic SMCs begin to exit the cell cycle and normally become quiescent. These investigators further showed that exit from the cell cycle *in vivo* correlated with loss of an autonomous growth phenotype characterized by the ability of SMCs cultured from embryonic rat aorta to proliferate in a defined serum-free medium without the addition of exogenous mitogens *in vitro*. Consistent with the possibility that development of the elastic fiber is coupled to SMC cell cycle exit *in vivo*, proliferating cell nuclear antigen (PCNA) staining revealed that subendothelial SMC proliferation rates were sustained at much higher levels in elastin (−/−) mice than in wild-type mice (88% vs. 35%). In mice made hemizygous for the tropoelastin gene, an increase of 35% in the number of elastic lamellae and SMC number was found.[62] Examination of humans with hemizygosity for the *ELN* (elastin) locus revealed a remarkable 2.5-fold increase in number of elastic lamellae in the aortic wall. Unlike what was previously assumed, this work showed that the number of elastic lamellae is neither genetically fixed nor species specific. Rather, these studies revealed a remarkable sensitivity of arterial SMCs to proper formation of elastic fibers for normal vessel wall structure and growth control. However, adaptation to normalize wall stress by increasing the number of elastic lamellae only occurs early in development. After birth, the artery wall responds to increased wall tension by SMC hypertrophy and wall thickening but with no change in the number of elastic lamellae. Thus, development of large conduit arteries is extremely responsive to reductions in elastin expression with compensatory changes to normalize wall stress and elasticity being evidenced by increases in the number of rings of elastic fibers and SMC mass during arterial development. These striking results suggest that acquisition of SMC quiescence and normal assembly of fibrillar collagens, proteoglycans, and elastic lamellae depend on production of the correct amounts of tropoelastin within the tunica media during embryogenesis. In addition, they raise the possibility that vascular injury, thrombosis, and inflammation later in life contribute to obstructive arterial disease by disrupting SMC–elastic fiber interactions. A similar argument can be made for fibrillar collagens in the artery wall. Koyama et al.[55] reported that SMCs become arrested in the G1 phase of the cell cycle when plated on polymerized type I collagen fibrils as a result of increased levels of the cdk2 inhibitors p27-Kip1 and p21-Cip1/Waf1. These effects on SMC cell cycle regulators were not seen if SMCs were plated on monomeric type I collagen.

Perhaps, in retrospect, it should not be surprising to find that ECM components of the vessel wall play an active, instructive role in control of tunica media formation and SMC growth control.[49,105] After all, vascular SMCs are surrounded by an extensively cross-linked ECM from the earliest times in their development.[49] Moreover, the ECM is a rich depository for many different morphogens and growth factors. Cell-matrix interactions have been well documented as playing essential roles in the development and differentiation of many cell and tissue types,[3] and the vessel wall is no exception.[17,112] Moreover, matrix degradation and remodeling is essential for organogenesis in general and vessel wall formation in particular.[19] Indeed, for large arteries at least, it could be argued that the main function of SMCs is to produce an ECM with an appropriate combination of elasticity and strength to function under high pulse pressures. The challenge for the future is to identify the signaling pathways that link constituents of the ECM to cell surface receptor–mediated changes in SMC gene expression.

Developmental Expression of Elastin, Fibulins and Fibrillins

Transcription of the elastin gene begins shortly after the first recognizable SMCs appear in vascular development.[48,90,93] Elastin is the primary ECM molecule in large conduit arteries and serves a structural role in nearly all vessels.[74] The arrangement of cross-linked elastic fibers differs significantly between vessels that contain neural crest–derived SMCs and those that contain mesoderm-derived SMCs.[89] Elastic fibers are composed of a number of important proteins in addition to elastin itself including fibulin-1, fibulin-2, fibrillins, emelin, and nidogen-2. For example, fibulin-1 is a core component of microfibrils that provide a scaffold for elastic fiber formation. Fibulin-1 is among the first ECM proteins to be expressed as mesenchymal cells become committed to a SMC fate and functions first as a provisional matrix for assembly of the tunica media and later as a template for deposition of elastic fibers.[50] Fibulins are well suited for a role as organizers of matrix formation since they can interact with a number of important ECM constituents including fibronectin, laminin, nidogen, aggrecan versican, and tropoelastin.[88] Moreover, inasmuch as fibulin-1 is strongly expressed at sites of epithelial to mesenchymal transformation throughout the embryo, it may also be important for transdifferentiation of endothelial cells into SMCs, for transformation of endocardial cells into valve interstitial cells, and for conversion of epicardial cells into coronary SMCs.[120]

Fibrillin-1 and -2 are additional microfibril components that play key roles in the development of the tunica media. Human mutations that map to fibrillin genes have been linked to connective tissue disorders with prominent vascular phenotypes. Marfan syndrome is caused by mutations in fibrillin-1,[31] and congenital contractural arachnodactyly maps to mutations in fibrillin-2[83] (for a recent review, see ref. 86). Studies of the developmental patterns of fibrillin gene expression during mammalian development led to the proposal that fibrillin-1 is primarily a load-bearing structural element in the vessel wall, whereas fibrillin-2 is largely involved with the control of elastic fiber assembly. Both fibrillin-1 and fibrillin-2 are expressed very early in vessel wall formation.[36,88] Moreover, fibrillin-2 is an early marker of vascular SMC heterogeneity in developing avian embryo.[50] Other important components of the elastic fiber include emilin, a 115-kd protein that is particularly abundant at the elastin-microfibril interface and thought to play a role in elastin fiber assembly,[14] and microfibril-associated glycoprotein-1, the major glyprotein antigen of elastin-associated microfibrils.

SMOOTH MUSCLE DIFFERENTIATION

Developmental Expression of SMC-Specific Contractile and Cytoskeletal Genes

Expression of SMαActin is the earliest marker currently known for vascular SMC differentiation.[81] Although it is not specific for vascular smooth muscle, expression of SMαActin is a useful and sensitive marker to identify the time and location at which signaling pathways that specify SMC-specific transcription have become activated. The first cells to express SMαActin during vascular development appear at the ventral surface of the dorsal aorta in apposition to the foregut endoderm.[50] As development of the aortic wall proceeds, additional markers of the mature SMC phenotype appear in temporal sequence. Duband et al.[32] showed that vascular SMC differentiation *in vivo* occurs in at least two steps defined by the timing of onset of vascular SMC marker expression. For example, SMαActin and desmin are representative early stage markers, whereas calponin and SM22α expression marks a later step in SMC differentiation. Smooth muscle myosin heavy chain (SM-MHC) is a highly selective marker for developing SMCs that it is not expressed in cardiac or skeletal muscle in the embryo.[75] Smooth muscle specific contractile and cytoskeletal proteins are produced not only by cell type–specific transcriptional controls but also by cell type–specific alternative splicing including α-tropomyosin, heavy caldesmon, meta-vinculin, SM1 and SM2 isoforms of SM-MHC, and smooth muscle α-actinin.[81]

Transcriptional Control of SMC Differentiation

In an effort to understand how transcription of SMC-specific contractile and cytoskeletal genes becomes activated in vascular development, several SMC-restricted promoters have been studied including SMαActin,[96] SM22α,[102] calponin,[76] SMγA,[15,85] telokin,[46] and SM-myosin heavy chain.[53,69] The common objective in these studies was to identify a *cis*-acting element(s) in the regulatory regions of the gene that mediates SMC-restricted gene expression. The vast majority of these studies have led to the identification of an evolutionarily conserved element called the CArG box (5′-CC(AT)$_6$GG-3′) as a critical motif for SMC-specific transcription. The CArG box forms the core binding sequence of the serum response element (SRE), a DNA sequence that binds homodimers of the serum response factor (SRF).[52] Although often regarded as a widely expressed transcription factor, in the developing embryo SRF expression is actually largely restricted to skeletal, cardiac, and smooth muscle–containing tissues.[24] Gene deletion studies have shown that SRF is required for very early stages in mesoderm formation and specification.[8] No defect in cell proliferation was found in SRF (−/−) ES cells,[8] consistent with the suggestion by Croissant et al.[24] that the primary role of SRF in developing embryos is transcriptional control of muscle differentiation including skeletal, cardiac, and smooth muscle lineages, rather than cell proliferation per se.

SRF is a highly conserved member of the MCM1, agamous, deficiens, SRF (MADS) box family of DNA-binding proteins.[97] It is composed of a central 60-amino acid MADS domain, responsible for DNA binding and dimerization, followed by a C-terminal transactivation domain.[110] A variety of studies have shown that MADS box factors play important roles in muscle-specific gene transcription in vertebrates. A null mutation of the *Drosophila* MADS box factor DMef-2 resulted in failure of somatic, cardiac, and visceral muscle differentiation.[65] Microinjection of antibodies to SRF blocked myogenic differentiation in two skeletal myoblast cell lines, mouse C2 and rat L6.[104] Multiple CArG elements are found in a number of muscle-specific promoters, including those that direct cell-specific transcription in vertebrate smooth muscle. A dependence on upstream CArG-elements for SMC specific transcription has been reported for SMαActin,[13] SMγA,[108] SM22α,[64] SM-myosin heavy chain,[121] caldesmon,[119] α$_1$-integrin,[79] and telokin.[46] Smooth muscle–rescricted transcription of the SM22α promoter in transgenic mice was also found to be dependent on an upstream CArG box found within the first 280 base pairs (bp) of the 5′ promoter sequence.[54] Thus factors that interact either directly or indirectly with the CArG box are likely to be important for maintenance of ongoing SMC-specific gene expression in cells that are already expressing a SMC phenotype. Moreover, Landerholm et al.[59] used two different dominant-negative forms of SRF to show that initial steps in transcriptional activation of SMC marker genes for coronary SMC differentiation from proepicardial cells also requires transcriptionally active SRF. Therefore, it seems likely that SRF is necessary for both activation of SMC-specific transcription during development,[8,59] and maintenance of SMC-specific transcription in adult vessels.[78,81]

While necessary for SMC differentiation, it is clear that SRF is not sufficient to activate SMC specific transcription. Increasing evidence suggests that interactions between SRF and other DNA-binding proteins including Nkx homeodomain factors, Mhox, GATA factors, and Kruppel-like Zn-finger proteins can act cooperatively to stimulate target gene transcription. Interactions of SRF with yet to be identified lineage-restricted factors may then promote SRF binding to the general transcription factors or the transcription factor II family. One *cis*-acting element that has been shown to function cooperatively with the SRF is the so-called TGF-β control element (TCE) adjacent to CArG box-A in the SMαActin promoter.[43] The TCE is a GC-rich element that was shown to bind the Krupple-like Zn-finger factor GKLF (gut *K*ruppel-*like* *f*actor) in a yeast one-hybrid screen.[2] However, cotransfection studies in cultured SMCs showed that GKLF acted as a repressor of TCE-dependent transcription in that GKLF inhibited SM22α and SMαActin promoter activity stimulated by TGF-β$_1$.[2] Moreover, GKLF was not highly expressed in differentiated SMC *in vivo*. Overexpression of the related Zn-finger factor BTEB-2 (*b*asic *t*ranscription *e*lement *b*inding *p*rotein-2), however, transactivated SM22α promoter activity in cultured SMCs. Additional evidence of a role for Kruppel-like factors in control of SMC gene expression comes from the work of Nagai's group,[116] which showed that a 15-bp sequence designated SE1 from the embryonic smooth muscle myosin heavy chain (*SMemb*) gene was essential for promoter activity in cultured SMCs and bound BTEB-2 in nuclear extracts from C2/2 rabbit vascular SMCs. In adult rabbits, BTEB-2 is highly expressed in bladder, uterus, and intestine but is developmentally downregulated in adult aorta. BTEB-2 expression is markedly induced in neointimal SMCs following balloon injury to rabbit aorta.[116] Taken together, these findings suggest that related Kruppel-like transcription factors may play reciprocal roles in the regulation of SMC-specific gene expression via interactions with TCE-like elements in regulatory regions of SMC marker genes.

A second class of DNA-binding proteins that may interact with CArG-box–dependent complexes in SMCs is the homeobox containing family of DNA binding proteins. Mutation of a homeodomain-like sequence (ATTA) located near the CArG-B element in the SMαActin promoter reduced promoter activity by half in rat aortic SMCs stimulated with angiotensin II (AII).[43] AII increases SRF binding to two proximal CArG boxes in the SMαActin promoter without increasing the overall levels of SRF in total cell extracts. Further studies showed that AII acted indirectly to promote SRF binding to CArG elements via its effects on the homeodomain factor Mhox.[43] AII increases MHox levels and Mhox binding to the ATTA sequence in SMCs, which acts cooperatively to enhance SRF affinity for CArG-B.[43] It is now becoming clear that coordinate and combinatorial interactions between multiple factors in addition to SRF is required to explain SMC-specific transcription. One explanation for why combinatorial interactions are so important is revealed by comparative analysis of the sequence of CArG elements in SMC specific promoters. Two common findings that emerge from this analysis are (1) multiple CArG boxes are often present, and (2) the sequence of these CArG elements frequently differs from the canonical $CC(AT)_6GG$ by having one or more G/C substitutions within the AT-rich core. Generally, the effect of these G/C substitutions is to reduce the affinity of SRF-CArG binding and render CArG-dependent transcription highly responsive to protein-protein interactions between SRF and other factors that increase the affinity of SRF binding to variant

CArG elements;[44] also see the references within ref. 44). This is an important feature of SMC-specific transcription via CArG elements that permits multiple levels of control involving both activators and inhibitors of SMC-specific transcription. Thus CArG-dependent transcription can be tailored to the developmental or physiologic context as a function of SRF binding partners. Given that the MADS box of SRF has been shown to interact with a large number of different nuclear factors including those that act to modify chromatin structure, and that both SRF and its partners can undergo stimulus-specific posttranslational modification, the range of control afforded by the variant CArG elements within SMC-specific promoters is large indeed. Future efforts will need to focus on characterization of the multiprotein complexes that are formed by combinatorial interactions with SRF to more completely explain both SMC differentiation in development and modulation of the differentiated phenotype for SMCs in settings of tissue injury, chronic inflammation, and disease.

Signal Transduction Pathways that Regulate SMC-Specific Gene Expression

To understand how SMC progenitor cells are directed to undergo SMC differentiation in development, it will be necessary to identify signaling pathways that couple extracellular factors secreted by neighboring cells to CArG-dependent transcription. One potentially important pathway in this regard is suggested by the recent report showing that activation of SRF-dependent transcription was closely linked to the process of cytoplasmic actin polymerization.[103] In this model, activation of the rhoA/p160 Rho kinase pathway led to phosphorylation and activation of LIM kinase, which in turn promoted stress fiber and focal adhesion formation via phosphorylation of the actin filament regulator cofilin. Polymerization of cytoplasmic actin into stress fibers was closely correlated with activation of SRF-dependent transcription.[103] Based on the results obtained with a variety of pharmacologic agents that promote or interfere with actin polymerization, a model was proposed that levels of free G-actin in the cell control SRF-dependent transcription perhaps via direct competitive interactions on the SRE itself.[103] A close correlation between rhoA/p160 rho-kinase activity, cytoplasmic actin reorganization, and focal adhesion assembly with SRF-dependent transcription is consistent with a requirement for epithelial to mesenchymal transformations in aortic SMC differentiation from neural crest cells, proepicardial cells, and endothelial cells. It is reasonable to assume that activation of rho family guanosine triphosphatase (GTPase) signaling pathways in SMCs will be found to be coupled to secreted factors derived from endothelial cells whose expression is upregulated by blood flow. The use of vertebrate models amenable to genetic analysis, coupled with high-throughput methods for analysis of gene expression and gene discovery, will accelerate the identification of essential genes involved in cell-cell signaling during tunica media formation.

SUMMARY

Lineage analysis studies of SMC origins in the embryo indicate that the vascular system is best viewed as a mosaic structure that is composed of unique parts that are pieced together from cells of different origins and developmental histories.

When viewed in this way, the diversity in functional properties and disease propensities of different blood vessels in the adult might not seem so mysterious. Future work will build upon the studies described above to (1) identify the critical genes and gene products in endothelial cells that direct recruitment and organziation of mesenchymal cells to form a tunica media, (2) characterize CArG box-dependent transcription in terms of cooperative and combinatorial interactions of SRF with both widely expressed and lineage-specific nuclear factors, and (3) define the important signal transduction pathways that couple endothelial-derived signals with SRF-dependent transcription in SMC differentiation.

ACKNOWLEDGEMENTS

The authors gratefully acknowledge helpful discussions with Thomas Landerholm, Robert Schwartz, Joseph Miano, Charles Little, John Schwarz, Adrianna Gittenberger-de Groot, and Gary Owens. Work in the authors laboratory was supported by grants from the National Institutes of Health (HL-41677), American Heart Association, and the Moran Foundation. M.W.M. is an Established Investigator of the American Heart Association.

REFERENCES

1. Abramovitch, R., Neeman, M., Reich, R., et al. (1998). Intercellular communication between vascular smooth muscle and endothelial cells mediated by heparin-binding epidermal growth factor-like growth factor and vascular endothelial growth factor. FEBS Lett 425:441–447.
2. Adam, P.J., Regan, C.P., Hautmann, M.B., Owens, G.K. (2000). Positive and negative acting Kruppel-like transcription factors bind a transforming growth factor beta control element required for expression of the smooth muscle cell differentiation marker SM22α *in vivo*. J Biol Chem 275:37798–37806.
3. Adams, J., Watt, F. (1993). Regulation of development and differentiation by the extracellular matrix. Development 117:1183–1198.
4. Adams, R., Wilkinson, G., Weiss, C., et al. (1999). Roles of ephrinB ligands and EphB receptors in cardiovascular development: demarcation of arterial/venous domains, vascular morphogenesis and sprouting angiogenesis. Genes Dev 13:295–306.
5. Altschul, R. (1944). Histologic analysis of arateriosclerosis. Arch Pathol 38:305–312.
6. Arciniegas, E., Ponce, L., Hartt, Y., Graterol, A., Carlini, R.G. (2000). Intimal thickening involves transdifferentiation of embryonic endothelial cells. Anat Rec 258:47–57.
7. Arciniegas, E., Sutton, A., Allen, T., Schor, A. (1992). Transforming growth factor beta-1 promotes the differentiation of endothelial cells into smooth muscle-like cells *in vitro*. J Cell Sci 103:521–529.
8. Arsenian, S., Weinhold, B., Oelgeschlager, M., Ruther, U., Nordheim, A. (1998). Serum response factor is essential for mesoderm formation during mouse embryogenesis. EMBO J 17:6289–6299.
9. Asahara, T., Chen, D., Takahashi, T., et al. (1998). Tie2 receptor ligands, angiopoietin-1 and angiopoietin-2, modulate VEGF-induced postnatal neovascularization. Circ Res 83:233–240.
10. Bazzoni, G., Dejana, E., Lampugnani, M.G. (1999). Endothelial adhesion molecules in the development of the vascular tree: the garden of forking paths. Curr Opin Cell Biol 11:573–581.

11. Benjamin, L., Golijanin, D., Itin, A., Pode, D., Keshet, E. (1999). Selective ablation of immature blood vessels in established human tumors follows vascular endothelial growth factor withdrawal. J Clin Invest 103:159–165.

12. Benjamin, L., Hemo, I., Keshet, E. (1998). A plasticity window for blood vessel remodeling is defined by pericyte coverage of the preformed endothelial network and is regulated by PDGF-B and VEGF. Development 125:1591–1598.

13. Blank, R., McQuinn, T., Yin, K., et al. (1992). Elements of the smooth muscle alpha-actin promoter required in cis for transcriptional activation in smooth muscle. Evidence for cell type-specific regulation. J Biol Chem 267:984–989.

14. Bressen, G., Daga-Gordini, D., Colombatti, A., Castellani, I., Marigo, V., Volpin, D. (1993). Emilin, a component of elastic fibers preferentially located at the elastin-microfibrils interface. J Cell Biol 121:201–212.

15. Browning, C., Culberson, D., Aragon, I., et al. (1998). The developmentally regulated expression of serum response factor plays a key role in the control of smooth muscle-specific genes. Dev Biol 194:18–37.

16. Camenisch, T., Spicer, A., Brehm-Gibson, T., et al. (2000). Disruption of hyaluronan synthase-2 abrogates normal cardiac morphogenesis and hyaluronan-mediated transformation of epithelium to mesenchyme. J Clin Invest 106:349–360.

17. Carey, D. (1991). Control of growth and differentiation of vascular cells by extracellular matrix proteins. Annu Rev Physiol 53:161–177.

18. Carmeliet, P. (2000). Mechanisms of angiogenesis and arteriogenesis. Nat Med 6:389–395.

19. Carmeliet, P., Collen, D. (1997). Genetic analysis of blood vessel formation. Trends Cardiovasc Med 7:271–281.

20. Carmeliet, P., Ferreira, V., Breier, G., et al. (1996). Abnormal blood vessel development and lethality in embryos lacking a single VEGF allele. Nature 380:435–439.

21. Clark, E., Clark, E. (1940). Microscopic observations on the extraendothelial cells of living mammalian blood vessels. Am J Anat 66:1–49.

22. Cook, C., Weiser, M., Schwartz, P., Jones, C., Majack, R. (1994). Developmentally timed expression of an embryonic growth phenotype in vascular smooth muscle cells. Circ Res 74:189–196.

23. Cox, C.M., Poole, T.J. (2000). Angioblast differentiation is influenced by the local environment: FGF-2 induces angioblasts and patterns vessel formation in the quail embryo. Dev Dyn 218:371–382.

24. Croissant, J., Kim, J., Eichele, G., et al. (1996). Avian serum response factor expression restricted primarily to muscle cell lineages is required for alpha-actin gene transcription. Dev Biol 177:250–264.

25. Crosby, J., Seifert, R., Soriano, P., Bowen-Pope, D.F. (1998). Chimaeric analysis reveals role of Pdgf receptors in all muscle lineages. Nat Genet 18:385–388.

26. Davies, P. (1995). Flow-mediated endothelial mechanotransduction. Physiol Rev 75:519–560.

27. De Angelis, L., Berghella, L., Coletta, M., et al. (1999). Skeletal myogenic progenitors originating from embryonic dorsal aorta coexpress endothelial and myogenic markers and contribute to postnatal muscle growth and regeneration. J Cell Biol 147:869–877.

28. de la Pompa, J.L., Timmerman, L.A., Takimoto, H., et al. (1998). Role of the NF-ATc transcription factor in morphogenesis of cardiac valves and septum. Nature 392:182–186.

29. DeRuiter, M., Poelmann, R., VanMunsteren, J., Mironov, V., Markwald, R., Gittenberger-de Groot, A.C. (1997). Embryonic endothelial cells transdifferentiate into mesenchymal cells expressing smooth muscle actins in vivo and in vitro. Circ Res 80:444–451.

30. Dickson, M., Martin, J., Cousins, F., Kulkarni, A., Karlsson, S., Akhurst, R. (1995). Defective haematopoiesis and vasculogenesis in transforming growth factor-β1 knock out mice. Development 121:1845–1854.

31. Dietz, H., Cutting, G., Pyeritz, R., et al. (1991). Marfan syndrome caused by a recurrent de novo missense mutation in the fibrillin gene. Nature 352:337–339.

32. Duband, J., Gimona, M., Scatena, M., Sartore, S., Small, J. (1993). Calponin and SM 22 as differentiation markers of smooth muscle: spatiotemporal distribution during avian embryonic development. Differentiation 55:1–11.

33. Eisenberg, L., Markwald, R. (1995). Molecular regulation of atrioventricular valvuloseptal morphogenesis. Circ Res 77:1–6.

34. Ferrara, N., Carver-Moore, K., Chen, H., et al. (1996). Heterozygous embryonic lethality induced by targeted inactivation of the VEGF gene. Nature 380:439–442.

35. Folkman, J., D'Amore, P. (1996). Blood vessel formation: what is its molecular basis? Cell 87:1153–1155.

36. Gallagher, B., Sakai, L., Little, C. (1993). Fibrillin delineates the primary axis of the early avian embryo. Dev Dyn 196:70–78.

37. Galvin, K., Donovan, M., Lynch, C., et al. (2000). A role for Smad6 in development and homeostasis of the cardiovascular system. Nat Genet 24:171–174.

38. Gao, Y., Li, M., Chen, W., Simons, M. (2000). Synectin, syndecan-4 cytoplasmic domain binding PDZ protein, inhibits cell migration. J Cell Physiol 184:373–379.

39. Gittenberger-de Groot, A., DeRuiter, M., Bergwerff, M., Poelmann, R. (1999). Smooth muscle cell origin and its relation to heterogeneity in development and disease. Arterioscler Thromb Vasc Biol 19:1589–1594.

40. Gittenberger-de Groot, A., Vrancken Peeters, M., Mentink, M., Gourdie, R., Poelmann, R. (1998). Epicardium-derived cells contribute a novel population to the myocardial wall and the atrioventricular cushions. Circ Res 82:1043–1052.

41. Gourdie, R., Mima, T., Thompson, R., Mikawa, T. (1995). Terminal diversification of the myocyte lineage generates Purkinje fibers of the cardiac conduction system. Development 121:1423–1431.

42. Gourdie, R., Wei, Y., Kim, D., Klatt, S., Mikawa, T. (1998). Endothelin-induced conversion of embryonic heart muscle cells into impulse-conducting Purkinje fibers. Proc Natl Acad Sci USA 95:6815–6818.

43. Hautmann, M., Thompson, M., Swarz, E., Olson, E., Owens, G. (1997). Angiotensin II-induced stimulation of smooth muscle α-actin expression by serum response factor and the homeodomain transcription factor MHox. Circ Res 81:600–610.

44. Hautmann, M.B., Madsen, C.S., Mack, C.P., Owens, G.K. (1998). Substitution of the degenerate smooth muscle (SM) alpha-actin CC(A/T-rich)6GG elements with c-fos serum response elements results in increased basal expression but relaxed SM cell specificity and reduced angiotensin II inducibility. J Biol Chem 273:8398–8406.

45. Hellstrom, M., Kalen, M., Lindahl, P., Abramsson, A., Betsholtz, C. (1999). Role of PDGF-B and PDGFR-β in recruitment of vascular smooth muscle cells and pericytes during embryonic blood vessel formation in the mouse. Development 126:3047–3055.

46. Herring, B., Smith, A. (1996). Telokin expression is mediated by a smooth muscle cell-specific promoter. Am J Physiol 270:C1656–1665.

47. Hirschi, K., D'Amore, P. (1996). Pericytes in the microvasculature. Cardiovasc Res 32:687–698.

48. Holzenberger, M., Lievre, C., Robert, L. (1993). Tropoelastin gene expression in the developing vascular system of the chicken: an in situ hybridization study. Anat Embryol (Berl) 188:481–492.

49. Hungerford, J., Little, C. (1999). Developmental biology of the vascular smooth muscle cell: building a multilayered vessel wall. J Vasc Res 36:2–27.

50. Hungerford, J., Owens, G., Aargraves, W., Little, C. (1996). Development of the aortic vessel wall as defined by vascular smooth muscle and extracellular markers. Dev Biol 178:375–392.

51. Hyer, J., Johansen, M., Prasad, A., et al. (1999). Induction of Purkinje fiber differentiation by coronary arterialization. Proc Natl Acad Sci USA 96:13214–13218.

52. Johansen, F., Prywes, R. (1995). Serum response factor: transcriptional regulation of genes induced by growth factors and differentiation. Biochim Biophys Acta 1242:1–10.

53. Katoh, Y., Loukianov, E., Kopras, E., Zilberman, A., Periasamy, M. (1994). Identification of functional promoter elements in the rabbit smooth muscle myosin heavy chain gene. J Biol Chem 269:30538–30545.

54. Kim, S., Ip, H., Lu, M., Clendenin, C., Parmacek, M. (1997). A serum response factor-dependent transcriptional regulatory program identifies distinct smooth muscle cell sublineages. Mol Cell Biol 17:2266–2278.

55. Koyama, H., Raines, E., Bornfeldt, K., Roberts, J., Ross, R. (1996). Fibrillar collagen inhibits arterial smooth muscle proliferation through regulation of Cdk2 inhibitors. Cell 87:1069–1078.

56. Krebs, L.T., Xue, Y., Norton, C.R., et al. (2000). Notch signaling is essential for vascular morphogenesis in mice. Genes Dev 14:1343–1352.

57. Kruger, O., Plum, A., Kim, J., et al. (2000). Defective vascular development in connexin 45-deficient mice. Development 127:4179–4193.

58. Kuo, C., Veselits, M., Barton, K., Lu, M., Clendenin, C., Leiden, J. (1997). The LKLF transcription factor is required for normal tunica media formation and blood vessel stabalization during murine embryogenesis. Genes Dev 11:2996–3006.

59. Landerholm, T., Dong, X.-R., Lu, J., Belaguli, N., Schwartz, R., Majesky, M. (1999). A role for serum response factor in coronary smooth muscle differentiation from proepicardial cells. Development 126:2053–2062.

60. LeLievre, C., Le Douarin, N. (1975). Mesenchymal derivatives of the neural crest: analysis of chimeric quail and chick embryos. J Embryol Exp Morphol 34:125–154.

61. Li, D., Brooke, B., Davis, E., et al. (1998). Elastin is an essential determinant of arterial morphogenesis. Nature 393:279–280.

62. Li, D., Faury, G., Talyor, D., et al. (1998). Novel arterial pathology in mice and humans hemizygous for elastin. J Clin Invest 102:1783–1787.

63. Li, D., Sorensen, L., Brooke, B., et al. (1999). Defective angiogenesis in mice lacking endoglin. Science 284:1534–1537.

64. Li, L., Liu, Z., Mercer, B., Overbeek, P., Olson, E. (1997). Evidence for serum response factor-mediated regulatory networks governing SM22α transcription in smooth, skeletal and cardiac muscle cells. Dev Biol 187:311–321.

65. Lilly, B., Zhao, B., Ranganayakulu, G., Paterson, B., Schulz, R., Olson, E. (1995). Requirement of MADS domain transcription factor D-MEF2 for muscle formation in Drosophila. Science 267:688–693.

66. Lindahl, P., Bostrom, H., Karlsson, L., Hellstrom, M., Kalen, M., Betsholtz, C. (1999). Role of platelet-derived growth factors in angiogenesis and alveogenesis. Curr Top Pathol 93:27–33.

67. Lindahl, P., Johansson, B., Leveen, P., Betsholtz, C. (1997). Pericyte loss and microanerysm formation in PDGF-B-deficient mice. Science 277:242–245.

68. Little, T., Beyer, E., Duling, B. (1995). Connexin 43 and connexin 40 gap junctional proteins are present in arteriolar smooth muscle and endothelium *in vivo*. Am J Physiol 268:H729–H739.

69. Madsen, C., Regan, C., Hungerford, J., White, S., Manabe, I., Owens, G. (1998). Smooth muscle-specific expression of the smooth muscle myosin heavy chain gene in transgenic mice requires 5′-flanking and first intronic DNA sequence. Circ Res 82:908–917.

70. Majesky, M., Schwartz, S. (1997). An origin for smooth muscle from endothelium? Circ Res 80:601–603.

71. Markwald, R., Mjaatvedt, C., Krug, E., Sinning, A. (1990). Inductive interactions in heart development. Role of cardiac adherons in cushion tissue formation. Ann NY Acad Sci 588:13–25.

72. Massague, J. (1998). TGF-β signal transduction. Annu Rev Biochem 67:753–791.

73. McAllister, K., Grogg, K., Johnson, D., et al. (1994). Endoglin, a TGF-beta binding protein of endothelial cells, is the gene for hereditary haemorrhagic telangiectasia type 1. Nat Genet 8:345–351.

74. Mecham, R., Stenmark, K., Parks, W. (1991). Connective tissue production by vascular smooth muscle in development and disease. Chest 99:43S–47S.

75. Miano, J., Cserjesi, P., Ligon, K., Periasamy, M., Olson, E. (1994). Smooth muscle myosin heavy chain exclusively marks the smooth muscle lineage during mouse embryogenesis. Circ Res 75:803–812.

76. Miano, J.M., Carlson, M.J., Spencer, J.A., Misra, R.P. (2000). Serum response factor-dependent regulation of the smooth muscle calponin gene. J Biol Chem 275:9814–9822.

77. Mikawa, T., Gourdie, R. (1996). Pericardial mesoderm generates a population of coronary smooth muscle cells migrating into the heart along with ingrowth of the epicardial organ. Dev Biol 174:221–232.

78. Moessler, H., Mericskay, M., Li, Z., Nagl, S., Paulin, D., Small, J. (1996). The SM 22 promoter directs tissue-specific expression in arterial but not in venous or visceral smooth muscle cells in transgenic mice. Development 122:2415–2425.

79. Obata, H., Hayashi, K., Nishida, W., et al. (1997). Smooth muscle cell phenotype-dependent transcriptional regulation of the alpha1 integrin gene. J Biol Chem 272: 26643–26651.

80. Ordahl, C. (1999). Myogenic shape-shifters. J Cell Biol 147:695–697.

81. Owens, G. (1995). Regulation of differentiation of vascular smooth muscle cells. Physiol Rev 75:487–517.

82. Pardanaud, L., Altmann, C., Kitos, P., Dieterlen-Lievre, F., Buck, C. (1987). Vasculogenesis in the early quail blastodisc as studied with a monoclonal antibody recognizing endothelial cells. Development 100:339–349.

83. Park, E., Putnam, E., Chitayat, D., Child, A., Milewicz, D. (1998). Clustering of FBN2 mutations in patients with congenital contractural arachnodactyly indicates an important role of the domains encoded by exons 24 through 34 during human development. Am J Med Genet 78:350–355.

84. Price, R., Owens, G., Skalak, T. (1994). Immunohistochemical identification of arteriolar development using markers of smooth muscle differentiation: evidence that capillary arterialization proceeds from terminal arterioles. Circ Res 75:520–527.

85. Qian, J., Kumar, A., Szucsik, J., Lessard, J. (1996). Tissue and developmental specific expression of murine smooth muscle γ-actin fusion genes in transgenic mice. Dev Dyn 207:135–144.

86. Ramirez, F., Pereira, L. (1999). The fibrillins. Int J Biochem Cell Biol 31:255–259.

87. Ranger, A.M., Grusby, M.J., Hodge, M.R., et al. (1998). The transcription factor NF-ATc is essential for cardiac valve formation. Nature 392:186–190.

88. Rongish, B., Drake, C., Argraves, W., Little, C. (1998). Identification of the developmental marker, JB3-antigen, as fibrillin-2 and its de novo organization into embryonic microfibrous arrays. Dev Dyn 212:461–471.

89. Rosenquist, T., Beall, A. (1990). Elastogenic cells in the developing cardiovascular system: Smooth muscle, nonmuscle and cardiac neural crest. Ann NY Acad Sci 588: 106–119.

90. Rosenquist, T., McCoy, J., Waldo, K., Kirby, M. (1988). Origin and propagation of elastogenesis in the developing cardiovascular system. Anat Rec 221:860–871.

91. Sarkisov, D., Kolokolchikova, E., Kaem, R., Paltsyn, A. (1988). Vascular changes in maturing granulation tissue. Bull Exp Biol Med 105:604–605.

92. Scott, N., Cipolla, G., Ross, C., et al. (1996). Identification of a potential role for the adventitia in vascular lesion formation after balloon overstretch injury of porcine coronary arteries. Circulation 93:2178–2187.

93. Selmin, O., Volpin, D., Bressen, G. (1991). Changes of cellular expression of mRNA for tropoelastin in the intraembryonic arterial vessels of developing chick by *in situ* hybridization. Matrix 11:347–358.

94. Shi, Y., O'Brien, J., Fard, A., Mannion, J., Zalewski, A. (1996). Adventitial myofibroblasts contribute to neointimal formation in injured porcine arteries. Circulation 94:1655–1664.

95. Shima, D.T., Mailhos, C. (2000). Vascular developmental biology: getting nervous. Curr Opin Genet Dev 10:536–542.

96. Shimizu, R., Blank, R., Jervis, R., Lawrenz-Smith, S., Owens, G. (1995). The smooth muscle alpha-actin gene promoter is differentially regulated in smooth muscle versus non-smooth muscle cells. J Biol Chem 270:7631–7643.

97. Shore, P., Sharrocks, A. (1994). The transcription factors Elk-1 and serum response factor interact by direct protein-protein contacts mediated by a short region of Elk-1. Mol Cell Biol 14:3283–3291.

98. Shutter, J.R., Scully, S., Fan, W., et al. (2000). Dll4, a novel Notch ligand expressed in arterial endothelium. Genes Dev 14:1313–1318.

99. Sims, D. (1991). Recent advances in pericyte biology—implications for health and disease. Can J Cardiol 7:431–443.

100. Skalak, T., Price, R., Zeller, P. (1998). Where do new arterioles come from? Mechanical forces and microvessel adaptation. Microcirculation 5:91–94.

101. Solway, J., Forsythe, S., Halayko, A., Vieira, J., Hershenson, M., Camoretti-Mercado, B. (1998). Transcriptional regulation of smooth muscle contractile apparatus expression. Am J Respir Crit Care Med 158:S100–S108.

102. Solway, J., Seltzer, J., Samaha, F., et al. (1995). Structure and expression of a smooth muscle cell-specific gene, SM22 alpha. J Biol Chem 270:13460–13469.

103. Sotiropoulos, A., Gineitis, D., Copeland, J., Triesman, R. (1999). Signal-regulated activation of serum response factor is mediated by changes in actin dynamics. Cell 98:159–169.

104. Soulez, M., Rouviere, C., Chafey, P., et al. (1996). Growth and differentiation of C2 myogenic cells are dependent on serum response factor. Mol Cell Biol 16:6065–6074.

105. Stenmark, K., Mecham, R. (1997). Cellular and molecular mechanisms of pulmonary vascular remodeling. Annu Rev Physiol 59:89–144.

106. Suri, C., Jones, P., Patan, S., et al. (1996). Requisite role of angiopoietin-1, a ligand for the tie2 receptor, during embryonic angiogenesis. Cell 87:1171–1180.

107. Suzuki, T., Nagai, R., Yazaki, Y. (1998). Mechanisms of transcriptional regulation of gene expression in smooth muscle cells. Circ Res 82:1238–1242.

108. Szucsik, J., Lessard, J. (1995). Cloning and sequence analysis of the mouse smooth muscle gamma-enteric actin gene. Genomics 28:154–162.

109. Topouzis, S., Majesky, M. (1996). Smooth muscle lineage diversity in the chick embryo: two types of aortic SMC differ in growth and receptor-mediated signaling responses to transforming growth factor-beta. Dev Biol 178:430–445.

110. Triesman, R. (1994). Ternary complex factors: growth factor regulated transcriptional activators. Curr Opin Genet Dev 4:96–101.

111. Tuder, R., Groves, B., Badesch, D., Voekel, N. (1994). Exuberant endothelial growth and elements of inflammation are present in plexiform lesions of pulmonary hypertension. Am J Pathol 144:275–285.

112. Vernon, R., Sage, E. (1995). Between molecules and morphology: extracellular matrix and creation of vascular form. Am J Pathol 147:873–882.

113. Vikkula, M., Boon, L., Carraway, K., et al. (1996). Vascular dysmorphogenesis caused by an activating mutation in the receptor tyrosine kinase TIE2. Cell 87:1181–1190.

114. Waldo, K., Kirby, M. (1993). Cardiac neural crest contribution to the pulmonary artery and sixth aortic arch artery complex in chick embryos aged 6 to 18 days. Anat Rec 237:385–399.

115. Wang, H., Chen, Z., Anderson, D. (1998). Molecular distinction and angiogenic interaction between embryonic arteries and veins revealed by ephrin-B2 and its receptor Eph-B4. Cell 93:741–753.
116. Watanabe, N., Kurabayashi, M., Shimomura, Y., et al. (1999). BTEB2, a Kruppel-like transcription factor, reguates expression of the SMemb/nonmuscle myosin heavy chain B (SMemb/NMHC-B) gene. Circ Res 85:182–191.
117. Wong, L.C., Langille, B.L. (1996). Developmental remodeling of the internal elastic lamina of rabbit arteries: effect of blood flow. Circ Res 78:799–805.
118. Yawashita, J., Itoh, H., Hirashima, M., Ogawa, M., Nishikawa, S., Yurugi, T., Naito, M., Nakao, K., Nishikawa, S. (2000) Flk-1-positive cells desired from embryonic stem cells serve as vascular progenitors. Nature 408:92–96.
119. Yano, H., Hayashi, K., Momiyama, T., Saga, H., Haruna, M., Sobue, K. (1995). Transcriptional regulation of the chicken caldesmon gene: activation of gizzard-type caldesmon promoter requires a CArG box-like motif. J Biol Chem 270:23661–23666.
120. Zhang, H., Timpl, R., Sasaki, T., Chu, M., Ekblom, P. (1996). Fibulin-1 and fibulin-2 expression during organogenesis in the developing mouse embryo. Dev Dyn 205: 348–364.
121. Zilberman, A., Dave, V., Miano, J., Olson, E., Periasamy, M. (1998). Evolutionarily conserved promoter region containing CArG*-like elements is crucial for smooth muscle myosin heavy chain gene expression. Circ Res 82:566–575.

CHAPTER 7

Vascular Development of the Heart

Robert J. Tomanek, Xinping Yue, and Wei Zheng

Myocardial vascularization during development is initiated through vasculogenesis, i.e., vascular tubes form from precursor cells (angioblasts). This phenomenon begins during the fourth week in humans, as the compact regions of the ventricular walls thicken and the diffusion distance for oxygen from the ventricular lumen lengthens. Subsequently, the tubes coalesce and continue to grow by sprouting (angiogenesis). Thus, a microvascular network, consisting of endothelial tubes, is formed. This network fuses at the sinus venosus to form veins. Subsequently a capillary plexus coalesces and penetrates the aorta and forms two major coronary arteries, which incorporate smooth muscle cells. From this point on, the coronary arteries expand and remodel, a process that continues into postnatal life.

Our understanding of coronary vasculogenesis and angiogenesis was, until recently, limited to descriptive data. During the past few years, our knowledge regarding stem cells, their migration, and growth factors, which may be important regulators of these processes, has begun to increase. Although the role of altered prenatal vascular development in congenital heart disease has not been established, there is now considerable evidence that coronary atherosclerosis has perinatal origins (Leistikow and Bolande, 1999). This evidence is based on historical, epidemiologic, and experimental observations.

The first part of this chapter examines spatial-temporal events leading to the establishment of a functional coronary circulation and its remodeling during the growth process. The second part examines the stimuli that initiate and regulate the vasculogenic and angiogenic events.

THE EPICARDIUM AND SUBEPICARDIAL MESENCHYME: CELL PRECURSORS FOR CORONARY VESSELS

Proepicardium

The proepicardium, or proepicardial organ, resembles a cluster of grapes and lies between the venous sinus horns and the liver primordium. This structure contains precursor cells that migrate to, and spread over, the surface of the heart via villous projections (Hiruma and Hirakow, 1989; Männer, 1992). First mesothelial cells envelop the heart and then the precursor cells for endothelium, smooth muscle, and fibroblasts migrate into the subepicardium (Viragh et al., 1993; Vrancken Peeters et al., 1995; Landerholm et al., 1999; Männer, 1999; Pérez-Pomares et al., 1997; Vrancken Peeters et al., 1999). Studies utilizing the electron microscope have demonstrated the migration pattern (Manasek, 1969; Virágh and Challice, 1973, 1981; Ho and Shimada 1978; Komiyama et al., 1987; Hiruma and Hirakow, 1989; Männer, 1992). As described in the chick embryo, at stage 17 multiple villus processes adhere to the dorsal wall of the atrioventricular canal to begin epicardial formation; the dorsal ventricular wall is covered first, and by stage 24 the entire ventricle is enveloped by the epicardium (Hiruma and Hirakow, 1989) (see Fig. 7.1). The formation and spreading of a subepicardial plexus was first demonstrated in embryonic pigs (Bennet, 1936). *In situ* hybridization has shown that epicardin, a member of the helix-loop-helix family of transcription factors, is strongly expressed in the proepicardium of mice at 9.5 days postcoitum and later in the epicardium as it is formed (Robb et al., 1998). These authors hypothesized that this gene functions in specification, proliferation, migration, and differentiation of the epicardium and its coronary derivatives. Epicardial progenitor cells also express α4 integrin which is believed to be essential for epicardial development (Pinco et al., 2001).

Subepicardial Mesenchyme

Progenitor cells (mesenchyme), which will differentiate into several myocardial cell types, have been traced in quail from the transverse septum to the proepicardium and finally to the subepicardium (Van den Eijnde et al., 1995), where they differentiate into three cell types: fibroblasts, endothelial, and smooth muscle. Mikawa and Fischman (1992) injected retrovirus label, *in ovo*, into the precardiac mesoderm (Hamburger-Hamliton stages 4 to 9) or into the myocardium (stages 10 to 18) of chickens. Prior to the formation of the epicardium all of the labeled cells were myocardial myocytes. However, injections after stage 17, especially those into the mesoderm, labeled endothelial, smooth muscle, and connective tissue cells. Poelmann et al. (1993), using the QH-1 antibody specific for quail endothelial cells and their precursors, traced the appearance of these cells first in the epicardium near the sinus venosus (stage HH 19) and then throughout the epicardium and subsequently into the myocardium. Mikawa and Gourdie (1996), using confocal microscopy and immunochemistry, identified both smooth muscle and endothelial cell populations in the proepicardial organ. Thus, the smooth muscle lineage is established prior to their migration of their precursors.

Pérez-Pomares et al. (1998a) used quail-chicken chimeras to demonstrate that at least a considerable part of the mesenchyme found in the proepicardium originated from the proepicardial mesothelium. Their conclusion that a cytoskeletal shift from epithelial to mesenchymal phenotype occurs is based on evidence that cytokeratin, an epithelial marker, and vimentin, a mesenchymal cell component, are both found in most of these cells. In addition, the squamous donor cells were found adhering to the myocardium of the atrial wall and entering the atrioventricular groove and outflow tract. Many of these donor-derived cells (both mesothelial and mesenchymal) were positive for the endothelial marker, QH-1.

This study suggests that the transformation of proepicardial and epicardial mesothelial cells contribute substantially to the subepicardial mesenchyme including the endothelial progenators. In another study (Pérez-Pomares et al., 1998b), the authors reported that, in the hamster, a few subepicardial cells become immunoreactive for the vascular endothelial growth factor (VEGF) receptor, flk-1, 10 days postcoitum with the number of reactive cells increasing markedly by day 11. Cells that were flk and cytokeratin immunoreactive developed into a capillary network. Accordingly, two cell types may serve as precursors for coronary endothelial cells. Dettman and colleagues (1998) found that quail epicardial monolayers produced invasive mesenchyme in collagen gels when stimulated with serum or growth factors. When quail epicardial cells labeled with LacZ were grafted into the pericardial space of embryonic day E2 chickens, they formed a chimeric epicardium, invaded the myocardial wall, and differentiated into smooth muscle, and perivascular and fibroblast cells. Similar experiments by Vrancken Peeters et al. (1999) confirmed the role of epicardial-mesenchymal transformation in providing myocardial vascular smooth muscle cells and fibroblasts.

A recent study by Landerholm et al. (1999) has shown that coronary smooth muscle cell differentiation from proepicardial cells requires transcriptionally active serum response factor. Their experiments revealed that explanted quail proepicardial cells were initially positive for cytokeratin (an epicardial marker) but showed no evidence of smooth muscle or endothelial cell markers. After 24 hours in culture, some cells were positive for smooth muscle α-actin, an event that coincided with the onset of serum response factor expression in virtually all epicardial cells. The appearance of smooth muscle (SM) markers such as calponin-SM22 and smooth muscle γ-actin were blocked by two different dominant-negative serum response factor constructs.

FORMATION OF ENDOTHELIAL-LINED TUBES

As summarized in Fig. 7.1, vascular tubes first appear in the epicardium and adjacent myocardium as the epicardium is formed. Endothelial precursor cells, sometimes positioned in clusters with erythrocytes constituting blood islands, have been observed in humans (Hirakow, 1983; Hutchins et al., 1988), rats (Rongish et al., 1994), and chickens (Hiruma and Hirakow, 1989). In humans, these structures first appear in the interventricular sulcus (Hutchins et al., 1988). These tubes coalesce and proliferate to form a microvascular network consisting of endothelium and associated pericytes.

The early myocardium is avascular. Its oxygen supply, which is from the ventricular lumen, is facilitated by the trabecular structure of the myocardium. With the expansion of the compact layer of the myocardium, microvessels appear. The left ventricle is vascularized slightly ahead of the right ventricle, with vascularization occurring between 48 and 59 days of gestation (Rychter et al., 1975). The relationship of coronary vasculogenesis to myocardial development is summarized in Fig. 7.2.

We have documented a vascular gradient from epicardium to endocardium in embryonic/fetal rats (Tomanek et al., 1996). In E15 hearts more than 60% of endothelial lined channels were found in the epicardial half of the myocardial compact zone; this gradient tended to disappear at the end of the gestational period (E22). The formation of vascular tubes (capillaries) and the growth of larger vessels

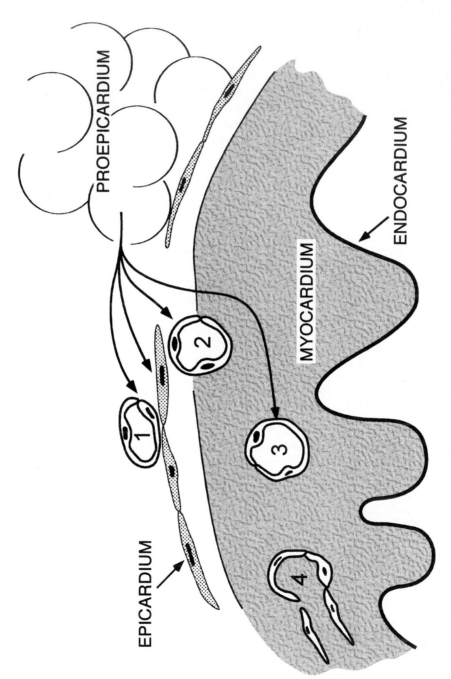

FIGURE 7.1. Angioblast migration and tube formation. With the formation of the epicardium, angioblasts migrate from the proepicardium and form vascular tubes on the epicardium (1), subepicardium-myocardium (2), and subsequently deeper in the myocardium (3). These structures coalesce and undergo sprouting or partitioning, i.e., angiogenesis (4).

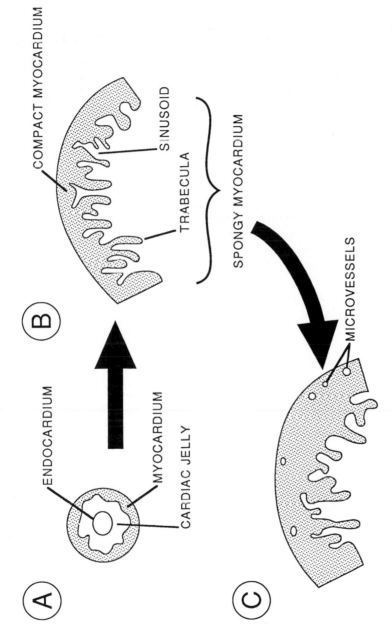

FIGURE 7.2. Development of the myocardium and the onset of coronary vasculogenesis. The early heart consists of endocardium, cardiac jelly, and myocardium (A). Myocardial growth results in a spongy layer (trabeculae with sinusoids lined by endocardium) and an out-compact region (B). As the compact region grows, microvessels appear near the myocardium closest to the epicardium (C).

shows a temporal-spatial relationship with the extracellular matrix (ECM). Immunohistochemical studies in our laboratory have documented the deposition of ECM components in relation to vascular tube formation in the rat (Rongish et al., 1996). (The ECM is discussed in a subsequent section of this chapter.) We have utilized electron microscopy to follow the events of coronary vasculogenesis and angiogenesis (Rongish et al., 1994, 1995). Tube formation is associated with blood islands, angioblast assembly, or canalization of individual angioblasts/endothelial cells (Fig. 7.3). Maturation of endothelial cells is characterized by development of cytoplasmic vesicles, thinning of the cytoplasm, formation of specialized junction, and the association of pericytes (Porter and Bankston, 1987a,b; Rongish et al., 1994). The formation of endothelial tubes, their coalescence and branching, prior to the formation of the two main coronary arteries and consequently coronary perfusion, occurs over a 3- or 4-day period in chickens (Rychter and Ostádal, 1971), mice (Virágh and Challice, 1981) and rats (Dbalý et al., 1968; Voboril and Schiebler, 1969; Tomanek et al., 1996), and about 2 weeks in humans (Hirakow, 1983).

The early postnatal period is characterized by rapid growth of the coronary vasculature in response to marked cardiomyocyte growth (Rakusan et al., 1965; Rakusan et al., 1994). In the rat, aggregate ventricular capillary length more than doubles during the second week of postnatal life (Olivetti et al., 1980), and total capillary length has been shown to correspond closely to ventricular weight during the postnatal period (Mattfeld and Mall, 1987). Rakusan and Turek (1985) estimated that 60% of the capillary bed is formed by the time weaning occurs. In sheep, intercapillary distance is maintained throughout postnatal development (Smolich et al., 1989). Postnatal capillary growth, as documented in rats, occurs by splitting of existing capillaries (van Groningen et al., 1991) as well as by sprouting.

FORMATION OF VEINS AND ARTERIES

The venous and arterial systems develop from the endothelial-lined tubes. By using quail-chicken chimeras, it was possible to show that the grafted quail endothelial cells, which are positive for the QH-1 antibody, were seen first as precursors in the myocardium and then incorporated into the sinus venosus, the site of the developing venous system (Poelmann et al., 1993). Prior to that study, it was believed that the veins, first seen in this region, were outgrowths of the sinus venosus (Grant, 1926; Goldsmith and Butler, 1937; Voboril and Schiebler, 1969; Rychter and Ostádal, 1971). Using the endothelial cell marker QH-1 in quail and markers of vascular smooth muscle, Vrancken Peeters and colleagues (1997b) noted that smooth muscle becomes apparent at the most proximal veins at stage HH (Hamburger Hamliton) 39. Contact of the capillary plexus with the coronary sinus is a relatively early event, occurring at 33 to 41 days of gestation in humans.

Formation of the right and left coronary arteries occurs as capillary plexi coalesce and penetrate the aortic root. Prior to this time the entire myocardial vasculature consists only of endothelial-lined tubes. In 1968 Dbalý and colleagues noted that they could not find evaginations from the aorta that would support outgrowth of coronary arteries, while they did note the closely apposed capillary plexus at the aortic root. Conclusive evidence that these coronary vessels undergo ingrowth into the aorta, rather than outgrowth, was not established until 20 years later.

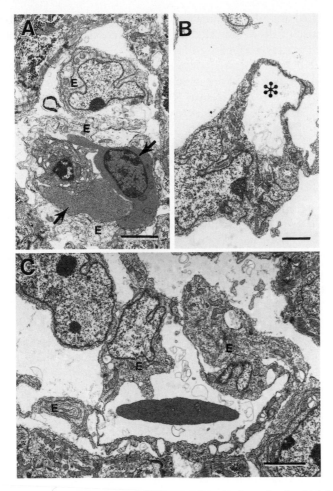

FIGURE 7.3.   Electron micrographs illustrating vascular tube formation in chick hearts. A: Clusters of blood cells (arrows) and angioblasts/endothelial cells (E) occur mainly in the subepicardium. B: Hollowing-out (canalization) of an endothelial cell via vacuole formation (*) is a common phenomenon. C: Assembly of endothelial cells (E) to form a vascular tube is also a characteristic of the developing myocardium. Bar = 2 μm.

Bogers et al. (1989) labeled endothelial cells with the QH-1 monoclonal antibody and were able to show that the capillary plexus fused and grew into the aorta. Waldo and colleagues (1990) provided similar evidence in chicks that had been injected with ink and the tissue then rendered translucent with a clearing agent. This ingrowth was also verified by using chick-quail chimeras (Poelmann et al., 1993). We have documented this phenomenon in rats by utilizing serial histologic sections (Tomanek et al., 1996). Virtually nothing is known about the factors that direct and regulate the penetration of the aorta by the vascular plexus. Interestingly, formation of the left coronary ostium prior to the right has been found to be consistent in humans (Hirakow, 1983; Mandarim-deLacerda, 1990).

Development of the media of the coronary arteries begins after the coronary channels become confluent with the aortic lumen. Perivascular cells appear around the newly formed vascular channel penetrating the aortic wall and differentiate into smooth muscle cells as indicated by smooth muscle α-actin staining (Ratajska and Fiejka, 1999). As described by Hood and Rosenquist (1992) smooth muscle α-actin in the chick is expressed after 8 days of incubation and spreads downstream, and progresses toward the apex only as far as parasympathetic neurons are found. Ablation of the neural crest in 30-hour embryos resulted in ectopic origin of the coronary arteries, e.g., subclavian. In every case only one coronary artery formed and then branched to give rise to the right and left coronary arteries. Subsequently, Waldo et al. (1994) used quail-to-chick chimeras to investigate the role of the cardiac neural crest in the development of coronary arteries. In these experiments the coronary neural crest cells in chicken embryos were replaced with quail coronary neural crest cells. The authors did not find cardiac neural crest cells within the tunica media of coronary arteries, but established the presence of clumps of neural crest cells disrupting the tunica media around the ostia of coronary arteries. Branches of coronary neural crest–derived parasympathetic ganglia and nerves entered the roots of coronary arteries at the points where neural crest cells were positioned. These data suggest that these neural crest cells along with the parasympathetic fibers may be essential for the development of coronary arteries.

The origin of vascular cells, i.e., smooth muscle and endothelial, from the proepicardium and the epi-subepicardium was noted in a previous section. Smooth muscle progenators must migrate to endothelial cell–lined channels. These channels transform into larger passageways and acquire smooth muscle as observed in rat (Dbalý et al., 1968; Rychter et al., 1972; Heintzberger, 1983) and chick (Rychter and Oštádal, 1971a,b). Although formation of the media begins in the main coronary arteries as the vascular tubes connect to the aorta (stage HH 32 in quail), the formation of the media of the veins occurs later, i.e. HH 39 (Vrancken Peters et al., 1997). These authors suggested that the origin of smooth muscle cells in veins differs from that of arteries, based on the finding that venous smooth muscle cells express β-myosin in heavy chain, which is also expressed in atrial myocardial cells. They hypothesized that smooth muscle cells in veins are recruited from atria. The discovery of a novel protein, bves (blood vessel/epicardial stubstand), has been used as an early marker of developing vascular smooth muscle cells (Reese et al., 1999). This protein was found expressed in the avian proepicardial organ, migrating epicardium, epicardial-derived mesenchyme, and smooth muscle of the developing myocardial arterial system.

Fetal coronary arteries do not show the tree-branching patterns noted in postnatal hearts. Rather, they contain more numerous anastomoses (Tomanek et al., 1996), which necessitates extensive remodeling. Due to the increase in diameter of the major vessels, remodeling must occur both prenatally and postnatally. Postnatal growth of arterioles is characterized by increases in wall thickness and segment length (Kurosawa et al., 1986; Matonoha and Zechmeister, 1992). After birth the left coronary artery and the left descending and circumflex branches rapidly increase in diameter, more rapidly than the right coronary artery (Ito et al., 1998). Dbalý (1973), who studied rat postnatal coronary artery development, found that anastomoses between arteries developed during the first few days of neonatal life. He found that the arterial system was well developed by postnatal day 12. The length of the large arteries increases in proportion to the linear expansion of the heart and the formation of new branches is limited beyond the third order arteries (Reinecke and Hort, 1992). Postnatal growth of the major coronary arteries is characterized by the tendency for external diameter and wall thickness to increase more in males during postnatal development than in age-matched females (Neufeld et al., 1962).

In humans the patterns of coronary vascular growth are similar to those in other species. The middle cardiac vein appears at 33 to 37 days of gestation, just 1 to 3 days after capillary plexuses are evident (Goldsmith and Butler, 1937; Hirakow, 1983; Hutchins et al., 1998). Establishment of the coronary arteries is forecast by endothelial buds at the root of the aorta, which give rise to the left coronary artery, usually first, and then the right at 44 to 50 days (Conte, 1984; Hutchins, et al., 1988; Mandarim-de-Lacerda, 1990). The development of the major branches follows closely; i.e., by 57 days all major branches are present (Conte, 1984). The development of the arterial system, as in other species, is from base to apex (Rychter et al., 1975).

EXTRACELLULAR MATRIX

The role of the extracellulr matrix (ECM) in development is well documented, and its contributions to heart development have been reviewed (Little and Rongish, 1995; Carver et al., 1997). The role of the ECM during vascular development is reviewed by Xu and Brooks in Chapter 4. This section is concerned only with findings that relate to the ECM of coronary vessel formation, a topic that has received relatively little attention.

Using immunohistochemistry, previous work in our laboratory explored the relationship between selected components of the ECM and vascularization of the developing rat heart (Rongish et al., 1996). When vascular tubes first appear at E13, fibronectin is already present and appears to provide a primary scaffolding for angioblast migration. Laminin deposition coincides with tube formation and is followed closely by the appearance of collagen IV, a major component of the basement membrane. These data fit with previous findings on other vascular beds that indicate that fibronectin facilitates endothelial cell migration and proliferation, while laminin secretion provides for vessel maturation (Risau and Lemmon, 1988). Collagens I and III are very sparse in the prenatal heart and accordingly do not play a major role in early formation of endothelial lined channels. These collagens increase in postnatal life and are mainly limited to (1) septa and (2) tunica adventitia of blood vessels. Our work also showed that the development of the ECM is independent of loading conditions, since hearts grafted *in oculo* experienced the same sequence of ECM deposition as hearts developing *in utero*. Our observation that platelet endothelial adhesion molecule is an early marker of endothelial cells fits with the notion that this adhesion molecule, which has functionally distinct isoforms, may play a major role in cell to cell adhesion and thus tube formation (Baldwin et al., 1994).

Tenascin-X (TN-X), a relatively new member of the tenascin family of ECM proteins, has been linked to migrating epicardial cells in the developing rat heart (Burch et al., 1995). TN-X is first expressed in the subepicardium in E12 hearts, just prior to the first evidence of tube formation. Its expression is associated with developing blood vessels and with nonmyocytes within the myocardium. Considering the spatial temporal expression of TN-X and the fact that tenascins are a family of glycoproteins that are expressed embryonically in distinct patterns, TN-

X may play an important role in cell migration of angioblasts and other cells originating in the epicardium/subepicardium.

Pericytes, as well as endothelial cells, contribute to ECM formation. *In vitro* experiments on cells from postnatal rats documented that both cell types have transcripts for fibronectin, and collagens I, IV, and VI (He and Spiro, 1995).

REGULATION OF CORONARY VASCULOGENESIS AND ANGIOGENESIS

Regulation by Growth Factors

VEGF and bFGF

Vascular endothelial growth factor (VEGF, or VEGF-A) and basic fibroblast growth factor (bFGF, or FGF-2) have been implicated in myocardial neovascularization in the adult, especially with regard to ischemia and hypoxia (Ladoux and Frelin, 1993; Banai et al., 1994; Cohen et al., 1994; Gu et al., 1997). Less is known about their precise roles in myocardial vasculogenesis and angiogenesis during development. VEGF is critical for vascular development since homozygous VEGF-deficient mice die in mid-gestation (Carmeliet et al., 1996). Similarly, mice deficient in VEGF receptor flk-1 die prior to day 10 postcoitum (Shalaby et al., 1995).

Embryonic and fetal hearts contain an abundance of bFGF (Joseph-Silverstein et al., 1989; Consigli and Joseph-Silverstein, 1991; Parlow et al., 1991; Spirito et al., 1991; Tomanek et al., 1996). We documented that bFGF transcripts are highest in E14–15 and neonatal rats (Tomanek et al., 1996). These time points correspond to the stages of early tube formation and rapid capillary growth, respectively. Although these data suggest a role for bFGF in coronary vasculogenesis and angiogenesis at these time points, this was not established by this study, since FGFs target other cell types and may be linked to their growth. However, bFGF has been shown to affect myocardial vascularization. Precocious expression of bFGF, due to retrovirus injection into the epicardial primordium of chicks, was shown to cause abnormal branching of coronary vessels (Mikawa, 1995). *In ovo* injection of bFGF just prior to the onset of myocardial capillary formation in chicks effected a marked increase in vascular volume density (Tomanek et al., 1998). In addition, the spongy/compact ratio in the ventricles was higher after bFGF injection. We recently provided evidence that early postnatal arteriolar growth is dependent on bFGF (Tomanek et al., 2001b). When neutralizing antibodies to bFGF were administered to neonatal rats during a 7-day period, arteriolar length density was markedly less than in the controls.

VEGF, which functions in targeting endothelial cells, has been found to be temporally and spatially related to early stages of myocardial vascularization. As noted previously, some hamster epicardial cells acquire the flk-1 receptor as early as E10 (Pérez-Pomares, et al., 1998b). VEGF transcripts are most abundant in the myocardial region near the epicardium in rat hearts before and during the early stages of microvessel formation (E13–16) and correlate with immunolocalization of VEGF protein (Tomanek et al., 1999). As tube formation occurs in the deeper areas of the myocardium, VEGF localization becomes more ubiquitous. Experimental studies in our laboratory have established a role for VEGF in the formation of coronary vascular structures. *In ovo* injections of the VEGF protein at stage HH 21 in the chicken increased vascular volume and density by 44% and 62%, respectively (Tomanek et al., 1998). Moreover, early postnatal capillarization of the heart is dependent on both VEGF and bFGF (Tomanek et al., 2001a). When monoclonal neutralizing antibodies to both VEGF and bFGF were administered for 1 week to rat neonates capillary and arteriolar diameters were increased, a finding that suggests that these growth factors facilitate maturation of the expanding vessel.

Messenger RNA (mRNA) encoding VEGF is very abundant in human fetal hearts, and VEGF protein is localized in cardiac myocytes and epithelial and vascular smooth muscle cells (Shifren et al., 1994). VEGF receptors are evident in the human fetal heart, but vary somewhat according to the vessel type (Partanen et al., 1999). Coronary arteries, capillaries, and veins express VEGFR-1 (flt-1) receptors, while VEGFR-2 (flk-1) is present in all of the vessel types except epicardial arteries. In the endothelium of the endocardium and endocardial blood vessels VEGFR-2 is more intensely expressed than VEGFR-1. In 5-week-old embryos capillaries show expression of VEGFR-3 (flt-4). Unlike VEGF-A knockout mice, which die during embryonic development, VEGF-B knockout mice survive and are fertile (Bellomo et al., 2000). However, their hearts are hypotrophic and their circulation is impaired as evidenced by poor recovery from myocardial ischemia. Thus, the VEGF-A gene is critical for normal coronary formation and development, while its VEGF-B counterpart plays a role in normal function of the coronary circulation. While null mutation of the VEGF-A gene is lethal, over expression of VEGF-A during embryogenesis causes aberrant coronary artery development and defective ventricular septation (Miqueral et al., 2000).

We have utilized explanted hearts cultured on collagen gels as an *ex vivo* model to study perturbations that influence tube formation. In this system, mesenchymal cells including angioblasts migrate from the embryonic heart explant and the latter, under the appropriate conditions, form tubes. Our data on embryonic rat heart explants showed that exogenous bFGF enhanced the number of cells that migrated onto the collagen gel, while VEGF stimulated vascular tube formation (Ratajska et al., 1995). The action of these growth factors was dependent on the stage of the explanted heart, i.e., bFGF was effective at E12, while VEGF was effective at E14. Our current studies, based on quail heart explants, indicate that vascular tube formation in this system is attenuated when anti-VEGF or anti-bFGF neutralizing antibodies are administered (Tomanek et al., 2001b). When neutralizing antibodies to both growth factors are given in combination, tube formation is nearly totally inhibited. Administration of a soluble Tie-2 receptor also attenuates tube formation; when combined with either bFGF or VEGF neutralizing antibodies, tube formation is reduced by 95%. These data indicate that tube formation is facilitated by more than one growth factor in this model. Our data also show that VEGF effects on tube formation are dependent on the pressure of bFGF and angiopoietins.

Other Growth Factors

The roles of angiopoietins and tie receptors are discussed in Chapter 3 by Suri and Yancopoulos. These investigators demonstrated that angiopoietin-1 and the Tie-2 receptor play a distinct role in embryonic angiogenesis, since deletion of the gene for either is lethal (Suri et al., 1996). Angiopoietin-1 or Tie-2 deletion in mice causes embryonic lethality as early as E9.5, and retardation of cardiac growth as evidenced by a lack of trabecular system development. Tie mRNA has been found to be prominent in the endocardium and endothelium of developing (E8.5) hearts, and accordingly appears to play a major role in morphogenesis (Korhonen et al., 1994). Human fetal coronary arteries, capillaries, and veins have recently been shown to contain both Tie-1 and Tie-2, as well as neuropilin receptors (Partanen et al., 1999).

We have shown that adding TGF-β protein to explanted quail embryonic hearts has an inhibitory effect on tube formation (Tomanek et al., 2001b). Consistent with this finding, we note that neutralizing antibodies to TGF-β promote tube formation. As noted by Pepper (1997), the stimulatory effects of this growth factor are likely dependent on local inflammation or the activation of other growth factors. However, the importance of TGF-β during development is supported by studies that show that its ancillary receptor, endoglin is required for extraembryonic angiogenesis (Arthur et al., 2000) and that TGF-β plays a critical role in glomerular capillary formation (Liu et al., 1999). Furthermore, TGF-β_3 expression in smooth muscle progenitor cells in arteries during development implicates this isoform in media formation (Yamagishi et al., 1999).

Although some other angiogenic growth factors have not been studied in the coronary vasculature, they undoubtedly play roles in the formation of the coronary vascular tree. Platelet-derived growth factor (PDGF) is expressed at sites of new blood vessel formation and is known to influence smooth muscle and fibroblast migration and proliferation, and to upregulate other growth factors and cytokines (Deuel, 1999). PDGF-A and PDGF-B null mice are unable to recruit smooth muscle cells. Accordingly, formation of the media in coronary vessels is undoubtedly dependent on this growth factor.

Metabolic and Mechanical Factors as Primary Stimuli

Growth factors, and consequently vessel formation, most likely require a primary stimulus for activation. Figure 7.4 outlines potential roles for metabolic (hypoxia) and mechanical (stretch) stimuli. The role of physical factors as stimulators of angiogenesis in adult heart and skeletal muscle have been reviewed (Hudlicka, 1988; Hudlicka and Brown, 1993). During heart development rapid growth of the myocardium and expansion of the ventricle may serve as a mechanical stimulus for the release of angiogenic growth factors. Once the coronary circulation is established, flow may act as a stimulus for vascular growth.

Hypoxia

Because hypoxia is a major trigger for upregulation of VEGF, we tested the role of hypoxic environment on vascular tube formation in our quail embryonic heart explant model (Yue and Tomanek, 1999). Explants were cultured under normoxic, hypoxic (2% to 10% O_2) and hyperoxic (95% O_2) conditions. Hypoxia enhanced tube formation, visualized by immunostaining with quail-specific endothelial cell marker (QH1), and upregulated three VEGF isoforms ($VEGF_{122}$, $VEGF_{166}$, and $VEGF_{190}$) in the quail hearts as detected by reverse-transcription polymerase chain

Coronary Vasculogenesis / Angiogenesis

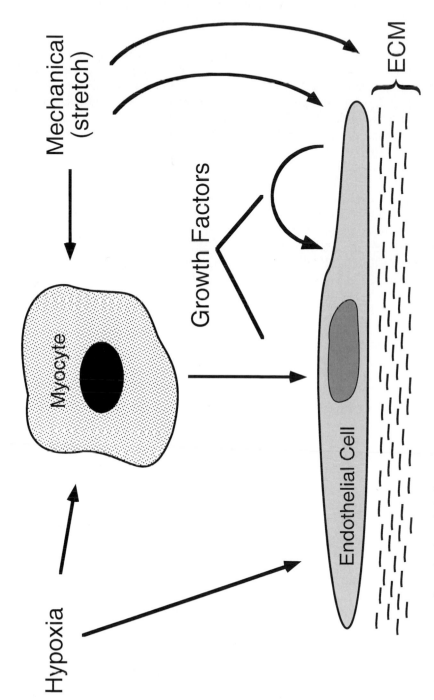

FIGURE 7.4. Mechanical and metabolic factors regulate coronary vasculogenesis. Hypoxia and mechanical factors, such as stretch, may stimulate growth factors in both myocytes and endothelial cells. Both paracrine and endocrine signaling can activate receptors in the latter. Stretch may also promote angiogenesis by its effects on the extracellular matrix (ECM).

reaction (RT-PCR). Hyperoxia had the opposite effect. To compare the functions of different VEGF isoforms induced by hypoxia, recombinant human $VEGF_{121}$ and $VEGF_{165}$ were added to the culture medium for 48 hours (Yue and Tomanek, 2001). $VEGF_{165}$ enhanced vascular growth in a dose-dependent manner: 5 ng/mL $VEGF_{165}$ increased the proliferation of endothelial cells; 10 ng/mL $VEGF_{165}$ increased both the number of endothelial cells and their incorporation into tubular structures; when the concentration was raised to 20 ng/mL, wider tubes with greater diameters were formed, and this growth pattern reached maximum at 50 ng/mL. No apparent growth pattern and enhancement of vascular growth were seen with the addition of $VEGF_{121}$ to the culture medium. We then compared the combined effects of hypoxia and exogenous $VEGF_{165}$. Tube formation from the heart explants treated with both hypoxia and $VEGF_{165}$ (50 ng/mL) had a morphology intermediate to those treated with hypoxia (thin and long tubes) or $VEGF_{165}$ (short and wide tubes) alone. Immuno-EM (QH-1) revealed endothelial cell lumenization (via vacuole formation) under all culture conditions. However, the addition of $VEGF_{165}$ stimulated the coalescence of endothelial cells to form larger vessels. We conclude (1) VEGF signaling is involved in hypoxia-induced coronary vasculogenesis/angiogenesis; (2) $VEGF_{121}$ and $VEGF_{165}$ induced by hypoxia have different functions on coronary vascular growth; and (3) endothelial cell lumenization (lumenization) is a normal process in coronary vasculogenesis.

Stretch

The role of stretch in mechanotransduction has been demonstrated in a number of cell types, for example: endothelium (Zhao et al., 1995; Murata et al., 1996), cardiac muscle (Cadre et al., 1998), and smooth muscle (Cheng et al., 1997; Park et al., 1998). Studies have shown that stretch triggers the upregulation and/or release of growth factors, for example TGF-β in cardiac muscle (Seko et al., 1999), mesangial (Riser et al., 1998) and smooth muscle (Li et al., 1998), cells. During heart development a relative stretch could be imposed on both cardiac myocytes and endothelial cells because of rapid growth throughout the ventricular wall. In addition, endothelial cells become subjected to hemodynamic stretch after coronary flow is established and longitudinal stretch increases as heart mass is enhanced. In view of the above, we subjected both cardiac myocytes and coronary microvascular cells to cyclic stretch (10% elongation at 30 cycles/min) using a Flexercell strain unit (Zheng et al., 2001). When conditioned media from stretched myocytes was added to nonstretched endothelial cell cultures, proliferation, migration, and tube formation were enhanced (Fig. 7.5). VEGF mRNA and protein were elevated in the stretched myocytes after 30 minutes. TGF-β mRNA was also enhanced. Neutralizing VEGF antibodies blocked the proliferative enhancement and markedly reduced the increase in migration and tube formation on a collagen gel. VEGF mRNA was also enhanced when the coronary microvascular cells were subjected to stretch. These data illustrate that stretch of either cardiac myocytes or coronary microvascular endothelial cells stimulates angiogenic factors. Therefore, coronary vessel formation may occur via both paracrine and autocrine pathways of VEGF.

The early postnatal period is a time of rapid cardiac growth and marked angiogenesis. In rats, for example, capillary aggregate length in the left ventricle increases from 9.5 to 197 m during the 10 days of life (Olivetti et al., 1980). As noted by

STRETCH AND PARACRINE EFFECTS OF VEGF

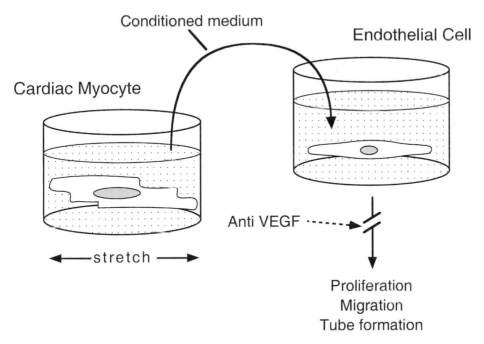

FIGURE 7.5.   Stretch and paracrine effects of vascular endothelial growth factor (VEGF). When conditioned medium from neonatal cardiac myocytes subjected to cyclic stretch is added to coronary microvascular endothelial cells, their proliferation, migration, and tube formation is markedly enhanced. These responses are nearly totally blocked when anti-VEGF neutralizing antibodies are added to the conditioned medium.

Rakusan (1999), neonatal "anemia" and low myoglobin content resulting from low iron due to a diet limited to milk may effect a relative cardiac hypoxia. In addition, blood pressure and coronary flow increase markedly during the early postnatal period (Rakusan, 1999). Thus, both hemodynamic and metabolic stimuli may serve as angiogenic triggers at this time.

Other Factors Influencing Myocardial Vascularization

In addition to the influences regulating coronary vascular growth discussed above, three other influences are considered in the following subsections.

Pericytes

Endothelial cells (ECs) and pericytes are anatomically closely related, as evidenced by their membranous contacts. Coculture systems have shown that these cell types interact via adhesion molecules, diffusible growth factors, and gap junctions (D'Amore, 1992; Hirschi and D'Amore, 1997). Pericytes control microvascular growth, as evidenced by their withdrawal as a requirement for EC proliferation

(Killingsworth, 1995). This withdrawal has been demonstrated during myocardial angiogenesis in adult pigs (Egginton et al., 1996). They may also guide endothelial sprouts during angiogenesis (Nehls et al., 1992). During embryonic and fetal development, pericytes in the heart appear some time after tube formation and thus are present only in more mature capillaries (Tomanek et al., 1996). In addition to influencing capillary growth, pericytes are considered to be pluripotent cells and may serve as precursors for smooth muscle cells (Hirschi and D'Amore, 1997).

Growth Rate and Ventricular Mass

To determine whether the rate of myocardial growth influences vascularization, we constricted the outflow tract of chicken hearts (stage HH 21), prior to the onset of myocardial vasculogenesis (Tomanek et al., 1999). At stage HH 29 ventricular mass and the left ventricular compact region were 64% and 43% greater, respectively, than in the shams. Vascular volume and numerical densities, however, were identical in the two groups, indicating that growth of the vasculature was proportional to the growth of the ventricle. By stage HH 36 ventricular mass was greater in the shams, due to a decelerated growth of the banded group. Vascular volume fraction, however, was virtually identical in the two groups. These data indicate that vascular growth is responsive to the magnitude of ventricular growth. We also noted that at stage HH 36 the banded hearts had stunted the growth of the main coronary arteries. These data are consistent with those obtained from embryonic rat hearts cultured *in oculo* (Rongish et al., 1996), which showed that unloaded hearts grow more slowly than those *in utero*, but vascular volume percent is similar in both groups. Thus, mass, rather than load per se, is the key factor governing neovasculariztion during embryonic/fetal development.

Innervation and Hormones

Our embryonic hearts grafted *in oculo* were also utilized to study the influence of sympathetic nerves and thyroxine. When hearts were grafted into sympathectomized eyes, vascular volume density was similar to the hearts grafted into eyes with intact sympathetic nerves (Torry et al., 1996). Since angiogenesis in young as well as senescent rat hearts is stimulated by exogenous thyroxine (Tomanek et al., 1995), we tested the hypothesis that thyroxine levels influence neovascularization in the embryonic heart (Rongish et al.). Thyroxine levels in host rats were either increased by injecting the hormone or virtually eliminated by injections of propylthiouracil. Neither of these interventions had any effect on vascular volume density of the grafted hearts. These studies indicate that the coronary vasculature in the embryonic heart is not sensitive to thyroxine levels or sympathetic innervation.

SUMMARY

Angioblasts from the proepicardium migrate to the subepicardium and subsequently to the myocardium where they differentiate into endothelial cells and form vascular tubes. This tubular (capillary) network precedes the development of coronary arteries, which facilitate a functional coronary circulation. Coalescence of these tubes constitutes the early formation of veins and arteries. The main coronary arteries are formed as a capillary plexus at the root of the aorta coalesces and penetrates the aortic wall. With the confluence of the tubular network with the aorta, smooth muscle cells are recruited to the root of the two major coronary arteries and a media is formed in a proximal to distal progression. Subsequently, smaller arteries and veins acquire a tunica media. Control of these processes is regulated by many factors including mechanical and metabolic influences, growth factors, and the extracellular matrix. Coronary vascular tube formation, as recently documented, is regulated not only by VEGF and its receptors, but also by bFGF and the angiopoietins. Thus, multiple growth factors and receptors play a role in establishing a coronary vasculature. Despite some important advances in coronary embryology during the last decade, our understanding of cell to cell signaling, temporal and spatial controls, and mechanisms underlying the growth and remodeling of the arterial system is quite limited.

ACKNOWLEDGMENTS

Work in the author's laboratory was supported by grants from the National Institutes of Health (HL 62178 and HL 48961).

REFERENCES

Aikawa, E., Kawant, J. (1982). Formation of coronary arteries sprouts from the primitive aortic sinus wall of the chick embryo. Experientia 38:816–818.

Arthur, H.M., Ure, J., Smith, A.J.H., et al. (2000). Endoglin, an ancillary TGF-β receptor is required for extreaembryonic angiogenesis and plays a key role in heart development. Dev Biol 217:42–53.

Baldwin, H.S., Shen, H.M., Yan, H-C., et al. (1994). Platelet endothelial cell adhesion molecule-1 (PECAM/CD31): alternatrively spliced, functionally distinct isoforms expressed during mammalian cardiovascular development. Development 120:2539–2553.

Banai, S., Jaklitsch, M.T., Shou, M., et al. (1994). Angiogenic-induced enhancement of collateral blood flow to ischemic myocardium by vascular endothelial growth factor in dogs. Circulation 89:2183–2189.

Bellomo, D., Headrick, J.P., Silins, G.U., et al. (2000). Mice lacking the vascular endothelial growth factor-B gene (Vegfb) have smaller hearts, dysfunctional coronary vasculature, and impaired recovery from cardiac ischemia. Circ Res 86:e29–e39.

Bennet, H.S. (1936). The development of the blood supply to the heart in the embryo pig. Am J Anat 60:27–53.

Bogers, A.J.J.C., Gittenberger-deGroot, A.C., Poelmann, R.E., Peault, B.M., Huysmans, H.A. (1989). Development of the origin of the coronary arteries, a matter of ingrowth or outgrowth? Anat Embryol 180:437–441.

Burch, G.H., Bedolli, M.A., McDonough, S., Rosenthal, S.M., Bristow J. (1995). Embryonic expression of tenascin-X suggests a role in limb, muscle, and heart development. Dev Dyn 203:491–504.

Cadre, B.M., Qi, M., Eble, D.M., Shannon, T.R., Bers, D.M., Samerel, A.M. (1998). Cyclic stretch down-regulates calcium transporter gene expression in neonatal rat ventricular myocytes. J Mol Cell Cardiol 30:2247–2259.

Carmeliet, P., Ferreira, V., Breier, G., et al. (1996). Abnormal blood vessel development and lethality in embryos lacking a single VEGF allele. Nature 380:435–439.

Carver, W., Terracio, L., Borg, T.K. (1997). Extracellular matrix maturation and heart formation. In: Development of cardiovascular systems. Molecules to organisms, pp. 43–56. Cambridge: Cambridge University Press.

Cheng, G.C., Briggs, W.H., Gerson, D.S., et al. (1997). Mechanical strain tightly controls fibroblast growth factor-2 release from cultured human vascular smooth muscle cells. Circ Res 80:28–36.

Cohen, M.V., Vernon, J., Yaghdjian, V., Hatcher, V.B. (1994). Longitudinal changes in myocardial basic fibroblast growth factor (FGF-2) activity following coronary artery ligation in the dog. J Mol Cell Cardiol 26:683–690.

Consigli, S.A., Joseph-Silverstein, J. (1991). Immunolocalization of basic fibroblast growth factor during chicken cardiac development. J Cell Physiol 146:379–385.

Conte, G. (1984). The embryology of the coronary arteries. In: Boucek, R.J., Morales, A.R., Romanelli, R., Judkins, M.P., eds. Coronary artery disease. Pathological and clinical assessment, pp. 38–65. Baltimore: Williams & Wilkins.

D'Amore, P.A. (1992). Capillary growth: a two-cell system. Sem Cancer Biol 3(2):49–56.

Dbalý, J, Ostádal, B., Rychter, Z. (1968). Development of the coronary arteries in rat embryos. Acta Anat 71:209–222.

Dbalý, J. (1973). Postnatal development of coronary arteries in the rat. Z Anat Entwickl-Gesch 141:89–101.

De Andres, A.V., Munoz-Chapuli, R., Sans-Coma, V. (1993). Development of the coronary arteries and cardiac veins in the dogfish. Anat Rec 235:436–442.

Dettman, R.W., Denetclaw, W., Jr., Ordahl, C.P., Bristow, J. (1998). Common epicardial origin of coronary vascular fibroblasts, and intermyocardial fibroblasts in the avian heart. Dev Biol 193:169–181.

Deuel, T. (1999). Platelet-derived growth factor and other modulators of angiogensis. In: Ware, J.A., Simons, M., eds. Angiogenesis and cardiovascular disease, pp. 128–140. New York: Oxford University Press.

Egginton, S., Hudlicka, O., Brown, M.D., Graciotti, L., Granata, A.L. (1996). *In vivo* pericyte-endothelial interaction during angiogenesis in adult cardiac and skeletal muscle. Microvasc Res 51:213–228.

Goldsmith, J.E., Butler, H.V. (1937). The development of the cardiac coronary ciruculation system. Am J Anat 60:185–201.

Grant, R.T. (1926). Development of the cardiac coronary vessels in the rabbit. Heart 13:261–271.

Gu, J.W., Santiago, D., Olowe, Y., Weinberger, J. (1997). Basic fibroblast growth factor as a biochemical marker of exercise-induced ischemia. Circulation 95:1165–1168.

He, Q., Spiro, M.J. (1995). Isolation of rat heart endothelial cells and pericytes: evaluation of their role in the formation of extracellular mtrix components. J Mol Cell Cardiol 27:1173–1183.

Heintzberger, C.F.M. (1983). Development of myocardial vascularization in the rat. Acta Morphol Neerl Scand 21:267–284.

Hirakow, R. (1983). Development of the cardiac blood vessels in staged human embryos. Acta Anat 115:220–230.

Hirschi, K.K., D'Amore, P.A. (1997). Control of angiogenesis by the pericyte: molecular mechanisms and significance. Exs 79:419–428.

Hiruma, T., Hirakow, R. (1989). Epicardial formation in embryonic chick heart: computer-aided reconstruction, scanning, and transmission electron microscopic studies. Am J Anat 184:129–138.

Ho, E., Shimada, Y. (1978). Formation of the epicardium studied with the scanning electron microscope. Dev Biol 66:579–585.

Hood, L.C., Rosenquist, T.H. (1992). Coronary artery development in the chick: origin and development of smooth muscle cells, and the effect of neural crest ablation. Anat Rec 234:291–300.

Hudlicka, O. (1988). Capillary growth: role of mechanical factors. News Physiol Sci 3:117–120.

Hudlicka, O., Brown, M.D. (1993). Physical forces and angiogenesis. In: Rubanyi, G.M., ed. Mechanoreption by the Vascular Wall, pp. 197–241. Mount Kisco, NY: Futura.

Hutchins, G.M., Kessler-Hanna, A., Moore, G.W. (1988). Development of the coronary arteries in the embryonic human heart. Circulation 77:1250–1257.

Ito, T., Harada, K. Tamura, M., Takada, G. (1998). *In situ* morphometric analysis of the coronry arterial growth in perintal rats. Early Human Growth 52:21–26.

Joseph-Silverstein, J., Consigli, S.A., Lyser, K.M., VerPault, C. (1989). Basic fibroblast gorwth factor in the chick embryo: immunolocalization to striated muscle cells and their precursors. J Cell Biol 108:2459–2466.

Killingsworth, M.C. (1995). Angiogenesis in early choroidal neovascularization secondary to age-related macular degeneration. Graefex Arch Clin Exp Ophthalmol 233:313–323.

Komiyama, M., Ito, K., Shimada, Y. (1987). Origin and development of the epicardium in the mouse embryo. Anat Embryol 176:183–189.

Korhonen, J., Polvi, A., Partanen, J., Altialo, K. (1994). The mouse Tie receptor tyrosine kinase gene: expression during embyronic angiogenesis. Oncogene 9:395–403.

Kurosawa, S., Kurosawa, H., Becker, A.E. (1986). The coronary arterioles in newborns, infants and children. A morphometric study of normal hearts and hearts with aortic atresia and complete transposition. Int J Cardiol 10:43–56.

Ladoux, A., Frelin, C. (1993). Hypoxia is a strong inducer of vascular endothelial growth factor mRNA expression in the heart. Biochem Biophys Res Commun 195:1005–1010.

Landerholm, T.E., Dong, X.R., Lu, J., Belaguli, N.S., Schwartz, R.J., Majesky, M.W. (1999). A role for serum response factor in coronary smooth muscle differentiation from proepicardial cells. Development 126:2053–2062.

Leistikow, E.A., Bolande, R.P. (1999). Perinatal origins of coronary atherosclerosis. Pediatr Dev Pathol 2:3–10.

Li, Q., Muragaki, T., Hatamura, I., Ueno, H., Ooshima, A. (1998). Stretch-induced collagen synthesis in cultured smooth muscle cells from rabbit aortic media and a possible involvement of angiotensin II and transforming growth factor-β. J Vasc Res 35:93–103.

Little, C.D., Rongish, B.J. (1995). The extracellular matrix during heart development. Experientia 51:873–882.

Liu, K.A., Darik, A., Ballerman, B.J. (1999). Neutralizing TGF-β1 antibody infusion in neonatal rats delays *in vivo* glomcrular capillary formation. Kidney Int 56:1334–1348.

Manasek, F.J. (1969). Embryonic development of the heart. II. Formation of the epicardium. J Embryol Exp Morphol 22:333–348.

Mandarim-de-Lacerda, C.A. (1990). Development of the coronary arteries in staged human embryos (the Paris Embryological Collection revisited). An Acad Bras Cienc 62:79–84.

Männer, J. (1992). The development of pericardial villi in the chick embryo. Anat Embryol 186:379–385.

Männer, J. (1999). Does the subepicardial mesenchyme contribute myocardioblasts to the myocardium of the chick embryo heart? A quail-chick chimera study tracing the fate of the epicardial primordium. Anat Rec 255:212–226.

Matonoha, P., Zechmeister, A. (1992). Structure of the coronary arteries during the prenatal period in man. Funct Dev Morphol 2:209–212.

Mattfeldt, T., Mall, G. (1987). Growth of capillaries and myocardial cells in the normal rat heart. J Mol Cell Cardiol 19:1237–1246.

Mikawa, T. (1995). Retroviral targeting of FGF and FGFR in cardiomyocytes and coronary vascular cells during heart development. Ann NY Acad Sci 752:506–575.

Mikawa, T., Borisov, A., Brown, A.M.C., Fischman, D.A. (1992). Clonal analysis of cardiac morphogenesis in the chicken embryo using a replication-defective retrovirus: I. Formation of the ventricular myocardium. Dev Dyn 193:11–23.

Mikawa, T., Fischman, D. (1992). Retroviral analysis of cardiac morphogenesis: discontinuous formation of coronary vessels. Proc Natl Acad Sci USA 89:9504–9508.

Mikawa, T., Gourdie, R.G. (1996). Pericardial mesoderm generates a population of coronary smooth muscle cells migrating into the heart along with ingrowth of the epicardial organ. Dev Biol 174:221–232.

Miquerol, L., Langille, B.L., Nagy, A. (2000). Embryonic development is disrupted by modest increases in vascular endothelial growth factor gene expression. Development 127:3941–3949.

Murata, K., Mills, I., Sumpio, B.E. (1996). Protein phosphatase 2A in stretch-induced endothelial cell proliferation. J Cell Biochem 63:311–319.

Nehls, V., Denzer, K., Drenckhahn, D. (1992). Pericyte involvement in capillary sprouting during angiogenesis *in situ*. Cell Tissue Res 270:469–474.

Neufeld, H.N., Wagenvoort, C.A., Edwards, J.E. (1962). Coronary arteries in fetuses, infants, juveniles and young adults. Lab Invest 11:837–844.

Olivetti, G., Anversa, P., Loud, A.V. (1980). Morphometric study of early postnatal development in the left and right ventricular myocardium of the rat. II. Tissue composition, capillary growth, and sarcoplasmic alterations. Circ Res 46:503–512.

Oštádal, B., Schiebler, T.H., Rychter, Z. (1975). Relations between the development of the capillary wall and myoarchitecture of the rat heart. Adv Exp Med Biol 53:375–388.

Park, J.M., Borer, J.G., Freeman, M.R., Peters, C.A. (1998). Stretch activates heparin-binding EGF-like growth factors expression in bladder smooth muscle cells. Am J Physiol 275:C1247–1254.

Parlow, M.H., Bolender, D.L., Kokan-Moore, N.P., Lough, J. (1991). Localization of bFGF-like proteins as punctate inclusions in the preseptation myocardium of the chicken embryo. Dev Biol 146:139–147.

Partanen, T.A., Makinen, T., Arola, J., Suda, T., Weich, H.A., Alitalo, K. (1999). Endothelial growth factor receptors in human fetal heart. Circulation 100:583–586.

Pepper, M.S. (1997). Transforming growth factor-beta: vasculogenesis, and vessel wall integrity. Cytokine Growth Factor Rev 8:21–43.

Pérez-Pomares, J.M., Macías, D., García-Garrido, L., Muñoz-Chápuli, R. (1997). Contribution of the primitive epicardium to the subepicardial mesenchyme in hamster and chick embryos. Dev Dyn 210:96–105.

Pérez-Pomares, J.M., Macias, D., García-Garrido, L., Muñoz-Chápuli, R. (1998a). The origin of the subepicardial mesenchyme in the avian embryo: an immunohistochemical and quail-chick chimera study. Dev Biol 200:57–68.

Pérez-Pomares, J.M., Macias, D., García-Garrido, L., Muñoz-Chápuli, R. (1998b). Immunolocalization of the vascular endothelial growth factor receptor-2 in the subepicardial mesenchyme of hamster embryos: identification of the coronary vessel precursors. Histochem J 30:627–634.

Pinco, K.A., Liu, S., Yang, J.T. (2001). α4 integrin is expressed in a subset of cranial neural crest cells and in epicardial progenitor cells during early mouse development. Mech Dev 100:99–103.

Poelmann, R.E., Gittenberger-deGroot, A.C., Metink, M.M.T., Bökenkamp, R., Hogers, B. (1993). Development of the cardiac vascular endothelium, studied with anti-endothelial antibodies in chicken-quail chimeras. Circ Res 73:559–568.

Porter, G.A., Bankston, P.W. (1987a). Maturation of myocardial capillaries in the fetal and neonatal rat: an ultrastructural study with a morphometric analysis of the vesicle populations. Am J Anat 178:116–125.

Porter, G.A., Bankston, P.W. (1987b). Functional maturation of the capillary wall in the fetal and neonatal rat heart: permeability characteristics of developing myocardial capillaries. Am J Anat 180:323–331.

Rakusan, K. (1999). Principles underlying vascular adaptation/angiogenesis. Adv Exp Med And Biol 471:439–443.

Rakusan, K., Cicutti, N., Flanagan, M.F. (1994). Changes in the cardiovascular network during cardiac growth, development and aging. Cell Mol Biol Res 40:117–122.

Rakusan, K., Jelinek, J., Korecky, B., Soukupova, M., Poupa, O. (1965). Postnatal development of muscle fibres and capillaries in the rat heart. Physiol Bohemoslov 14:32–37.

Rakusan, K., Turek, Z. (1985). Protamine inhibits capillary formation in growing rat hearts. Circ Res 57:393–398.

Ratajska, A., Fiejka, E. (1999). Prenatal development of coronary arteries in the rat: morphologic patterns. Anatomy and Embryology 200(5):533–540.

Ratajska, A., Torry, R.J., Kitten, G.T., Kolker, S.J., Tomanek, R.J. (1995). Modulation of cell migration and vessel formation by vascular endothelial growth factor and basic fibroblast growth factor in cultured embryonic heart. Dev Dyn 203:399–407.

Reese, D.E., Zavaljevski, M., Streiff, N.L., Bader, D. (1999). bves: a novel gene expressed during coronary blood vessel development. Dev Biol 209:159–171.

Reinecke, P., Hort, W. (1992). The growth of coronary artery branches in man under physiological conditions. Morphological studies of corrosion casts of the anterior interventricular branch of the coronary artery. (article in Geman) Z Kardiol 8:100–115.

Risau, W., Lemmon, V. (1988). Changes in the vascular extracellular matrix during embryonic vasculogenesis and angiogenesis. Dev Biol 125:441–450.

Riser, B.L., Cortes, P., Yee, J., et al. (1997). Mechanical strain- and high glucose-induced alterations in mesangial cell collagen metabolism: role of TGF β. J Am Soc Nephrol 10:827–836.

Riser, B.L., Cortes, P., Yee, J., Sharba, A.K., Asano, K., Rodriquez-Barbero, A., Narins, R.G. (1998). Mechanical strain- and high glucose-induced alterations in mesangial cell collagen metabolism: role of TGF-β. J Am Soc Nephrol 9:827–836.

Robb, L., Mifsud, L., Hartley, L., et al. (1998). Epicardin: a novel basic helix-loop-helix transcription factor gene expressed in epicardium, branchial arch, myoblasts, and mesenchyme of developing lung, gut, kidney and gonads. Dev Dyn 213:105–113.

Rongish, B.J., Hinchman, G., Doty, M.K., Baldwin, H.S., Tomanek, R.J. (1996). Relationship of extracellular matrix to coronary neovascularization during development. J Mol Cell Cardiol 28:2203–2215.

Rongish, B.J., Torry, R.J., Tomanek, R.J. (1995). Coronary neovascularization of embryonic rat hearts cultured in oculo is independent of thyroid hormones. Am J Physiol 268:H811–H816.

Rongish, B.J., Torry, R.J., Tucker, D.C., Tomanek, R.J. (1994). Neovascularization of embryonic rat hearts cultured in oculo closely mimics in utero coronary vessel development. J Vasc Res 31:205–215.

Rychter, Z., Jelinek, R., Marhan, O. (1972). Shape and location of non-vascularized area of ventricular myocardium in rat embryos during terminal phase of heart vascularization. Folia Morphol (Praha) 20:21–28.

Rychter, Z., Jirasek, J.E., Rychterova, V., Uher, J. (1975). Vascularization of heart in human embryo: location and shape of non-vascularized part of cardiac wall. Folia Morphol (Praha) 23:88–96.

Rychter, Z., Ostádal, B. (1971a). Mechanisms of the development of coronary arteries in chick embryo. Folia Morphol (Praha) 19:113–124.

Rychter, Z., Ostádal, B. (1971b). Fate of "sinusoidal" intrabecular spaces of the cardiac wall after development of the coronary vascular bed in chick embryo. Folia Morphol (Praha) 19:31–44.

Seko, Y., Seko, Y., Takahashi, N., Tobe, K., Kadowaki, T., Yazaki, Y. (1999). Pulsatile stretch activates mitrogen-activated protein kinase (MAPK) family members and focal adhesion kinase (p125[FAK]) in cultured rat cardiac myocytes. Biochem Biophys Res Commun 259:8–14.

Shalaby, F., Rossant, J., Yamaguchi, T.P., et al. (1995). Failure of blood island formation and vasculogenesis in Flk-1-deficient mice. Nature 376:62–66.

Shifren, J.L., Doldi, N., Ferrara, N., Mesiano, S., Jaffe, R.B. (1994). In the human fetus, vascular endothelial growth factor is expressed in epithelial cells an myocytes, but not vascular endothelium. J Clin Endocrinol Metab 79:316–322.

Smolich, J.J., Walker, A.M., Campbell, G.R., Adamson, T.M. (1989). Left and right ventricular myocardial morphometry in fetal, neonatal, and adult sheep. Am J Physiol 257:H1–H9.

Spirito, P., Fu, Y.-M., Yu, Z.-X., Epstein, S.E., Casscells, W. (1991). Immunohistochemical localization of basic and acidic fibroblast growth factors in the developing rat heart. Circulation 84:322–332.

Sugi, Y., Sasse, J., Barron, M., Lough, J. (1995). Developmental expression of fibroblast growth factor receptor-1 (cek-I; Fig) during heart development. Dev Dyn 202:115–125.

Suri, C., Jones, P.F., Paten, S., et al. (1996). Requisite role of angiopoietin-1, a ligand for the tie2 receptor, during embryonic angiogenesis. Cell 87:1171–1180.

Tomanek, R.J. (1996). Formation of the coronary vasculature: a brief review. Cardiovasc Res 31:E46–E51.

Tomanek, R.J., Connell, P.M., Butters, C.A., Torry, R.J. (1995). Compensated coronary microvascular growth in senescent rats with thyroxine-induced cardiac hypertrophy. Am J Physiol 268:H419–H425.

Tomanek, R.J., Haung, L., Suvarna, P.R., O'Brien, L.C., Ratajska, A., Sandra, A. (1996). Coronary vascularization during development in the rat and its relationship to basic fibroblast growth factor. Cardiovasc Res 31:E116–E126.

Tomanek, R.J., Hu, N., Phan, B., Clark, E.B. (1999). Rate of coronary vascularization during embryonic chicken development is influenced by the rate of myocardial growth. Cardiovasc Res 41:663–671.

Tomanek, R.J., Lotun, K., Clark, E.B., Suvarna, P.R., Hu, N. (1998). VEGF and bFGF stimulate myocardial vascularization in embryonic chick. Am J Physiol 274:H1620–H1626.

Tomanek, R.J., Ratajska, A. (1997). Vasculogenesis and angiogenesis in developing heart. In: Burggren, W.W., Keller, B.B., eds. Development of cardiovascular systems, pp. 35–42. Cambridge: Cambridge University Press.

Tomanek, R.J., Ratajska, A., Kitten, G.T., Yue, X., Sandra, A. (1999). Vascular endothelial growth factor coincides with coronary vasculogenesis and angiogenesis. Dev Dyn 215: 54–61.

Tomanek, R.J., Sandra, A., Zheng, W., Brock, T., Bjercke, R.J. (2001a) Vascular endothelial growth factor and basic fibroblast growth factor differentially modulate early postnatal coronary angiogenesis Circ Res 88:1135–1141.

Tomanek, R.J., Torry, R.J. (1994). Growth of the coronary vasculature in hypertrophy: mechanisms and model dependence. Cell Mol Biol Res 40:129–136.

Tomanek, R.J., Zheng, W., Peters, K.G., Lin, P., Holifield, J.S., Suvarna, P.R. (2001b) Multiple growth factors regulate coronary embryonic vasculogenesis. Dev Dyn 221: 265–273.

Torry, R.J., Rongish, B.J., Tucker, D.C., Kostreva, D.R., Tomanek, R.J. (1996). Influence of graft innervation on neovascularization of embryonic heart tissue grafted in oculo. Am J Physiol 270:H33–H37.

Unger, E.F., Banai, S., Shou, M., (1994). Basic fibroblast growth factor enhances myocardial collateral blood flow in a canine model. Am J Physiol 266:H1588–H1595.

Van den Eijnde, S.M., Wenink, A.C., Vermeij-Keers, C. (1995). Origin of subepicardial cells in rat embryos. Anat Rec 242:96–102.

Van Groningen, J.P., Weninck, A.C.G., Testers, L.H.M. (1991). Myocardial capillaries: increase in number by splitting of existing vessels. Anat Embryol 184:65–70.

Virágh, S., Challice, C.E. (1973). Origin and differentiation of cardiac muscle cells in the mouse. J Ultrastruct Res 43:1–24.

Virágh, S., Challice, C.E. (1981). The origin of the epicardium and the embryoic myocardial circulation in the mouse. Anat Rec 201:157–168.

Virágh, S., Gittengerger-deGroot, A.C., Poelmann, R.E., Kalman, F. (1993). Early development of quail heart epicardium and associated vascular and glandular structures. Anat Embryol 188:381–393.

Virágh, S., Kálmán, F., Gittengcrger-deGroot, A.C., Poelmann, R.E., Moorman, A.F.M. (1990). Angiogenesis and hematopoiesis in the epicardium of the vertebrate embryo heart. Ann NY Acad Sci 588:455–458.

Virágh, S., Szabo, E., Challice, C.E. (1989). Formation of the primitive myo- and endocardial tubes in the chicken embryo. J Mol Cardiol 21:123–137.

Voboril, A., Schiebler, T.H. (1969). Uber die Entwicklung des Grefässversorgung des Rattenherzen. Z Anat Entw Gesch 129:24–40.

Vrancken Peters, M.-P.F.M., Gittenberger-deGroot, A.C., Mentink, M.M.T., Hungerford, J.E., Little, C.D., Poelmann, R.E. (1997a). Differences in development of coronary arteries and veins. Cardiovasc Res 36:101–110.

Vrancken Peeters, M.-P.F.M., Gittenberger-de Groot, A.C., Mentink, M.M.T., Hungerford, J.E., Little, C.D., Poelmann, R.E. (1997b). The development of the coronary vessels and their differentiation into arteries and veins in the embryonic quail heart. Dev Dyn 208: 338–348.

Vrancken Peeters, M.-P.F.M., Gittenberger-de Groot, A.C., Mentink, M.M.T., Poelmann, R.E. (1999). Smooth muscle cells and fibroblasts of the coronary arteries derive from epithelial-mesenchymal transformation of the epicardium. Anat Embryol 199:367–378.

Vrancken Peeters, M.-P.F.M., Mentink, M.M.T., Poelmann, R.E., Gittenberger-de Groot, A.C. (1995). Cytokeratins as a marker for epicardial formation in the quail embryo. Anat Embryol 191:503–508.

Wagner, R.C. (1980). Endothelial cell embryology and growth. Adv Microcirc 9:45–75.

Waldo, K.L., Kumski, D.H., Kirby, M.L. (1994). Association of the cardiac neural crest with development of the coronary arteries in the chick embryo. Anat Rec 239:315–331.

Waldo, K.L., Willner, W., Kirby, M.L. (1990). Origin of the proximal coronary artery stems and a review of ventricular vascularization in the chick embryo. Am J Anat 188:109–120.

Yamagishi, T., Nakajima, Y., Nakamura, H. (1999). Expression of TGFβ3 RNA during chick embryogenesis: a possible important role in cardiovascular development. Cell Tissue Res 298:85–93.

Yamaguchi, T.P., Dumont, D.J., Conlon, R.A., Breitman, M.L., Rossant, J. (1993). FlK-1, an flt-related receptor tyrosine kinase is an early marker for endothelial cell precursors. Development 118:489–498.

Yue, X., Tomanek, R.J. (1999). Stimulation of coronary vasculogenesis/angiogenesis by hypoxia in cultured embryonic hearts. Dev Dyn 216:28–36.

Yue, X., Tomanek, R.J. (2001) Effects of $VEGF_{165}$ and $VEGF_{121}$ on vasculogenesis and angiogenesis in cultured embryonic guail hearts. Am J Physiol 280:H2240–H2247.

Zheng, W., Seftor, E.A., Meininger, C.J., Hendrix, M.J.C., Tomanek, R.J. (2001) Mechanisms of coronary angiogenesis in response to stretch: role of VEGF and TGF-β. Am J Physiol, 280:H909–H917.

Zhao, S., Suciu, A., Ziegler, T., et al. (1995). Synergistic effects of fluid shear stress and cyclic circumferential stretch on vascular endothelial cell morphology and cytoskeleton. Atheroscler. Thromb Vasc Biol 15:1781–1786.

CHAPTER 8

Vascular Development of the Brain and Spinal Cord

Haymo Kurz and Bodo Christ

Understanding blood vessel formation in the central nervous system (CNS), that is, the brain and spinal cord, is of interest to both developmental biologists and medical researchers. Embryologists address the fundamental question—Which primordial tissues provide cells that build the vascular wall?—and ask by which mechanisms vascular tubes within the compact epithelium of the early CNS are established. Doctors are interested in therapeutic tools that either promote blood vessel growth in the differentiated brain, following, for example, ischemia, or prevent angiogenesis, for example, in brain tumors.

This chapter discusses the embryology of CNS angiogenesis, but mentions the medical aspects only briefly. Recent advances in studying neural (Goldman et al., 1997; McKay, 1997; Levison and Goldman, 1997; Brüstle et al., 1998), mesenchymal (Young et al., 1995), and bone marrow (Eglitis and Mezey, 1997; Dieterlen-Lièvre, 1998) stem cells may have important implications for both fundamental and applied research on vascular development of the CNS.

After providing a basic nomenclature of CNS development, we bring together divergent lines of research and speculate on the implications of recent findings. We emphasize the issues of cell lineage and differentiation of vascular and perivascular cells. Fundamental mechanisms and growth factor signaling have been reviewed elsewhere (Wilting et al., 1995b, 1998; LaBonne and Bronner-Fraser, 1998; Little et al., 1998; Risau, 1998; Neufeld et al., 1999; Ortega et al., 1999; Weinstein, 1999).

This chapter provides a vivid impression of cell morphology during CNS and meningeal vascularization through conventional imaging and confocal laser scanning microscopy (CLSM) of sections of avian embryos (Japanese quail, *Coturnix coturnix japonica*, and white leghorn chicken, *Gallus gallus domesticus*) at various stages of development. Chimeras of these species have been extensively used for cell lineage studies that were based on the endogenous nucleolar marker (Le Douarin, 1973) or monoclonal antibodies like MB1/QH1 (Peault et al., 1983). Based on the generally high conservation of fundamental developmental mechanisms, as demonstrated repeatedly by concordance of the quail-chick system with mammalian development, we are confident that our results can be extrapolated to, and combined with, observations in mammalian species (mouse, rat, rabbit) and humans.

BUILDING MATERIALS

Vertebrate embryos are formed from three germ layers that arise during gastrulation. The *ectoderm* gives rise to the neural tube (NT) that is surrounded by *mesoderm* after neurulation, whereas the *endoderm* is displaced further to the ventral side. The endoderm is essential for angiopoietic commitment of mesodermal cells (Pardanaud and Dieterlen-Liévre, 1999), but we will not take it into account here because it does not influence CNS vascularization. We focus on the developmental potentials of the ectoderm and mesoderm. With respect to blood vessel development, we distinguish between the head region, where the NT is intrinsically segmented into rhombomeres, but the mesoderm is not, and the trunk region, where the mesoderm is segmented into somites but the NT is not (Köntges and Lumsden, 1996; Bronner-Fraser and Fraser, 1997; Le Douarin et al., 1998). The neural crest (NC) in the head region has a far greater potential to differentiate into various tissues than trunk NC (Groves and Bronner-Fraser, 1999); Interestingly, the border between the head and trunk regions is immediately caudal to the otic placode for the mesoderm, but further caudally located (5th somite) for the NC. The term *mesectoderm* has been assigned to emigrated head NC because this cell population constitutes most of the head and branchial arch mesenchyme.

Neuroectoderm and Neural Crest

The neural tube (NT) is formed by ectoderm during the process of neurulation (Le Douarin et al., 1998), which takes place at about the same time the paired heart anlagen, the dorsal aortae, as well as related vascular intraembryonic structures and the blood islands of the yolk sac are forming and circulation starts (between 20 and 36 hours in chicken embryos, corresponding to embryonic day E8.5 in the mouse). The NT is a purely epithelial hollow tube with a primitive basement membrane. The NT internally becomes organized into the unpaired ventral floor plate and dorsal roof plate, and on either side ventral basal and dorsal alar plates. The roof plate provides the neural crest cells that undergo epitheliomesenchymal transition and emigrate from the dorsal NT. Some of these NC cells stay near the NT and become neurons in the spinal ganglia that project their axons back into the NT alar plate. These axon bundles are called dorsal root and the pertaining spinal ganglia are hence called dorsal root ganglia (DRG). The DRG-forming NC cells have to perforate the NT basement membrane twice: once upon emigration and once upon reentrance of their axons into the NT (Kalcheim and Le Douarin, 1986; Teillet et al., 1987; Poelmann et al., 1990). In contrast, ventral motoneurons project their axons outward through the basement membrane.

   While the roof plate provides NC cells, the floor plate remains epithelial. The most apparent changes occur in the basal and alar plate, where three zones become clearly distinguishable: in the innermost layer, the ependymal zone forms the surface of the central canal, and its neuroblast nuclei are densely packed and undergo mitosis; in the intermediate layer, the mantle zone, neuroblasts are elongated and eventually detach from the inner surface, a process known as the "birth" of a neuron; in the outer layer, the marginal zone, the projections (dendrites, axons) of developing neurons accummulate. In this way, the alar plate differentiates into the dorsolateral sensory column, and the basal plate into the ventrolateral motor column (gray matter), both enveloped by white fiber tracts. Together with neurons, glial cells are produced from the same progenitors that appear as radial glia initially and later differentiate into macroglia, that is astroglia and the myelin-forming oligodendroglia (Galileo et al., 1990; Artinger et al., 1995). In the ependymal zone, some neuronal stem cells seem to persist throughout a lifetime (Goldman et al., 1997; McKay, 1997). We will use the term *neuroblast* in the broad sense of both an embryonic neural or glial progenitor or a persisting potential neuronal stem cell.

The early patterning of the NT as just described essentially takes place between 40 and 100 hours of chick development (corresponding to E9 to E11.5 in mice, or to the third to fourth week in humans) and progresses in a rostrocaudal direction over the entire length of the embryo. In this process, complex signaling pathways differentially regulate gene expression, thus establishing the craniocaudal and dorsoventral axes, and gross bilateral symmetry in the NT. Since the details are beyond the scope of this chapter, readers who are interested in the mutual influence of neural and nonneural ectoderm and axial and paraxial mesoderm are referred to the literature (Kalcheim and Le Douarin, 1986; Brand-Saberi et al., 1996; Baker and Bronner-Fraser, 1997a; Bronner-Fraser and Fraser, 1997; Sosic et al., 1997; Christ et al., 1998; LaBonne and Bronner-Fraser, 1998).

Mesoderm and Mesectoderm

The NT is initially surrounded by axial (notochord) and paraxial mesoderm (somites, paraxial head mesoderm), but not yet by blood vessels. However, jointly with the process of somite differentiation into sclerotome and dermomyotome, a large number of endothelial cells (ECs) are derived from all somitic compartments (Huang et al., 1994, 1997; Wilting et al., 1995a). These contribute to the dorsal aortae, and form segmental vessels and the spinal perineural vascular plexus (PNVP). In the paraxial mesoderm of the head, ECs differentiate (without somite formation) and contribute to the branchial arch arteries and cerebral PNVP (Noden, 1990; Couly et al., 1995). Interestingly, the vascular plexuses on either side remain unconnected for a long time in the trunk, but connect early in the midline of the head region. It should be noted that mesoderm is the only source of all ECs, no matter what their later location and modification will be (Noden, 1988). In addition, trunk mesoderm, but not head mesoderm, can give rise to vascular smooth muscle cells (vSMCs) and most components of the dermis (Couly et al., 1992; Korn et al., 2001).

In the head region, the NC provides a large mesenchymal cell population, the aforementioned mesectoderm, that partially intermingles with the branchial mesoderm-derived mesenchyme (producing nearly exclusively ECs and striated muscle fibers), and that gives rise to most of the connective tissues of the head and branchial arches. Moreover, the vSMCs of branchial arch arteries and of meningeal vessels have been shown to be of mesectodermal origin (Couly and Le Douarin, 1987), whereas mesectoderm-derived ECs have never been observed. The potential of the ectodermal placodes of the head (ear, lens, and olfactory) in the formation of mesectoderm has not yet been firmly established (D'Amico-Martel and Noden, 1983; Couly and Le Douarin, 1985; Noden, 1988; Baker and Bronner-Fraser, 1997b; LaBonne and Bronner-Fraser, 1998).

Segmentation of the head NT, and thus the positional identity of head NC is established through homeobox genes. The multitude of NC fates in the mesectoderm are regulated via various members of the transforming growth factor-β (TGF-β) superfamily and their receptors (Shah et al., 1996) and by endothelin-1/endothelin A receptor (Nataf et al., 1998). Moreover, matrix molecules like fibronectin, laminin, tenascin, and versican are important regulators of NC migration and differentiation (Poelmann et al., 1990; Lallier et al., 1992). The molecular basis for the building of perineural vessels, that is, the interaction of mesoderm- or mesectoderm-derived vSMC and pericytes with mesoderm-derived ECs is an area of ongoing research (D'Amore and Smith, 1993; Nehls and Drenckhahn, 1993; Hirschi and D'Amore, 1996, 1997; Pepper, 1997; Adam et al., 1998; Drake et al., 1998; Etchevers et al., 1999, 2001; Jiang et al., 2000).

ENDOTHELIAL CELL INVASION, INTERACTION, AND DIFFERENTIATION

Invasion of Endothelial Cells

Spinal Cord

Despite extensive blood vessel growth in the mesenchyme surrounding the early CNS, the NT remains completely avascular until the end of the third incubation day in avian embryos (E9.5 in mice, fourth week in humans). By then, a well-developed PNVP is supplied with arterial blood from segmental branches of the aorta in the so-called primitive arterial tracts (Strong, 1961) that are located ventrally on either side of the floor plate (Fig. 8.1A). The lateral portion of the PNVP drains to the cardinal veins.

The first endothelial sprouts into the NT originate from the primitive arterial tracts in the cervical region at about 72 hours of incubation (Fig. 8.1B). At this time, the floor plate and to a lesser degree the ependymal layer express vascular endothelial growth factor-A (VEGF-A) (Breier et al., 1992; Aitkenhead et al., 1998). Moreover, the process of invasion appears to be Tie-2/Angiopoietin-1 dependent (Suri et al., 1996, 1998; Koblizek et al., 1998). The role of ephrinB/EphB signaling during arteriovenous demarcation around and sprouting into the CNS needs to be studied further (Adams et al., 1999).

The ECs adjacent to the basal membrane of the NT produce long filopodia that protrude between the neuroepithelium immediately lateral to the floor plate (Fig. 8.1C). It appears as if the filopodia first open spaces and then pull the rest of the ECs into the epithelium of the NT. The number and length of the filopodia are unmatched by any vascular sprouting phenomena outside the CNS. The location of sprout invasion is exactly defined in the mediolateral direction, but appears random along the craniocaudal axis and bilaterally asymmetric (Kurz et al., 1996). In any case, the variable axial distance between subsequent sprouts (10 to 50 µm) is not related to the segmental vascular pattern in the mesoderm (about 200 µm) (Kurz et al., 1996), and the left and right sides develop independently (Klessinger and Christ, 1996). These ventral sprouts bifurcate and form arcades in an axial direction, located in the region between the ependymal and the mantle layer. About 12 hours after the initial ventral sprouting, the ECs in the lateral (venous)

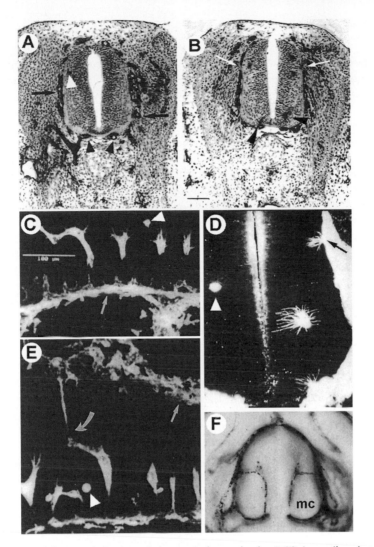

FIGURE 8.1. Early vascularization of the cervical neural tube (NT) in quail embryos. QH1-immunoperoxidase of transverse paraffin sections (A,B); QH1-immunofluorescence using confocal laser scanning microscopy (CLSM) of longitudinal (C,E) or transverse (D) vibratome sections. Transverse section of ink-injected, cleared embryo (F). A: At the end of embryonic day E3, the NT is surrounded by the perineural vascular plexus (PNVP, black arrows) that receives blood from the primitive arterial tracts (black arrowheads). The NT is devoid of sprouts, but an isolated QH1+ cell (white arrowhead; presumably a macrophage) is found in its dorsal region. B: Early at E4, invasion of endothelial sprouts starts from the primitive arterial tracts lateral to the floor plate (black arrowhead). Note several isolated QH1+ cells near the dorsal roots (white arrows). C: During E4, tips of ventral sprouts bifurcate and form longitudinal arcades with their neighbors. Note presence of isolated QH1+ cell (arrowhead) near tips. The segmental vessels lateral to the PNVP (arrow) are spaced at about 250-μm intervals, whereas the intraneural sprouts are spaced at 30 to 50 μm (bar = 100 μm). D: Later at E4, additional sprouts from the lateral PNVP invade the NT (arrow). At this higher resolution, the extremely long and numerous filopodia at the tips of sprouts become visible on the right side, whereas a single round cell (arrowhead) is present on the left side, indicating bilateral asymmetry of invasion. E: Following ventral arcade formation and lateral sprouting from the PNVP (arrow), the first contacts between endothelial cells are formed (curved arrow). Note isolated QH1+ cell (arrowhead) between ventral sprouts. F: At the end of E4, perfusion starts in a capillary loop around the motor column (mc). Note that vascular systems of both sides appear rather symmetric, but are connected only via the dorsal PNVP, not in the ventral region or inside the NT. Modified from Physiology of angiogenesis. Kurz, H. J Neuro-Oncol (2001) 50:17–35, with permission of Kluwer Academic.

PNVP start sprouting into the NT (Fig. 8.1D). No structural peculiarity of the NT has been observed yet at this lateral site; hence, we assume that a dorsoventral gradient of signaling molecules may be interpreted by the ECs. Eventually, the tips of the ventral and lateral sprouts establish contacts and the ECs from both entrance regions form cords without a lumen (Fig. 8.1E). Finally, a lumen is opened between the ECs, and perfusion starts in a capillary loop around the motor column (Fig. 8.1F).

We have never observed DNA synthesis in the leading ECs of intraneural sprouts. However, ECs do proliferate inside the CNS once they have formed patent capillaries, albeit at a much lower rate than in the plexuses outside the NT (Cossmann et al., 1997), and despite their expression of VEGFR-2 and the presence of its mitogenic ligand VEGF-A in the NT (Aitkenhead et al., 1998; Neufeld et al., 1999; Plate, 1999; see also Chapter 2). This somewhat reduced proliferative activity (and the peculiar filopodia) most likely are due to the lack of molecules like mitosis- (and migration-) enhancing fibronectin (Britsch et al., 1989) in the early ventral CNS matrix (Poelmann et al., 1990). This will be discussed below in more detail.

Brain

Essentially the same sequence of sprouting, first from ventral and then from lateral parts of the PNVP, is observed in the brain (Fig. 8.2B). In contrast to the rostro-caudal progression in the spinal cord, however, EC invasion of the brain anlage proceeds from rhombencephalic to more rostral levels (Strong, 1964). Moreover, the brain anlage deviates considerably from the basic NT patterning, most obviously by its size and ventricle formation, but also by formation of special sense organs (retina, olfactory bulb), and by internal segmentation. Interestingly, we were not able to detect any segmental relationship of the vascular sprouts to the early rhombomeric pattern.

The PNVP vessels anastomose in the ventral region between the floor plate and notochord, but as in the spinal cord, intraneural vascular plexus formation is bilaterally independent (Fig. 8.2C). In analogy to the aforementioned longitudinal arcades of the trunk NT, an extensive subependymal plexus is formed in the brain anlage that is connected to the PNVP via a large number of radially oriented capillaries (Fig. 8.3A).

It is beyond the scope of this chapter to describe in detail the variable vascular architecture in all parts of the brain during their growth and remodeling. Instead, we and others (Galileo et al., 1990; Bertossi et al., 1993; Virgintino et al., 1993; Yamagata et al., 1995; Roncali et al., 1996) mostly focus on the avian optic tectum (mesencephalon) and the mammalian telencephalic cortex as paradigms for brain vascular development.

Interactions of Endothelial Cells

All blood vessels are formed via interactions of ECs with each other, with other cell types, and with the extracellular matrix (ECM). Preceding invasion into the NT, EC precursors aggregate to form the PNVP in a fibronectin-rich matrix and expand the plexus in the primitive meninges, most likely via a combination of sprouting and nonsprouting, so-called intussusceptive (Burri and Tarek, 1990;

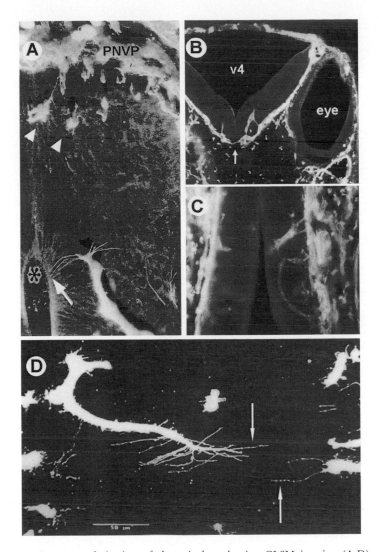

FIGURE 8.2.   Later vascularization of the spinal cord using CLSM imaging (A,D), and early vascularization of the hindbrain using conventional fluorescence (B,C) of QH1 immunostained vibratome sections. A: At the end of E6, ventral sprouts have proceeded dorsally to the level of the roof of the central canal (asterisk), where their filopodia, but never their cell bodies, reach the inner-most region of the ependymal zone (arrow). In the dorsalmost region of the NT, the population of polymorphic QH1+ cells (arrowheads) is steadily growing (pseudo–three-dimensional (3D) visualiza-tion using IMARIS software). B: Transverse section of the E4 rhombencephalon (fourth ventricle: v4) shows a sprout on the left, and a capillary loop on the right side of the floor plate. Note that the PNVPs of both sides are connected between floor plate and notochord (arrow). C: Longitudinal section of the E4 hindbrain, similar to the NT, shows an asymmetric pattern of sprouting, longitudinal arcade for-mation, and macrophage invasion. D: In the dorsal E7 medulla, long filopodia with varicosities approach the midline from both left and right sides (arrows), thus establishing connections between the intraneural vascular plexuses of both sides. Modified from Physiology of angiogenesis. Kurz, H. J Neuro-Oncol (2001) 50:17–35, with permission of Kluwer Academic.

164   H. Kurz and B. Christ

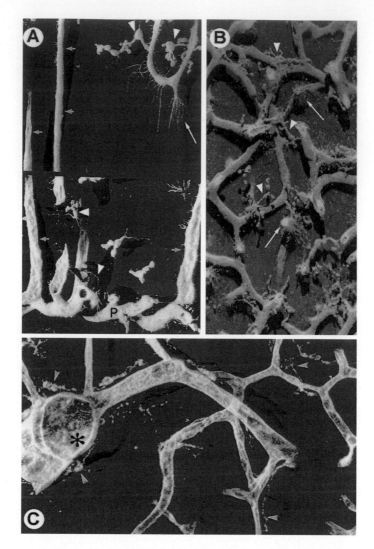

FIGURE 8.3.   CLSM-Pseudo–3D visualization of QH1-immunostained vibratome sections through
E5 mesencephalon (optic tectum: A) and E14 telencephalon (B,C). A: From the pial surface (outside
upper edge), radial vessels (small arrows) supply the subependymal vascular plexus (P). Note the U-
shaped, nonlumenized endothelial cord with its long, radially oriented filopodia (arrow). Weakly ramified macrophages sometimes are found in contact with sprouts (upper arrowheads) or detached from
vascular structures (lower arrowheads). They usually have several broad lamellipodia but may also
possess one or two filopodia. B: A dense capillary network is formed via intensive intraneural sprouting (arrows). Most, but not all, individual QH1+ cells (microglia) are found in a perivascular position
(arrowheads). C: Higher magnification of a small artery (asterisk) reveals perivascular microglia near
a bifurcation and along the capillary network (arrowheads). Modified from Physiology of angiogenesis. Kurz, H. J Neuro-Oncol (2001) 50:17–35, with permission of Kluwer Academic.

Schlatter et al., 1997) angiogenesis. They furthermore interact with meningeal
fibroblasts that are derived from trunk mesoderm (Halata et al., 1990) or from
head mesectoderm (Couly and Le Douarin, 1987). However, the precise course of
vascular development in the meninges has received little attention as yet. From our

observations on operated chick embryos lacking the calvaria, but with apparently intact cerebral perfusion, we conclude that the interactions between ECs and vascular supporting cells in the perineural vasculature are largely independent of the development of the pachymeninx and subarachnoidal spaces.

The conditions for vascular morphogenesis change dramatically once ECs enter the NT. Instead of a loose mesenchyme with ample extracellular spaces, invasive ECs face a tightly packed epithelium and the peculiar CNS matrix. Hyaluronic acid (HA) and HA-binding molecules (Poelmann et al., 1990; Turley et al., 1994; Jaworski et al., 1995; Margolis et al., 1996; Gary et al., 1998) are concentrated in the ependymal layer. Vitronectin (Seiffert et al., 1995) can be found in the floor plate, but all regions are devoid of fibronectin. In this situation, extremely long filopodia are formed and extended between the neuroblasts and radial glia (Fig. 8.2A) before the EC body moves into and further through the NT. Along these filopodia and at their tips, varicosities are observed (Figs. 8.2D and 8.4B) that might contain matrix-degrading enzymes (Forsyth et al., 1999), similar to "invadopodia" (Chen, 1996; Kelly et al., 1998). Moreover, it has been demonstrated that VEGF-A can upregulate matrix metalloproteinase (MMP) activity in ECs (Lamoreaux et al., 1998; Zucker et al., 1998), and that ECs can produce filopodia containing vesicles with active MMP-2 (Nguyen et al., 1999), but conclusive evidence that this is the case during normal CNS vascularization is still lacking.

We propose that the mechanical problem of moving the ECs through an epithelium lacking typical adhesion molecules is solved by the filopodia in two ways: they find their way to the inner surface of the NT, where they can attach between the tight junctions of the ependymal neuroblasts (Fig. 8.2A), and they find the filopodia of neighboring ventral and lateral sprouts that are anchored in the NT basement membrane and the PNVP (Fig. 8.1E). The enormous number, length, and apparently random orientation of the filopodia suggest that the first intraneural contacts between sprouting ECs are generated via a trial-and-error mechanism. Cell membrane–bound molecules, however, that make ECs recognize each other and form cords at this early stage have not been identified yet.

Using conventional and confocal microscopy, we consistently saw at least two ECs come in contact before a vascular lumen was formed (Fig. 8.4A), similar to *in vitro* observations (Nehls et al., 1998). Our observations do not contradict directly the conclusions by Wolff and Bär (1972), who proposed that in the CNS (and elsewhere) single ECs can form a lumen by internal vacuolization (Bär et al., 1984), but there is no evidence that the latter process contributes much to the process of rapid vascular growth. Likewise, the proposal (Bär and Wolff, 1972; Bär, 1980) that some sort of meningeal EC displacement into the NT, rather than sprouting, represented the initial step of CNS vascularization appears highly unlikely. Moreover, by exchanging NT between chick and quail embryos, we showed that putative ectoderm-derived EC precursors (Strong, 1961, 1964) do not exist (Kurz et al., 1996). The possibility of invasive, nonsprouting angioblasts that contribute to the aforementioned sequence of sprouting and capillary loop formation is discussed below in the context of macrophage invasion.

The process of vascular growth is more complex than we have described until now. In particular, the formation of a lumen and the expansion of the endothelial lining during vessel growth need more detailed consideration.

With respect to the process of lumen formation, we have shown that fibronectin can first be detected around nonlumenized EC cords and is strongly expressed

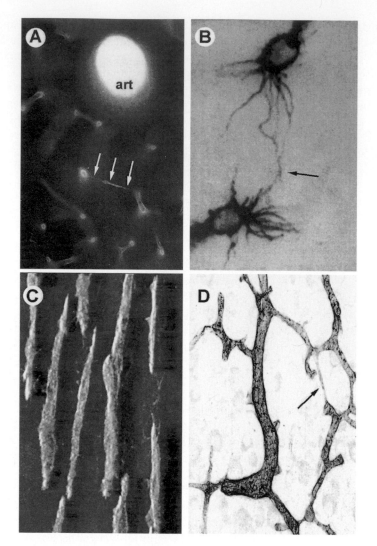

FIGURE 8.4.   Conventional (A) and CLSM (B–D) imaging of β_1 integrin (A), QH1 (B,C), or HT7 (D) in vibratome sections of E14 quail (A,B) or chick (C,D) brain. A: The integrin signal is very intense in the wall of an artery (art), and around the lumenized capillaries. Note absence of labeled sprouts, and the faint connection between two neighboring capillaries (arrows), presumably indicating the process of lumen formation. B: High magnification (single confocal section) of two endothelial cells at sprout tips with filopodia in contact (arrow). The "empty" region in each cell corresponds to the nucleus. C: Following orthotopic transplantation of quail paraxial head mesoderm into a chick, the endothelial lining of a large artery in the host is chimeric. Note that graft-derived QH1+ endothelial cells are oriented in direction of flow, and that "empty" spaces correspond to chick endothelial cells. From Embryonic central nervous system angiogenesis does not involve blood-borne endothelial progenitors. Kurz, H. et al. J Comp Neurol (2001) 436:263–274. © Wiley-Liss. D: Expression of HT7 (neurothelin) starts after E12, is weakly expressed in the smallest lumenized capillaries (arrow), but never in sprouts or nonlumenized endothelial cords. Note the strong, sometimes punctuate expression in medium-sized vessels. From Induction of the blood-brain barrier marker neurothelin/HT7 in endothelial cells by a variety of tumors in chick embryos. Papoutsi, M., et al. (2000). Histochem Cell Biol 113:105–113, with permission of Springer-Verlag.

around established vessels, and that β_1-integrin expression is almost exclusively seen around lumenized capillaries (Fig. 8.4B). Later in development, strong vitronectin expression around CNS capillaries has been described (Seiffert et al., 1995). This leads to the conclusion that the initial apical cell-cell adhesion between the future luminal sides of adjacent EC is replaced by basal cell-matrix adhesion on the abluminal side, thus permitting blood pressure to open the capillary lumen (Kurz, 1997).

With respect to the expansion of the established endothelial lining of perfused vessels, it is noteworthy that, following mesoderm transplantation (Wilting et al., 1995a; Huang et al., 1997), chimeric vessels can be observed (Fig. 8.4C). Not only intercalated growth via locally proliferating EC but also integration of migrating ECs takes place during expansion of the intraneural vascular network. In particular, ECs have been shown to migrate along the luminal surface of blood vessels (Christ et al., 1990; Wilms et al., 1991). In contrast, the proposed contribution of blood-borne angioblasts during angiogenesis (Caprioli et al., 1998; Hatzopoulos et al., 1998; Yamashita et al., 2001) does not play a significant role during embryonic CNS vascularization, as indicated by the absence of blood-derived ECs in chick-quail parabiosis or after transplantation of hematopoietic tissues (Kurz and Christ, 1998; Kurz et al., 2001). Our view does not exclude that circulating EC precursors may contribute to blood vessel growth during, for example, pathologic angiogenesis (Asahara et al., 1997; Goldbrunner et al., 1999).

Macrophages and Microglia

It is generally accepted that macrophages (MΦ), besides their ability to digest cellular debris and matrix components, are involved in many normal and pathologic processes, including angiogenesis (Sunderkötter et al., 1994). Moreover, monocytes in adults have been shown to express VEGFR-1, which is thought to mediate VEGF-A– [and possibly VEGF-B– and placenta growth factor (PlGF-)] dependent motility (Clauss et al., 1996). But is there a role for MΦ (comprising in this context also blood monocytes and microglia) during embryonic blood vessel formation, in particular in the CNS? We cannot answer this question conclusively but feel that this multifunctional cell population might warrant more attention, and hence will report here on the invasion, distribution, proliferation, and differentiation of embryonic MΦ in the developing spinal cord and brain.

The early embryonic MΦs stem from the yolk sac and soon spread through almost all embryonic tissues as has been conclusively demonstrated with chick-quail yolk sac chimeras (Cuadros et al., 1992; Cuadros and Navascues, 1998). Soon after neurulation, they invade the dorsal NT, i.e., roof plate and alar plates (Cuadros et al., 1993). This MΦ invasion precedes the ventral EC invasion by several hours, that is, it takes place where and when NC cells have emigrated, or DRG neurons have projected their central processes back into the NT. During differentiation of the NT, the immigrated MΦs populate the marginal layer but are rarely found in the inner mantle layer, where the intraneural vascular plexus forms. They proliferate, migrate along the developing fiber tracts toward the ventral side, and develop an increasingly polymorphic, often ramified cell shape (Cossmann et al., 1997, 2000). Interestingly, they approach the newly formed blood vessels and move in the perivascular space, or sometimes establish connections to early, nonlumenized EC cords (Kurz, 1997). We have shown that these cells often have large vacuoles, may ingest matrix components like fibronectin or laminin, and express the common leukocyte antigen CD45 (Kurz et al., 2001).

On the other hand, there is a possibility that not all of these dorsal immigrants are determined to become MΦs. From our observation that they occasionally become situated between the tips of ventral EC sprouts, we concluded that they might serve as angioblasts that will integrate into the early EC cords and thus directly participate in intraneural angiogenesis (Kurz et al., 1996). Given that EC and MΦ share many molecular markers from the earliest stages of development (Peault et al., 1983), and that EC regularly transform into hemopoietic cells in the floor of the aorta (Jaffredo et al., 1998), thus giving rise also to MΦs, a (reversible) switch between the EC and the hematopoietic MΦ fate appears possible (Pardanaud and Dieterlen-Liévre, 1999). However, with *in situ* hybridization, no VEGFR-2–positive cells with a distribution similar to that of our putative intraneural angioblasts have been detected yet (J. Wilting, personal communication).

Clearly, embryonic macrophages invade the CNS from the mesenchyme at the pial surface, and are not directly derived from circulating monocyte-like precursors that penetrate the wall of intraneural blood vessels (Kurz and Christ, 1998). Their appearance is only loosely linked to developmental neuron death (Cuadros and Navascues, 1998; Zhang and Galileo, 1998). An influence by macrophage inhibitory factor remains to be shown (Suzuki et al., 1999). Macrophage migration in the fiber tracts is independent from blood vessels, but in the gray matter MΦs frequently migrate along the abluminal side of microvessels. They gradually develop the morphology and distribution of mature microglia. In particular, a ramified shape is frequent in MΦ that are positioned farther away from blood vessels or in contact with sprouting ECs (Fig. 8.3A), whereas those in a perivascular position around more mature vessels often have an ameboid, elongated, or fusiform appearance (Fig. 8.3B,C).

We found that such perivascular MΦs are located relatively often at vessel bifurcations and gather around the growing vessel wall of arteries (Fig. 8.3C). While their precise role during development of the brain remains enigmatic, one should perhaps consider the possibility that they are also involved in maturation of the vascular system, that is, in the recruitment and assembly of pericytes and smooth muscle cells. Presently, increasing evidence indicates that the MΦ/microglia population does play a role in neovascularization (Roncali et al., 1996; Pennell and Streit, 1997; Earle and Mitrofanis, 1998) and is functionally and morphologically heterogeneous from the beginning of development (Streit and Graeber, 1993; Andjelkovic et al., 1998). Given the enormous importance of adult microglia in both antigen presentation and intercellular signaling via chemokines and growth factors in the course of practically every lesion of the CNS (Morioka et al., 1993; Streit, 1993, 1996; Angelov et al., 1996, 1998; Moore and Thanos, 1996; Lehrmann et al., 1998; Hurley et al., 1999), understanding the development of perivascular MΦs and their relationship to other periendothelial cells appears to be an important field of investigation.

Pericytes and Smooth Muscle Cells

The maturation of primordial vascular tubes is accompanied by the emergence of periendothelial cells that participate in matrix formation, and that eventually become pericytes (PCs defined as cells located in an immediate periendothelial position and in the same basement membrane as the ECs), or vascular smooth muscle cells (vSMCs are defined as cells outside the vascular basement membrane, and equipped with specific actin-myosin contractile filaments). Recent findings support the almost forgotten view (Draeger and England, 1998) that special intimal vSMCs may exist (Kohler et al., 1999), which we include in our summary of all vSMCs.

Whereas several early and specific markers for ECs have been established since long for many species, the situation is different for PCs and vSMCs. Most vSMCs may be characterized by the expression of (almost) smooth muscle–specific α-actin (αSMA) and myosin heavy chain isoforms and by their perivascular location, but αSMA may also be present in some PC populations. The PCs, whose morphologic and functional characteristics depend on the various vascular beds, have been characterized by a variety of markers like αSMA, desmin, or vimentin, none of which is specific. Apparently, the developmental history and the functional state of both PCs and vSMCs may be rather heterogeneous (Herman and D'Amore, 1985; Nehls and Drenckhahn, 1991; Yablonka-Reuveni et al., 1995, 1998; Bergwerff et al., 1998; Drake et al., 1998; Hungerford and Little, 1999). Despite recent progress (Alliot et al., 1996b), notion also holds for CNS vessels (Ehler et al., 1995), where even less is known about cell lineage and diversification during development.

It has been shown that pericyte investment (Bär and Budi, 1984) is a very important step in vascular maturation, in which TGF-β signaling (D'Amore and Smith, 1993; Pepper, 1997) as well as the angiopoietin-1/Tie-2 and related systems (Suri et al., 1996, 1998; Asahara et al., 1998; Koblizek et al., 1998) play a major role (see Chapter 3 for details). Recently, the transmembrane molecule endoglin has been identified as an essential regulator of PC and vSMC differentiation (Li et al., 1999) that modulates the interaction of TGF-β superfamily members and their receptors (Gougos and Letarte, 1990; Cheifetz et al., 1992; Lastres et al., 1996; Rius et al., 1998) and is even expressed in vSMCs (Adam et al., 1998).

In the CNS, platelet-derived growth factor (PDGF) B/PDGF-Rβ signaling seems to be of particular importance, since mice lacking either ligand or receptor in some brain regions fail to recruit PCs and vSMCs and thus develop cerebral hemorrhage (Lindahl et al., 1997, 1998; Crosby et al., 1998; Hellstrom et al., 1999), whereas exogenous PDGF-B promoted vSMC differentiation and vascular maturation in the chorioallantoic membrane assay (Oh et al., 1998). Perhaps the best-understood example of PDGF-related EC-PC interactions (Hirschi et al., 1999) in the CNS is the retina, where the capillary growth- and regression-preventing function of PCs has been demonstrated (Benjamin et al., 1998). Moreover, PCs may even perform the tasks of antigen presentation and phagocytosis in the mature CNS (Balabanov et al., 1999). With respect to pathologic or therapeutic angiogenesis, the key phenomenon of PC destabilization (Benjamin et al., 1999) via angiopoietin-2, for example (Maisonpierre et al., 1997; Asahara et al., 1998) becomes more evident.

But what are the precursors for the somewhat enigmatic brain periendothelial cells? During our proliferation studies on ECs, we always questioned why there was practically no labeling of nuclei in periendothelial cells (Cossmann et al., 1997, 2000). In the developing brain, and in contrast to extraneural vessels (Lee et al., 1997), we never saw proliferation of αSMA-expressing cells, which indicates that vSMCs are (conditionally) postmitotic during CNS growth. So, in contrast to the proliferative immigrant ECs and MΦs, the PC and vSMCs population apparently is not amplified by cell division. This thesis is compatible with five alternative hypotheses, that PCs and vSMCs in the CNS are produced via (1) EC transdifferentiation, as described for the aorta (DeRuiter et al., 1997); (2) recruitment and differentiation of blood-borne precursors or of intraneural MΦs; (3) continuous perivascular immigration from extraneural mesoderm; (4) reimmigration from emigrated NC cells; or (5) recruitment of local or internally migratory neuroectoderm. In a series of chick-quail transplantation experiments, we recently investigated this question of PC and vSMC lineage. Our results are strongly favoring hypothesis 5, are compatible with 4, and exclude hypotheses 1 through 3. While Etchevers et al. (2001) demonstrated forebrain PCs from NC, we have shown that in addition to NC, internally migratory neuroectoderm in fact can differentiate into brain vSMCs, but head or limb bad mesoderm can not (Korn et al., 2000, 2001). This may seem unlikely to some investigators, but one should consider that ventral NT cells can emigrate and give rise to intestinal SMCs (Bockman and Sohal, 1998), and that the smooth dilatator pupillae muscle is derived from the retina anlagen. Evidence both from the potential of cranial NCs to form the vSMCs in the meningeal, head, and branchial vessels (Couly and Le Douarin, 1987; Bergwerff et al., 1998) and from embryoid body culture that easily transform from a neural to a smooth muscle phenotype (Drab et al., 1997) supports the view that vSMCs can be generated directly from neuroblasts during development. In any case, differentiation of local neuroectoderm to PCs and vSMCs could recruit precursors (in addition to those from NC-derived head mesenchyme) from the highly proliferative neuroblast pool in the ependymal layer (known to contain multipotent neuronal stem cells), thus resolving the enigma of nonproliferative periendothelial cells.

Periendothelial cells make their appearance in the developing CNS only shortly after the initial EC invasion. Starting during the fifth incubation day in chicks (E12 in mice), an increasing number of PCs around capillaries (Bertossi et al., 1995) and of vSMCs around the major vessels were observed. Interestingly, N-Cadherin is required for EC-PC interactions (Gerhardt et al., 2001). While there is little doubt that both cell types are needed for stabilizing CNS capillaries, veins, and arteries, some uncertainty exists regarding PC function in establishing and maintaining the blood-brain barrier.

Blood-Brain Barrier Maturation

A prominent feature of mature CNS blood vessels is their extremely low permeability for polar substances and higher molecular weight molecules (Cornford and Hyman, 1999). The "checkpoint" between the blood and the neural parenchyma is realized mainly by vascular cells and is called the blood-brain barrier (BBB), a term comprising also the blood-retina barrier (Liebner et al., 1997). It is supplemented by the blood-liquor barrier of pial (Allt and Lawrenson, 1997; Cassella et al., 1997a) and choroid plexus (Bertossi et al., 1988; Wilting and Christ, 1989) vessels. The tightness of the BBB is principally achieved through structural features of the ECs, like the lack of pinocytosis and fenestrations, together with numerous inter-EC tight junctions (Rubin, 1992; Farrell and Risau, 1994; Staddon and Rubin, 1996; Risau et al., 1998) and high occludin expression (Hirase et al., 1997; Nico et al., 1999; Rubin and Staddon, 1999).

The specific permeability of the BBB for metabolites results from a set of transport proteins, like the insulin-independent glucose transporter 1 (GLUT1) (Boado, 1998; Virgintino et al., 1998), the transferrin receptor, γ-glutamyl transpeptidase, and alkaline phosphatase (Risau et al., 1986; Lawrenson et al., 1999), and the permeability glycoproteins (PGPs) that act as multidrug transporters and are transcribed from the mdr genes in mice, rats, and humans (Qin and Sato, 1995; Matsuoka et al., 1999).

Several other markers have been described, whose functions in the BBB are less well understood, like the endothelial barrier antigen (EBA) (Cassella et al., 1996, 1997b; Orte et al., 1999), the matrix component agrin (Barber and Lieth, 1997; Cotman et al., 1999), or the integral membrane glycoprotein neurothelin (Schlosshauer et al., 1995), also known as HT7 in chick (Fig. 8.4D), OX-47 in rat, and basigin in mouse (Albrecht et al., 1990; Miyauchi et al., 1990; Seulberger et al., 1990, 1992; Ikeda et al., 1996; Fan et al., 1998).

In addition to these EC features, the PCs contribute important BBB properties via their wide metabolic repertoire, phagocytic potential, and immunomodulatory function (Risau et al., 1992; Balabanov and Dore-Duffy, 1998; Ramsauer et al., 1998; Balabanov et al., 1999; Muldoon et al., 1999; Alliot et al., 1999b).

From the developmental point of view, we want to know how and when ECs (and PCs) develop their peculiar properties. But despite considerable experimental effort (and medical interest), the question of BBB differentiation is far from settled yet. We will briefly address the problem of BBB induction, and the development of structural and molecular (as defined by PGP/mdr and HT7) BBB properties.

Several events related to signal transduction and cell adhesion have been described that precede BBB formation (Achen et al., 1995; Lossinsky and Wisniewski, 1998; Lossinsky et al., 1999), of which the transient expression of N-cadherin in ECs (Yamagata et al., 1995; Gerhardt et al., 1999) is of particular interest. But to the best of our knowledge, no specific signal has been identified yet that induces BBB properties during embryonic CNS angiogenesis. From descriptive studies (Bertossi et al., 1993) or from experiments using astrocyte-conditioned medium, some authors postulated astrocyte-derived factors (Neuhaus et al., 1991; Janzer et al., 1993; Kuchler-Bopp et al., 1999) but were never able to isolate such agents. In contrast, accumulating evidence supports the view that BBB properties develop in the absence of astrocytes (Holash et al., 1993), because the EC phenotype is altered soon after invasion into the NT (Bauer et al., 1993), at a time when neuroblasts but not astrocytes are present (Krum, 1996; Krum et al., 1997; Orte et al., 1999). Moreover, the presence of PCs appears to be an effective inducer of BBB properties (Balabanov and Dore-Duffy, 1998), whereas astrocytes may be needed for maintaining a functional BBB after its formation, as indicated by recent findings in angiotensinogen-deficient mice (Alliot et al., 1999a). However, the regulatory function of the perivascular angiotensin system in the CNS (Kakinuma et al., 1998), like that of vasoactive intestinal reptide-like molecules (Benagiano et al., 1996) remains to be analyzed.

Ultrastructural studies in the mouse revealed that EC, immediately after formation of the first intraneural capillaries (E10), no longer have fenestrations, in contrast to perineural ECs, and remain unfenestrated during development (Bauer et al., 1993). The number of pinocytotic vesicles decreases, and the number of tight junctional complexes increases over the next few days. Interestingly, the intimate relationship between ECs and neuroblasts on the one hand, and between ECs and PCs on the other hand, has been described at early stages of CNS vascularization (Bauer et al., 1993; Zerlin and Goldman, 1997; Balabanov and Dore-Duffy, 1998; Earle and Mitrofanis, 1998; Gerhardt et al., 1999).

While most authors did not address the origin of PCs, Bauer et al. (1993) and reported (but did not prove) early immigration of PCs from perineural mesenchyme, which appears to resemble NC-derived mesectoderm. A recent study on forebrain viability after NC removal supports this concept (Etchevers et al., 1999). Our findings on periendothelial cell recruitment favor the view that BBB is induced early (Møllgard et al., 1988) by neuroblasts (Stewart and Wiley, 1981) and (directly neuroblast-derived) PCs and VSMCs, and that in turn ECs (and PCs/VSMCs) may instruct astrocyte precursors (Zerlin and Goldman, 1997). This is in some contrast to both the traditional view about astrocyte-dependent BBB induction, and the reports by Bär and Wolff (1972), who postulated (in the rat) both late differentiation of PCs, and late vascular basement membrane and BBB maturation.

Although some discrepancies may be attributed to species differences, the observation of early (E10.5 in mice, 8 weeks in humans) and specific expression of the multidrug resistance gene *mdr1* in CNS capillaries indicates the rapid development of important BBB features (Qin and Sato, 1995; Schumacher and Møllgard, 1997). Hence, *mdr1* is the earliest specific BBB marker and is expressed at a time when astrocytes have not differentiated yet. Moreover, both the overall vascular pattern and *mdr1* expression are unaffected by inversion of cortical layers in the mouse mutant *reeler* (Qin and Sato, 1995).

Neurothelin/HT7 is observed later than PGP/*mdr1* expression, but still well in advance (E12 in chick) to glial fibrillary acidic protein (GFAP)-positive astrocyte differentiation (H. Kurz, J. Korn, unpublished). Chick HT7 has been shown to be inducible by grafted quail brain (Ikeda et al., 1996) or glioma cells (Papoutsi et al., 2000) in chick mesenteric or chorioallantoic vessels. On the other hand, basigin-deficient mouse embryos rarely survive implantation, and those surviving lack a distinct BBB phenotype (Igakura et al., 1996, 1998; Naruhashi et al., 1997). Moreover, the non-BBB function of this immunoglobulin superfamily molecule that is widely expressed in epithelia and in some neurons (Seulberger et al., 1992), and was also detected in developing heart capillaries (H. Kurz, unpublished), has been further characterized in human tumors (Biswas et al., 1995). Interestingly, we found that it could be induced in chick ECs by a variety of mammalian tumor cell lines (Papoutsi et al., 2000).

A functional test of the BBB was made by injecting horseradish peroxidase (HRP) or dyes into the circulation and tracing their distribution in the CNS. Using HRP, the existence of a functional BBB was demonstrated in E13 mice for most of the CNS, with the exception of the forebrain region where vessels are not impermeable before E15 (Risau et al., 1986). Likewise, the dye exclusion studies indicated that BBB maturation is a continuous process that lasts for several days and depends on the brain region, with some regions always remaining permeable.

Experiments involving ectopic transplantation of prospective choroid plexus epithelium once more supported the view that it is the neuroectoderm that determines EC permeability properties (Wilting and Christ, 1989), as indicated by progressive thinning and fenestration of ECs (Bertossi et al., 1988). Perhaps persistent VEGF-A expression is the signal for some CNS capillaries not to form BBB characteristics (Breier et al., 1992). As could be expected, VEGF signaling has been identified as an important regulator of vascular permeability, whose overexpression hinders BBB maturation (Dobrogowska et al., 1998; Rosenstein et al., 1998; Zhao et al., 1998; Fischer et al., 1999).

OPEN QUESTIONS AND MEDICAL APPLICATIONS

Research in the neurosciences is strongly focused on neurons and glia. While typical developmental problems like those of neuronal cell lineage, axon pathfinding, synapse formation, and circuit wiring appear to be quite independent of CNS vascularization, a growing body of data shows that neurons, glia, and vascular cells share molecular markers and are, more than perhaps expected, developmentally and functionally interrelated. We have discussed above examples for this perspective, like the recruitment of vSMC from neuroectoderm, the expression of the adhesion molecule N-cadherin in both neuroblasts and EC, and the contribution of the synapse-organizing molecule agrin to the BBB.

We will now briefly address the VEGF-A$_{165}$– and PlGF-2–binding activity of the axon-guiding semaphorin receptor neuropilin, expression of VEGFs and of VEGF receptors in cells other than ECs, and the question of VEGF overexpression and of hypoxia in the (developing) CNS, encompassing also nitric oxide (NO) signaling. Since there is increasing awareness that blood vessels, their matrix, and perivascular cells play a crucial role in practically all lesions and during regeneration of the CNS, we include in this review the recent literature on pathologic and therapeutic angiogenesis and antiangiogenesis. It will become clear that, while novel therapeutic approaches are promising in brain tumors or in some ischemic disorders, a full understanding of angiogenesis in the CNS will require much more work with special emphasis on embryonic development.

Neuropilin

The cell surface glycoprotein neuropilin was initially described as an important receptor for axonal pathfinding via semaphorin or collapsin signaling (Kawakami et al., 1996; Feiner et al., 1997; Fujisawa and Kitsukawa, 1998). Neuropilin is also expressed by migrating NC cells that consequently avoid regions rich in collapsin (Eickholt et al., 1999), and is downregulated in dHAND-deficient embryos that apparently lack proper EC-vSMC interactions (Yamagishi et al., 2000). Overexpression of neuropilin surprisingly indicated an influence of neuropilin on the cardiovascular system (Kitsukawa et al., 1995), and neuropilin-deficient mice showed (among other defects) impaired CNS vascularization (Kawasaki et al., 1999). Two more ligands (besides semaphorin and collapsin) for neuropilin have been identified as VEGF-A$_{165}$ (Soker et al., 1998) and placenta growth factor-2 (PlGF-2) (Migdal et al., 1998). It therefore appears possible that the well-known macroscopic association of nerves (axons) and blood vessels (ECs) in some regions is guided by similar signals. However, as described above, and supported by the probably independent distribution of neuropilin and ECs in the optic tectum (Yamagata et al., 1995), ECs in the CNS prefer the developing gray matter over the fiber tracts and hence the role of neuropilin/VEGF-A/PlGF-2/semaphorin interactions in the CNS remains to be determined (Bagnard et al., 2001).

VEGFs and VEGF Receptors

Among the growth factors of the VEGF family (see also Chapter 2), the highly EC-specific VEGF-A/VEGFR-2 signaling has received the most attention. Recently, however, it has been demonstrated that NC-derived cells (DRG neurons, Schwann cells) can also use the same survival and mitogenic signal in an autocrine fashion (Sondell et al., 1999, 2000; Sondell and Kanje, 2001). In the CNS, however, it apparently is only the ECs that first upregulate, and with maturation, down-regulate VEGFR-2, and it is only non-ECs (neurons, macroglia, microglia, PCs, vSMCs) that produce VEGF-A (Neufeld et al., 1999; Ortega et al., 1999). Important regulators of VEGF sinaling may be the Id1 and Id3 transcription factors, whose absence is associated with low VEGF-A and VEGFR-2 levels and vascular malformation in the early CNS (Lyden et al., 1999).

In contrast to VEGFR-2, VEGFR-1 can be found not only on the earliest hemangioblasts (Fong et al., 1999a) and ECs, but also on monocytes (Clauss et al., 1996), reactive astrocytes (Krum and Rosenstein, 1998), and even on vSMCs (Wang and Keiser, 1998). It is not yet known whether and when VEGFR-1 expression starts in these non-ECs in the embryonic CNS, but one could speculate that growth factors like PlGF-1 and VEGF-B may influence not only ECs (Olofsson et al., 1998) but also recruitment, proliferation, and differentiation of periendothelial cells (see above) via VEGFR-1 in the brain and spinal cord. In particular, a high VEGF-B expression has been found in developing brain (Lagercrantz et al., 1998), but is absent from brain vSMCs, not from extraneural vSMCs (Aase et al., 1999).

Hypoxia

While VEGF expression in the early NT does not appear to be hypoxia-dependent (see above), oxygen tension certainly influences remodeling of the vascular system via the VEGF and other systems throughout later life. However, controversial findings concerning CNS vascularization have been reported following exogenous VEGF-A application or long-term hypoxia. In concordance with the observation of extended extraneural vascular plexuses after extraneural injection (Drake and Little, 1995), intracerebral application of VEGF leads to grossly dilated perineural/pial vessels (and malformation of the heart), but not to alterations of the vascular pattern of the brain (Feucht et al., 1997). In neural embryonic explant cultures, however, VEGF administration led to enhanced complexity of the brain vasculature. In contrast, continuous VEGF infusion in adult mice resulted in dilated, but not more complex, CNS vessels, impaired BBB function, and VEGFR-1 expression in astrocytes (Rosenstein et al., 1998). Long-term hypoxia of adult rats and mice led to enhanced VEGF expression in glia and neurons, and VEGFR-1 upregulation in ECs (Marti and Risau, 1998; Patt et al., 1998), but to variably enhanced or unaltered capillary density (Boero et al., 1999). These discrepancies indicate that VEGFs and their growth-promoting receptors are not the only significant players during CNS angiogenesis or hypoxia, but that periendothelial cells or growth-preventing molecules may be equally important (Carmeliet et al., 1998; Kotch et al., 1999; Steinbrech et al., 1999).

As stated before, a special case of and perhaps the best studied system for CNS vascularization is the mammalian retina (Chan-Ling et al., 1990). In classical experiments, the effects of VEGF and PDGF on ECs and PCs, the sensitivity of vascular remodeling to oxygen tension, and the regulation of growth factors through neuroglia and matrix were demonstrated (Alon et al., 1995; Stein et al., 1995; Stone et al., 1995; Provis et al., 1997; Benjamin et al., 1998; Mousa et al., 1999). We are confident that further insights will be derived from studying ocular angiogenesis, in particular with respect to the modulation of the VEGF response and EC behavior by extracellular matrix and integrins (Soldi et al., 1999). Perivascular and CNS matrix components are still not well understood, but apparently play significant roles during both vessel maturation (Antonelli and D'Amore, 1991; Eggli and Graber, 1995) and CNS lesions (Ellison et al., 1998; Mendis et al., 1998; Zuo et al., 1998; Jaworski et al., 1999).

Ischemic CNS Lesions

In the case of CNS lesions leading to ischemia (arterial occlusion, spinal crush), two opposite strategies have been proposed. While no doubt exists about upregulation of VEGF and vascular changes in or around the ischemic area (Bartholdi et al., 1997; Hayashi et al., 1999; Issa et al., 1999; Vaquero et al., 1999), and some authors prefer proangiogenic treatment to reduce hypoxic damage (Hayashi et al., 1998a; Plate, 1999), others propose antiangiogenic strategies to reduce edema and matrix degradation (Wamil et al., 1998; Vaquero et al., 1999). Perhaps the future therapy of choice will also depend on other inducible factors like hepatocyte growth factor (HGF)/SF (scatter factor) (Hayashi et al., 1998b) and TGF-β (Lehrmann et al., 1998), and the recruitment of immunocompetent cells from the blood (Wekerle et al., 1991), the vascular wall (Graeber et al., 1989; Angelov et al., 1996), or the brain parenchyma (Moore and Thanos, 1996; Streit, 1996).

Nitric oxide (NO) has been identified as a molecule that is able to mediate or antagonize VEGF signals (Tsurumi et al., 1997; van der Zee et al., 1997). Most important, the expression of endothelial constitutive NO synthase has been demonstrated in early (rat) embryonic CNS blood vessels (Topel et al., 1998). While the permeability enhancing effect of NO at the BBB was shown (Fischer et al., 1999; Mayhan, 1999), we can only speculate whether NO is migration promoting (Murohara et al., 1999) also for brain ECs. In any case, NO appears to be an important player in normal development and in all hypoxic (and hypoglycemic) conditions (Faller 1999).

CNS Tumors

A variety of tumors in or around the brain and spinal cord have been found to be particularly angiogenesis-dependent, and numerous growth-promoting molecules have been identified that will, or already have, become targets for antiangiogenic strategies. Subsequent to the description of VEGF and PDGF in malignant gliomas (Plate et al., 1992), enhanced expression of TGF-β (Yamada et al., 1995) and endoglin (Bodey et al., 1998) have also been observed. The integrity of blood vessels in tumors was shown to be particularly VEGF dependent (Benjamin et al., 1999), and approaches inhibiting VEGF have produced promising results (Bernsen et al., 1998; Kirsch et al., 1998; Fong et al., 1999b; Im et al., 1999). Even an anti-tumor activity of endogenous soluble neuropilin as VEGF antagonist has recently been suggested (Gagnon et al., 2000). Since neuronal tumors also overexpress

VEGF and related molecules (Chan et al., 1998; Ribatti et al., 1998), it can be hoped that successful glioma therapies can soon be adapted for other CNS tumors.

A special case is the hemangioblastoma of the CNS that is related to mutations in the von Hippel–Lindau (VHL) gene (Wizigmann-Voos and Plate, 1996). These hemangioblastomas, in addition to expressing VEGF, also produce other hypoxia-inducible factors like erythropoietin (Krieg et al., 1998). It would be interesting to learn when in (embryonic) development these malformations arise, why they are preferentially located in the hindbrain and spinal region, and which specific therapy would be useful.

SUMMARY

The brain and spinal cord primordia are ectodermal epithelial structures that remain devoid of blood vessels during generation of the primordial cardiovascular network in the embryonic and yolk sac mesoderm. Lagging behind the process of somite formation, a perineural vascular plexus emerges that later remodels into meningeal vessels. From this plexus, and starting in the craniocervical transition zone, endothelial sprouts invade the neural tissue from the ventral and lateral sides. These endothelial cells remain in contact with the perineural plexus, and they use extremely long and numerous filopodia for pathfinding and for establishing first intercellular contacts. A capillary lumen is usually opened between two endothelial cells, following matrix deposition and adhesion molecule activation. VEGF in the neuroectoderm and VEGF receptors on endothelial cells are upregulated before and remain highly expressed during CNS angiogenesis, but are downregulated as the CNS matures.

Macrophages and microglia precursors enter the CNS before angiogenesis starts, migrate partially in the fiber tracts, partially along blood vessels, and may participate in regulating embryonic angiogenesis, pericyte and smooth muscle recruitment, and blood-brain barrier formation. Pericytes are found soon after endothelial immigration, and may be derived from extraneural or intraneural precursors. They develop partly inresponse to PDGF-B, stabilize CNS blood vessels, and contribute to BBB function. The appearance of contractile vascular smooth muscle cells lags several days behind the initial steps of CNS angiogenesis.

Morphologic features of the BBB are observed very early in the first intraneural capillaries, but BBB maturation lasts throughout embryonic and early postnatal life. While strong evidence indicates that induction of BBB properties in endothelial cells is independent of astrocytes, maintenance of its function apparently depends on both pericytes and astrocytes. The first known specific BBB markers are the *mdr* genes, whereas other putative markers like neurothelin were shown to be nonspecific. Permeability of the BBB is regulated via VEGF and NO.

While the pivotal role of VEGF-A/VEGFR-2 interaction during CNS angiogenesis is well established, the actual complexity of the regulatory network has become evident by demonstrating additional signaling via other VEGFs and their receptors, including neuropilin. Variable findings of CNS vascularization following exogenous VEGF application or hypoxia stress the point that additional influences (pericytes, matrix) are important and need to be investigated further. This corresponds to the trial of both pro- and antiangiogenic therapeutic approaches to treat ischemic conditions in the CNS. Tumors of the brain and spinal cord often overexpress VEGF and a multitude of other growth factors, thus mimicking embryonic angiogenesis to a variable degree. Antiangiogenic strategies appear particularly promising in these conditions.

ACKNOWLEDGMENTS

We thank Dr. Johannes Korn for his contribution of experimental work and image preparation. Confocal imaging was performed using the CLSM facilities at the Institute of Neuropathology at Freiburg (Dr. B. Volk, director) and at the Institute of Anatomy at Bern. In particular, we are obliged to Dr. Peter S. Eggli, Bern, for his expert support in producing some of the confocal images and three-dimensional visualizations. For stimulating discussions, we thank Drs. F. Dieterlen-Liévre, C. D. Little, and J. Wilting.

REFERENCES

Aase, K., Lymboussaki, A., Kaipainen, A., Olofsson, B., Alitalo, K., Eriksson, U. (1999). Localization of VEGF-B in the mouse embryo suggests a paracrine role of the growth factor in the developing vasculature. Dev Dyn 215:12–25.

Achen, M.G., Clauss, M., Schnürch, H., Risau, W. (1995). The non-receptor tyrosine kinase Lyn is localised in the developing murine blood-brain barrier. Differentiation 59:15–24.

Adam, P.J., Clesham, G.J., Weissberg, P.L. (1998). Expression of endoglin mRNA and protein in human vascular smooth muscle cells. Biochem Biophys Res Commun 247: 33–37.

Adams, R.H., Wilkinson, G.A., Weiss, C., et al. (1999). Roles of ephrinB ligands and EphB receptors in cardiovascular development: demarcation of arterial/venous domains, vascular morphogenesis, and sprouting angiogenesis. Genes Dev 13:295–306.

Aitkenhead, M., Christ, B., Eichmann, A., Feucht, M., Wilson, D.J., Wilting, J. (1998). Paracrine and autocrine regulation of vascular endothelial growth factor during tissue differentiation in the quail. Dev Dyn 212:1–13.

Albrecht, U., Seulberger, H., Schwarz, H., Risau, W. (1990). Correlation of blood-brain barrier function and HT7 protein distribution in chick brain circumventricular organs. Brain Res 535:49–61.

Alliot, F., Rutin, J., Leenen, P.J.M., Pessac, B. (1999a). Brain parenchyma vessels and the angiotensin system. Brain Res 830:101–112.

Alliot, F., Rutin, J., Leenen, P.J., Pessac, B. (1999b). Pericytes and periendothelial cells of brain parenchyma vessels co-express aminopeptidase N, aminopeptidase A, and nestin. J Neurosci Res 58:367–378.

Allt, G., Lawrenson, J.G. (1997). Is the pial microvessel a good model for blood-brain barrier studies? Brain Res Brain Res Rev 24:67–76.

Alon, T., Hemo, I., Itin, A., Pe'er, J., Stone, J., Keshet, E. (1995). Vascular endothelial growth factor acts as a survival factor for newly formed retinal vessels and has implications for retinopathy of prematurity. Nature Med 1:1024–1028.

Andjelkovic, A.V., Nikolic, B., Pachter, J.S., Zecevic, N. (1998). Macrophages/microglial cells in human central nervous system during development: an immunohistochemical study. Brain Res 814:13–25.

Angelov, D.N., Neiss, W.F., Streppel, M., Walther, M., Guntinas-Lichius, O., Stennert, E. (1996). ED2-positive perivascular cells act as neuronophages during delayed neuronal loss in the facial nucleus of the rat. Glia 16:129–139.

Angelov, D.N., Walther, M., Streppel, M., Guntinas-Lichius, O., Neiss, W.F. (1998). The cerebral perivascular cells. Adv Anat Embryol Cell Biol 147:1–87.

Antonelli, A., D'Amore, P.A. (1991). Density-dependent expression of hyaluronic acid binding to vascular cells *in vitro*. Microvasc Res 41:239–251.

Artinger, K.B., Fraser, S., Bronner-Fraser, M. (1995). Dorsal and ventral cell types can arise from common neural tube progenitors. Dev Biol 172:591–601.

Asahara, T., Chen, D., Takahashi, T., et al. (1998). Tie2 receptor ligands, angiopoietin-1 and angiopoietin-2, modulate VEGF-induced postnatal neovascularization. Circ Res 83:233–240.

Asahara, T., Murohara, T., Sullivan, A., et al. (1997). Isolation of putative progenitor endothelial cells for angiogenesis. Science 275:964–967.

Bagnard, D., Vaillant, C., Khuth, S.T., Dufay, N., Lohrum, M., Puschel, A.W., Belin, M.F., Bolz, J., Thomasset, N. (2001). Semaphorin 3A-vascular endothelial growth factor-165 balance mediates migration and apoptosis of neural progenitor cells by the recruitment of shared receptor. J Neurosci 21:3332–3341.

Baker, C.V., Bronner-Fraser, M. (1997a). The origins of the neural crest. Part I: embryonic induction. Mech Dev 69:3–11.

Baker, C.V., Bronner-Fraser, M. (1997b). The origins of the neural crest. Part II: an evolutionary perspective. Mech Dev 69:13–29.

Balabanov, R., Beaumont, T., Dore-Duffy, P. (1999). Role of central nervous system microvascular pericytes in activation of antigen-primed splenic T-lymphocytes. J Neurosci Res 55:578–587.

Balabanov, R., Dore-Duffy, P. (1998). Role of the CNS microvascular pericyte in the blood-brain barrier. J Neurosci Res 53:637–644.

Bär, T. (1980). The vascular system of the cerebral cortex. Adv Anat Embryol Cell Biol 59:1–62.

Bär, T., Budi, S.A. (1984). Identification of pericytes in the central nervous system by silver staining of the basal lamina. Cell Tissue Res 236:491–493.

Bär, T., Guldner, F.H., Wolff, J.R. (1984). "Seamless" endothelial cells of blood capillaries. Cell Tissue Res 235:99–106.

Bär, T., Wolff, J.R. (1972). The formation of capillary basement membranes during internal vascularization of the rat's cerebral cortex. Z Zellforsch 133:231–248.

Barber, A.J., Lieth, E. (1997). Agrin accumulates in the brain microvascular basal lamina during development of the blood-brain barrier. Dev Dyn 208:62–74.

Bartholdi, D., Rubin, B.P., Schwab, M.E. (1997). VEGF mRNA induction correlates with changes in the vascular architecture upon spinal cord damage in the rat. Eur J Neurosci 9:2549–2560.

Bauer, H.C., Bauer, H., Lametschwandtner, A., Amberger, A., Ruiz, P., Steiner, M. (1993). Neovascularization and the appearance of morphological characteristics of the blood-brain barrier in the embryonic mouse central nervous system. Dev Brain Res 75:269–278.

Benagiano, V., Virgintino, D., Maiorano, E., et al. (1996). VIP-like immunoreactivity within neurons and perivascular neuronal processes of the human cerebral cortex. Eur J Histochem 40:53–56.

Benjamin, L.E., Golijanin, D., Itin, A., Pode, D., Keshet, E. (1999). Selective ablation of immature blood vessels in established human tumors follows vascular endothelial growth factor withdrawal. J Clin Invest 103:159–165.

Benjamin, L.E., Hemo, I., Keshet, E. (1998). A plasticity window for blood vessel remodelling is defined by pericyte coverage of the preformed endothelial network and is regulated by PDGF-B and VEGF. Development 125:1591–1598.

Bergwerff, M., Verberne, M.E., DeRuiter, M.C., Poelmann, R.E., Gittenberger-de Groot, A.C. (1998). Neural crest cell contribution to the developing circulatory system: implications for vascular morphology? Circ Res 82:221–231.

Bernsen, H.J.J.A., Rijken, P.F.J.W., Peters, J.P.W., Bakker, H., van der Kogel, A.J. (1998). Delayed vascular changes after antiangiogenic therapy with antivascular endothelial growth factor antibodies in human glioma xenografts in nude mice. Neurosurgery 43: 570–575.

Bertossi, M., Ribatti, D., Nico, B., Mancini, Lozupone, E., Roncali, L. (1988). The barrier systems in the choroidal plexuses of the chick embryo studied by means of horseradish peroxidase. J Submicrosc Cytol Pathol 20:385–395.

Bertossi, M., Riva, A., Congiu, T., Virgintino, D., Nico, B., Roncali, L. (1995). A compared TEM/SEM investigation on the pericytic investment in developing microvasculature of the chick optic tectum. J Submicrosc Cytol Pathol 27:349–358.

Bertossi, M., Roncali, L., Nico, B., et al. (1993). Perivascular astrocytes and endothelium in the development of the blood-brain barrier in the optic tectum of the chick embryo. Anat Embryol 188:21–29.

Biswas, C., Zhang, Y., DeCastro, R., et al. (1995). The human tumor cell-derived collagenase stimulatory factor (renamed EMMPRIN) is a member of the immunoglobulin superfamily. Cancer Res 55:434–439.

Boado, R.J. (2000). Molecular regulation of the blood-brain barrier GLUT1 glucose transporter by brain derived factors. J Neural Transm Suppl 59:255–261.

Bockman, D.E., Sohal, G.S. (1998). A new source of cells contributing to the developing gastrointestinal tract demonstrated in chick embryos. Gastroenterology 114:878–882.

Bodey, B., Siegel, S.E., Kaiser, H.E. (1998). Upregulation of endoglin (CD105) expression during childhood brain tumor-related angiogenesis. Anti-angiogenic therapy. Anticancer Res 18:1485–1500.

Boero, J.A., Ascher, J., Arregui, A., Rovainen, C., Woolsey, T.A. (1999). Increased brain capillaries in chronic hypoxia. J Appl Physiol 86:1211–1219.

Brand-Saberi, B., Wilting, Ebensperger, C., Christ, B. (1996). The formation of somite compartments in the avian embryo. Int J Dev Biol 40:411–420.

Breier, G., Albrecht, U., Sterrer, S., Risau, W. (1992). Expression of vascular endothelial growth factor during embryonic angiogenesis and endothelial cell differentiation. Development 114:521–532.

Britsch, S., Christ, B., Jacob, H.J. (1989). The influence of cell-matrix interactions on the development of quail chorioallantoic vascular system. Anat Embryol 180:479–484.

Bronner-Fraser, M., Fraser, S.E. (1997). Differentiation of the vertebrate neural tube. Curr Opin Cell Biol 9:885–891.

Brüstle, O., Choudhary, K., Karram, K., et al. (1998). Chimeric brains generated by intra-ventricular transplantation of fetal human brain cells into embryonic rats. Nature Biotechnol 16:1040–1044.

Burri, P.H., Tarek, M.R. (1990). A novel mechanism of capillary growth in the rat pulmonary microcirculation. Anat Rec 228:35–45.

Caprioli, A., Jaffredo, T., Gautier, R., Dubourg, C., Dieterlen-Lievre, F. (1998). Blood-borne seeding by hematopoietic and endothelial precursors from the allantois. Proc Natl Acad Sci USA 95:1641–1646.

Carmeliet, P., Dor, Y., Herbert, J.M., et al. (1998). Role of HIF-1α in hypoxia-mediated apoptosis, cell proliferation and tumour angiogenesis. Nature 394:485–490.

Cassella, J.P., Lawrenson, J.G., Allt, G., Firth, J.A. (1996). Ontogeny of four blood-brain barrier markers: an immunocytochemical comparison of pial and cerebral cortical microvessels. J Anat 189:407–415.

Cassella, J.P., Lawrenson, J.G., Firth, J.A. (1997a). Development of endothelial paracellular clefts and their tight junctions in the pial microvessels of the rat. J Neurocytol 26:567–575.

Cassella, J.P., Lawrenson, J.G., Lawrence, L., Firth, J.A. (1997b). Differential distribution of an endothelial barrier antigen between the pial and cortical microvessels of the rat. Brain Res 744:335–338.

Chan, A.S.Y., Leung, S.Y., Wong, M.P., et al. (1998). Expression of vascular endothelial growth factor and its receptors in the anaplastic progression of astrocytoma, oligodendroglioma, and ependymoma. Am J Surg Pathol 22:816–826.

Chan-Ling, T.L., Halasz, P., Stone, J. (1990). Development of retinal vasculature in the cat: processes and mechanisms. Curr Eye Res 9:459–478.

Cheifetz, S., Bellon, T., Cales, C., et al. (1992). Endoglin is a component of the transforming growth factor-β receptor system in human endothelial cells. J Biol Chem 267: 19027–19030.

Chen, W.T. (1996). Proteases associated with invadopodia, and their role in degradation of extracellular matrix. Enzyme Prot 49:59–71.

Christ, B., Poelmann, R.E., Mentink, M.M., Gittenberger-deGroot, A.C. (1990). Vascular endothelial cells migrate centripetally within embryonic arteries. Anat Embryol 181: 333–339.

Christ, B., Schmidt, C., Huang, R., Wilting, J., Brand-Saberi, B. (1998). Segmentation of the vertebrate body. Anat Embryol 197:1–8.

Clauss, M., Weich, H., Breier, G., et al. (1996). The vascular endothelial growth factor receptor Flt-1 mediates biological activities. Implications for a functional role of placenta growth factor in monocyte activation and chemotaxis. J Biol Chem 271:17629–17634.

Cornford, E.M., Hyman, S. (1999). Blood-brain barrier permeability to small and large molecules. Adv Drug Deliv Rev 36:145–163.

Cossmann, P.H., Eggli, P.S., Christ, B., Kurz, H. (1997). Mesoderm-derived cells proliferate in the embryonic central nervous system: confocal microscopy and three-dimensional visualization. Histochem Cell Biol 107:205–213.

Cossmann, P.H., Eggli, P.S., Kurz, H. (2000). Three-dimensional analysis of DNA replication foci: a comparative study on species and cell type in situ. Histochem Cell Biol 113:195–205.

Cotman, S.L., Halfter, W., Cole, G.J. (1999). Identification of extracellular matrix ligands for the heparan sulfate proteoglycan agrin. Exp Cell Res 249:54–64.

Couly, G., Coltey, P., Eichmann, A., Le Douarin, N.M. (1995). The angiogenic potentials of the cephalic mesoderm and the origin of brain and head blood vessels. Mech Dev 53:97–112.

Couly, G.F., Coltey, P.M., Le Douarin, N.M. (1992). The developmental fate of the cephalic mesoderm in quail-chick chimeras. Development 114:1–15.

Couly, G.F., Le Douarin, N.M. (1985). Mapping of the early neural primordium in quail-chick chimeras. I. Developmental relationships between placodes, facial ectoderm, and prosencephalon. Dev Biol 110:422–439.

Couly, G.F., Le Douarin, N.M. (1987). Mapping of the early neural primordium in quail-chick chimeras. II. The prosencephalic neural plate and neural folds: implications for the genesis of cephalic human congenital abnormalities. Dev Biol 120:198–214.

Crosby, J.R., Seifert, R.A., Soriano, P., Bowen-Pope, D.F. (1998). Chimaeric analysis reveals role of Pdgf receptors in all muscle lineages. Nature Gen 18:385–388.

Cuadros, M.A., Coltey, P., Carmen, N.M., Martin, C. (1992). Demonstration of a phagocytic cell system belonging to the hemopoietic lineage and originating from the yolk sac in the early avian embryo. Development 115:157–168.

Cuadros, M.A., Martin, C., Coltey, P., Almendros, A., Navascues, J. (1993). First appearance, distribution, and origin of macrophages in the early development of the avian central nervous system. J Comp Neurol 330:113–129.

Cuadros, M.A., Navascues, J. (1998). The origin and differentiation of microglial cells during development. Prog Neurobiol 56:173–189.

D'Amico-Martel, A., Noden, D.M. (1983). Contributions of placodal and neural crest cells to avian cranial peripheral ganglia. Am J Anat 166:445–468.

D'Amore, P.A., Smith, S.R. (1993). Growth factor effects on cells of the vascular wall: a survey. Growth Factors 8:61–75.

DeRuiter, M.C., Poelmann, R.E., VanMunsteren, J.C., Mironov, V., Markwald, R.R., Gittenberger-de Groot, A.C. (1997). Embryonic endothelial cells transdifferentiate into mesenchymal cells expressing smooth muscle actins in vivo and in vitro. Circ Res 80: 444–451.

Dieterlen-Liévre, F. (1998). Hematopoiesis: progenitors and their genetic program. Curr Biol 8:R727–R730.

Dobrogowska, D.H., Lossinsky, A.S., Tarnawski, M., Vorbrodt, A.W. (1998). Increased blood-brain barrier permeability and endothelial abnormalities induced by vascular endothelial growth factor. J Neurocytol 27:163–173.

Drab, M., Haller, H., Bychkov, R., et al. (1997). From totipotent embryonic stem cells to spontaneously contracting smooth muscle cells: a retinoic acid and db-cAMP *in vitro* differentiation model. FASEB J 11:905–915.

Draeger, A., England, C. (1998). The intima: historic literature revisited. Anat Anz 180: 189–192.

Drake, C.J., Hungerford, J.E., Little, C.D. (1998). Morphogenesis of the first blood vessels. Ann NY Acad Sci 857:155–179.

Drake, C.J., Little, C.D. (1995). Exogenous vascular endothelial growth factor induces malformed and hyperfused vessels during embryonic neovascularization. Proc Natl Acad Sci USA 92:7657–7661.

Earle, K.L., Mitrofanis, J. (1998). Development of glia and blood vessels in the internal capsule of rats. J Neurocytol 27:127–139.

Eggli, P.S., Graber, W. (1995). Association of hyaluronan with rat vascular endothelial and smooth muscle cells. J Histochem Cytochem 43:689–697.

Eglitis, M.A., Mezey, E. (1997). Hematopoietic cells differentiate into both microglia and macroglia in the brains of adult mice. Proc Natl Acad Sci USA 94:4080–4085.

Ehler, E., Karlhuber, G., Bauer, H.C., Draeger, A. (1995). Heterogeneity of smooth muscle-associated proteins in mammalian brain microvasculature. Cell Tissue Res 279:393–403.

Eickholt, B.J., Mackenzie, S.L., Graham, A., Walsh, F.S., Doherty, P. (1999). Evidence for collapsin-1 functioning in the control of neural crest migration in both trunk and hind-brain regions. Development 126:2181–2189.

Ellison, J.A., Velier, J.J., Spera, P., et al. (1998). Osteopontin and its integrin receptor $\alpha_v\beta_3$ are upregulated during formation of the glial scar after focal stroke. Stroke 29:1698–1706.

Etchevers, H.C., Couly, G., Vincent, C., Le Douarin, N.M. (1999). Anterior cephalic neural crest is required for forebrain viability. Development 126:3533–3543.

Etchevers, H.C., Vincent, C., Le Douarin, N.M., Couly, G.F. (2001). The cephalic neural crest provides pericytes and smooth muscle cells to all blood vessels of the face and forebrain. Development 128:1059–1068.

Faller, D.V. (1999). Endothelial cell responses to hypoxic stress. Clin Exp Pharmacol Physiol 26:74–84.

Fan, Q.W., Yuasa, S., Kuno, N., et al. (1998). Expression of basigin, a member of the immunoglobulin superfamily, in the mouse central nervous system. Neurosci Res 30: 53–63.

Farrell, C.L., Risau, W. (1994). Normal and abnormal development of the blood-brain barrier. Microsc Res Technol 27:495–506.

Feiner, L., Koppel, A.M., Kobayashi, H., Raper, J.A. (1997). Secreted chick semaphorins bind recombinant neuropilin with similar affinities but bind different subsets of neurons *in situ*. Neuron 19:539–545.

Feucht, M., Christ, B., Wilting, J. (1997). VEGF induces cardiovascular malformation and embryonic lethality. Am J Pathol 151:1407–1416.

Fischer, S., Clauss, M., Wiesnet, M., Renz, D., Schaper, W., Karliczek, G.F. (1999). Hypoxia induces permeability in brain microvessel endothelial cells via VEGF and NO. Am J Physiol Cell Physiol 45:C812–C820.

Fong, G.-H., Zhang, L., Bryce, D.-M., Peng, J. (1999a). Increased hemangioblast commitment, not vascular disorganization, is the primary defect in *flt-1* knock-out mice. Development 126:3015–3025.

Fong, T.A., Shawver, L.K., Sun, L., et al. (1999b). SU5416 is a potent and selective inhibitor of the vascular endothelial growth factor receptor (Flk-1/KDR) that inhibits tyrosine kinase catalysis, tumor vascularization, and growth of multiple tumor types. Cancer Res 59:99–106.

Forsyth, P.A., Wong, H., Laing, T.D., et al. (1999). Gelatinase-A (MMP-2), gelatinase-B (MMP-9) and membrane type matrix metalloproteinase-1 (MT1-MMP) are involved in different aspects of the pathophysiology of malignant gliomas. Br J Cancer 79:1828–1835.

Fujisawa, H., Kitsukawa, T. (1998). Receptors for collapsin/semaphorins. Curr Opin Neurobiol 8:587–592.

Gagnon, M.L., Bielenberg, D.R., Gechtman, Z., et al. (2000). Identification of a natural soluble neuropilin-1 that binds vascular endothelial growth factor: *in vivo* expression and antitumor activity. Proc Natl Acad Sci USA 97:2573–2578.

Galileo, D.S., Gray, G.E., Owens, G.C., Majors, J., Sanes, J.R. (1990). Neurons and glia arise from a common progenitor in chicken optic tectum: demonstration with two retroviruses and cell type-specific antibodies. Proc Natl Acad Sci USA 87:458–462.

Gary, S.C., Kelly, G.M., Hockfield, S. (1998). BEHAB/brevican: a brain-specific lectican implicated in gliomas and glial cell motility. Curr Opin Neurobiol 8:576–581.

Gerhardt, H., Liebner, S., Redies, C., Wolburg, H. (1999). N-cadherin expression in endothelial cells during early angiogenesis in the eye and brain of the chicken: relation to blood-retina and blood-brain barrier development. Eur J Neurosci 11:1191–1201.

Gerhardt, H., Wolburg, H., Redies, C. (2000). N-cadherin mediates pericytic-endothelial interaction during brain angiogenesis in the chicken. Dev Dyn 218:472–479.

Goldbrunner, R.H., Bernstein, J.J., Plate, K.H., Vince, G.H., Roosen, K., Tonn, J.C. (1999). Vascularization of human glioma spheroids implanted into rat cortex is conferred by two distinct mechanisms. J Neurosci Res 55:486–495.

Goldman, J.E., Zerlin, M., Newman, S., Zhang, L., Gensert, J. (1997). Fate determination and migration of progenitors in the postnatal mammalian CNS. Dev Neurosci 19:42–48.

Gougos, A., Letarte, M. (1990). Primary structure of endoglin, an RGD-containing glycoprotein of human endothelial cells. J Biol Chem 265:8361–8364.

Graeber, M.B., Streit, W.J., Kreutzberg, G.W. (1989). Identity of ED2-positive perivascular cells in rat brain. J Neurosci Res 22:103–106.

Groves, A.K., Bronner-Fraser, M. (1999). Neural crest diversification. Curr Top Dev Biol 43:221–258.

Halata, Z., Grim, M., Christ, B. (1990). Origin of spinal cord meninges, sheaths of peripheral nerves, and cutaneous receptors including Merkel cells. An experimental and ultrastructural study with avian chimeras. Anat Embryol 182:529–537.

Hatzopoulos, A.K., Folkman, J., Vasile, E., Eiselen, G.K., Rosenberg, R.D. (1998). Isolation and characterization of endothelial progenitor cells from mouse embryos. Development 125:1457–1468.

Hayashi, T., Abe, K., Itoyama, Y. (1998a). Reduction of ischemic damage by application of vascular endothelial growth factor in rat brain after transient ischemia. J Cereb Blood Flow Metab 18:887–895.

Hayashi, T., Abe, K., Sakurai, M., Itoyama, Y. (1998b). Inductions of hepatocyte growth factor and its activator in rat brain with permanent middle cerebral artery occlusion. Brain Res 799:311–316.

Hayashi, T., Sakurai, M., Abe, K., Sadahiro, M., Tabayashi, K., Itoyama, Y. (1999). Expression of angiogenic factors in rabbit spinal cord after transient ischaemia. Neuropathol Appl Neurobiol 25:63–71.

Hellstrom, M., Kal, Lindahl, P., Abramsson, A., Betsholtz, C. (1999). Role of PDGF-B and PDGFR-β in recruitment of vascular smooth muscle cells and pericytes during embryonic blood vessel formation in the mouse. Development 126:3047–3055.

Herman, I.M., D'Amore, P.A. (1985). Microvascular pericytes contain muscle and nonmuscle actins. J Cell Biol 101:43–52.

Hirase, T., Staddon, J.M., Saitou, M., et al. (1997). Occludin as a possible determinant of tight junction permeability in endothelial cells. J Cell Sci 110:1603–1613.

Hirschi, K.K., D'Amore, P.A. (1996). Pericytes in the microvasculature. Cardiovasc Res 32:687–698.

Hirschi, K.K., D'Amore, P.A. (1997). Control of angiogenesis by the pericyte: molecular mechanisms and significance. EXS 79:419 428.

Hirschi, K.K., Rohovsky, S.A., Beck, L.H., Smith, S.R., D'Amore, P.A. (1999). Endothelial cells modulate the proliferation of mural cell precursors via platelet-derived growth factor-BB and heterotypic cell contact. Circ Res 84:298–305.

Holash, J.A., Noden, D.M., Stewart, P.A. (1993). Re-evaluating the role of astrocytes in blood-brain barrier induction. Dev Dyn 197:14–25.

Huang, R., Zhi, Q., Ordahl, C.P., Christ, B. (1997). The fate of the first avian somite. Anat Embryol 195:435–449.

Huang, R., Zhi, Q., Wilting, J., Christ, B. (1994). The fate of somitocoele cells in avian embryos. Anat Embryol 190:243–250.

Hungerford, J.E., Little, C.D. (1999). Developmental biology of the vascular smooth muscle cell: building a multilayered vessel wall. J Vasc Res 36:2–27.

Hurley, S.D., Walter, S.A., Semple-Rowland, S.L., Streit, W.J. (1999). Cytokine transcripts expressed by microglia *in vitro* are not expressed by ameboid microglia of the developing rat central nervous system. Glia 25:304–309.

Igakura, T., Kadomatsu, K., Kaname, T., et al. (1998). A null mutation in basigin, an immunoglobulin superfamily member, indicates its important roles in peri-implantation development and spermatogenesis. Dev Biol 194:152–165.

Igakura, T., Kadomatsu, K., Taguchi, O., et al. (1996). Roles of basigin, a member of the immunoglobulin superfamily, in behavior as to an irritating odor, lymphocyte response, and blood-brain barrier. Biochem Biophys Res Commun 224:33–36.

Ikeda, E., Flamme, I., Risau, W. (1996). Developing brain cells produce factors capable of inducing the HT7 antigen, a blood-brain barrier-specific molecule, in chick endothelial cells. Neurosci Lett 209:149–152.

Im, S.A., Gomez-Manzano, C., Fueyo, J., et al. (1999). Antiangiogenesis treatment for gliomas: transfer of antisense vascular endothelial growth factor inhibits tumor growth *in vivo*. Cancer Res 59:895–900.

Issa, R., Krupinski, J., Bujny, T., Kumar, S., Kaluza, J., Kumar, P. (1999). Vascular endothelial growth factor and its receptor, KDR, in human brain tissue after ischemic stroke. Lab Invest 79:417–425.

Jaffredo, T., Gautier, R., Eichmann, A., Dieterlen-Liévre, F. (1998). Intraaortic hemopoietic cells are derived from endothelial cells during ontogeny. Development 125:4575–4583.

Janzer, R.C., Lobrinus, J.A., Darekar, P., Juillerat, L. (1993). Astrocytes secrete a factor inducing the expression of HT7-protein and neurothelin in endothelial cells of chorioallantoic vessels. Adv Exp Med Biol 331:217–221.

Jaworski, D.M., Kelly, G.M., Hockfield, S. (1995). The CNS-specific hyaluronan-binding protein BEHAB is expressed in ventricular zones coincident with gliogenesis. J Neurosci 15:1352–1362.

Jaworski, D.M., Kelly, G.M., Hockfield, S. (1999). Intracranial injury acutely induces the expression of the secreted isoform of the CNS-specific hyaluronan-binding protein BEHAB brevican. Exp Neurol 157:327–337.

Jiang, X.B., Rowitch, D.H., Soriano, P., McMahon, A.P., Sucov, H.M. (2000). Fate of the mammalian cardiac neural crest. Development 127:1607–1616.

Kakinuma, Y., Hama, H., Sugiyama, F., et al. (1998). Impaired blood-brain barrier function in angiotensinogen-deficient mice. Nature Med 4:1078–1080.

Kalcheim, C., Le Douarin, N.M. (1986). Requirement of a neural tube signal for the differentiation of neural crest cells into dorsal root ganglia. Dev Biol 116:451–466.

Kawakami, A., Kitsukawa, T., Takagi, S., Fujisawa, H. (1996). Developmentally regulated expression of a cell surface protein, neuropilin, in the mouse nervous system. J Neurobiol 29:1–17.

Kawasaki, T., Kitsukawa, T., Bekku, Y., et al. (1999). A requirement for neuropilin-1 in embryonic vessel formation. Development 126:4895–4902.

Kelly, T., Yan, Y., Osborne, R.L., et al. (1998). Proteolysis of extracellular matrix by invadopodia facilitates human breast cancer cell invasion and is mediated by matrix metalloproteinases. Clin Exp Metastasis 16:501–512.

Kirsch, M., Strasser, J., Allende, R., Bello, L., Zhang, J.P., Black, P.M. (1998). Angiostatin suppresses malignant glioma growth *in vivo*. Cancer Res 58:4654–4659.

Kitsukawa, T., Shimono, A., Kawakami, A., Kondoh, H., Fujisawa, H. (1995). Over-expression of a membrane protein, neuropilin, in chimeric mice causes anomalies in the cardiovascular system, nervous system and limbs. Development 121:4309–4318.

Klessinger, S., Christ, B. (1996). Axial structures control laterality in the distribution pattern of endothelial cells. Anat Embryol 193:319–330.

Koblizek, T.I., Weiss, C., Yancopoulos, G.D., Deutsch, U., Risau. (1998). Angiopoietin-1 induces sprouting angiogenesis *in vitro*. Curr Biol 8:529–532.

Kohler, A., Jostarndt-Fögen, K., Alliegro, M.C., Rottner, C., Draeger, A. (1999). Intima-like smooth muscle cells: developmental link between endothelium and media? Anat Embryol 200:313–323.

Köntges, G., Lumsden, A. (1996). Rhombencephalic neural crest segmentation is preserved throughout craniofacial ontogeny. Development 122:3229–3242.

Korn, J., Christ, B., Kurz, H. (2000). Brain vascular smooth muscle cells from neuro-ectoderm. Ann Anat 182:132.

Korn, J., Christ, B., Kurz, H. (2002). Neuroectodermal origin of brain pericytes and vascular smooth muscle cells. J Comp Neurol 442:78–88.

Kotch, L., Narayan, V.I., Laughner, E., Semenza, G.L. (1999). Defective vascularization of HIF–1α-null embryos is not associated with VEGF deficiency but with mesenchymal cell death. Dev Biol 209:254–267.

Krieg, M., Marti, H.H., Plate, K.H. (1998). Coexpression of erythropoietin and vascular endothelial growth factor in nervous system tumors associated with von Hippel-Lindau tumor suppressor gene loss of function. Blood 92:3388–3393.

Krum, J.M. (1996). Effect of astroglial degeneration on neonatal blood-brain barrier marker expression. Exp Neurol 142:29–35.

Krum, J.M., Kenyon, K.L., Rosenstein, J.M. (1997). Expression of blood-brain barrier characteristics following neuronal loss and astroglial damage after administration of anti-Thy-1 immunotoxin. Exp Neurol 146:33–45.

Krum, J.M., Rosenstein, J.M. (1998). VEGF mRNA and its receptor flt-1 are expressed in reactive astrocytes following neural grafting and tumor cell implantation in the adult CNS. Exp Neurol 154:57–65.

Kuchler-Bopp, S., Delaunoy, J.P., Artault, J.C., Zaepfel, M., Dietrich, J.B. (1999). Astro-cytes induce several blood-brain barrier properties in non-neural endothelial cells. Neuroreport 10:1347–1353.

Kurz, H. (1997). Über die Verzweigungen der Blutgefäße—Embryonale Musterbildung zwischen Ordnung, Zufall und Chaos. Habilitation, Universität Freiburg.

Kurz, H. (2000). Physiology of angiogenesis. J Neuro-Oncol 50:17–35.

Kurz, H., Christ, B. (1998). Embryonic CNS macrophages and microglia do not stem from circulating, but from extravascular precursors. Glia 22:98–102.

Kurz, H., Gärtner, T., Eggli, P.S., Christ, B. (1996). First blood vessels in the avian neural tube are formed by a combination of dorsal angioblast immigration and ventral sprout-ing of endothelial cells. Dev Biol 173:133–147.

Kurz, H., Korn, J., Eggli, P.S., Huang, R., Christ, B. (2001). Embryonic central nervous system angiogenesis does not involve blood-borne endothelial progenitors. J Comp Neurol 436:263–274.

LaBonne, C., Bronner-Fraser, M. (1998). Induction and patterning of the neural crest, a stem cell-like precursor population. J Neurobiol 36:175–189.

Lagercrantz, J., Farnebo, F., Larsson, C., Tvrdik, T., Weber, G., Piehl, F. (1998). A comparative study of the expression patterns for vegf, vegf-b/vrf and vegf-c in the developing and adult mouse. Biochim Biophys Acta 1398:157–163.

Lallier, T., Leblanc, G., Artinger, K.B., Bronner-Fraser, M. (1992). Cranial and trunk neural crest cells use different mechanisms for attachment to extracellular matrices. Development 116:531–541.

Lamoreaux, W.J., Fitzgerald, M.E., Reiner, A., Hasty, K.A., Charles, S.T. (1998). Vascular endothelial growth factor increases release of gelatinase A and decreases release of tissue inhibitor of metalloproteinases by microvascular endothelial cells in vitro. Microvasc Res 55:29–42.

Lastres, P., Letamendia, A., Zhang, H., et al. (1996). Endoglin modulates cellular responses to TGF-β1. J Cell Biol 133:1109–1121.

Lawrenson, J.G., Reid, A.R., Finn, T.M., Orte, C., Allt, G. (1999). Cerebral and pial microvessels: differential expression of gamma-glutamyl transpeptidase and alkaline phosphatase. Anat Embryol 199:29–34.

Le Douarin, N.M. (1973). A Feulgen-positive nucleolus. Exp Cell Res 77:459–468.

Le Douarin, N.M., Teillet, M.A., Catala, M. (1998). Neurulation in amniote vertebrates: a novel view deduced from the use of quail-chick chimeras. Int J Dev Biol 42:909–916.

Lee, S.H., Hungerford, J.E., Little, C.D., Iruela-Arispe, M.L. (1997). Proliferation and differentiation of smooth muscle cell precursors occurs simultaneously during the development of the vessel wall. Dev Dyn 209:342–352.

Lehrmann, E., Kiefer, R., Christensen, T., et al. (1998). Microglia and macrophages are major sources of locally produced transforming growth factor-β1 after transient middle cerebral artery occlusion in rats. Glia 24:437–448.

Levison, S.W., Goldman, J.E. (1997). Multipotential and lineage restricted precursors coexist in the mammalian perinatal subventricular zone. J Neurosci Res 48:83–94.

Li, D.Y., Sorensen, L.K., Brooke, B.S., et al. (1999). Defective angiogenesis in mice lacking endoglin. Science 284:1534–1537.

Liebner, S., Gerhardt, H., Wolburg, H. (1997). Maturation of the blood-retina barrier in the developing pecten oculi of the chicken. Brain Res Dev Brain Res 100:205–219.

Lindahl, P., Hellstrom, M., Kalen, M., Betsholtz, C. (1998). Endothelial-perivascular cell signaling in vascular development: lessons from knockout mice. Curr Opin Lipidol 9:407–411.

Lindahl, P., Johansson, B.R., Leveen, P., Betsholtz, C. (1997). Pericyte loss and microaneurysm formation in PDGF-B-deficient mice. Science 277:242–245.

Little, C.D., Mironov, V., Sage, E.H. (1998). Vascular morphogenesis: in vivo, in vitro, in mente. Boston: Birkhäuser.

Lossinsky, A.S., Buttle, K.F., Pluta, R., Mossakowski, M.J., Wisniewski, H.M. (1999). Immunoultrastructural expression of intercellular adhesion molecule 1 in endothelial cell vesiculotubular structures and vesiculovacuolar organelles in blood-brain barrier development and injury. Cell Tissue Res 295:77–88.

Lossinsky, A.S., Wisniewski, H.M. (1998). Immunoultrastructural expression of ICAM-1 and PECAM-1 occurs prior to structural maturity of the murine blood-brain barrier. Dev Neurosci 20:518–524.

Lyden, D., Young, A.Z., Zagzag, D., et al. (1999). Id1 and Id3 are required for neurogenesis, angiogenesis and vascularization of tumour xenografts. Nature 401:670–677.

Maisonpierre, P.C., Suri, C., Jones, P.F., et al. (1997). Angiopoietin-2, a natural antagonist for Tie2 that disrupts in vivo angiogenesis. Science 277:55–60.

Margolis, R.K., Rauch, U., Maurel, P., Margolis, R.U. (1996). Neurocan and phosphacan: two major nervous tissue-specific chondroitin sulfate proteoglycans. Perspect Dev Neurobiol 3:273–290.

Marti, H.H., Risau, W. (1998). Systemic hypoxia changes the organ-specific distribution of vascular endothelial growth factor and its receptors. Proc Natl Acad Sci USA 95: 15809–15814.

Matsuoka, Y., Okazaki, M., Kitamura, Y., Taniguchi, T. (1999). Developmental expression of P-glycoprotein (multidrug resistance gene product) in the rat brain. J Neurobiol 39: 383–392.

Mayhan, W.G. (1999). VEGF increases permeability of the blood-brain barrier via a nitric oxide synthase/cGMP-dependent pathway. Am J Physiol Cell Physiol 45:C1148–C1153.

McKay, R.D.G. (1997). Stem cells in the central nervous system. Science 276:66–71.

Mendis, D.B., Ivy, G.O., Brown, I.R. (1998). SPARC/osteonectin mRNA is induced in blood vessels following injury to the adult rat cerebral cortex. Neurochem Res 23: 1117–1123.

Migdal, M., Huppertz, B., Tessler, S., et al. (1998). Neuropilin-1 is a placenta growth factor-2 receptor. J Biol Chem 273:22272–22278.

Miyauchi, T., Kanekura, T., Yamaoka, A., Ozawa, M., Miyazawa, S., Muramatsu, T. (1990). Basigin, a new, broadly distributed member of the immunoglobulin superfamily, has strong homology with both the immunoglobulin V domain and the β-chain of major histocompatibility complex class II antigen. J Biochem 107:316–323.

Møllgard, K., Dziegielewska, K.M., Saunders, N.R., Zakut, H., Soreq, H. (1988). Synthesis and localization of plasma proteins in the developing human brain. Dev Biol 128: 207–221.

Moore, S., Thanos, S. (1996). The concept of microglia in relation to central nervous system disease and regeneration. Prog Neurobiol 48:441–460.

Morioka, T., Kalehua, A.N., Streit, W.J. (1993). Characterization of microglial reaction after middle cerebral artery occlusion in rat brain. J Comp Neurol 327:123–132.

Mousa, S.A., Lorelli, W., Campochiaro, P.A. (1999). Role of hypoxia and extracellular matrix-integrin binding in the modulation of angiogenic growth factors secretion by retinal pigmented epithelial cells. J Cell Biochem 74:135–143.

Muldoon, L.L., Pagel, M.A., Kroll, R.A., Roman-Goldstein, S., Jones, R.S., Neuwelt, E.A. (1999). A physiological barrier distal to the anatomic blood-brain barrier in a model of transvascular delivery. Am J Neuroradiol 20:217–222.

Murohara, T., Witzenbichler, B., Spyridopoulos, I., et al. (1999). Role of endothelial nitric oxide synthase in endothelial cell migration. Arterioscler Thromb Vasc Biol 19:1156–1161.

Naruhashi, K., Kadomatsu, K., Igakura, T., et al. (1997). Abnormalities of sensory and memory functions in mice lacking Bsg gene. Biochem Biophys Res Commun 236:733–737.

Nataf, V., Grapin-Botton, A., Champeval, D., Amemiya, A., Yanagisawa, M., Le Douarin, N.M. (1998). The expression patterns of endothelin-A receptor and endothelin 1 in the avian embryo. Mech Dev 75:145–149.

Nehls, V., Drenckhahn, D. (1991). Heterogeneity of microvascular pericytes for smooth muscle type α-actin. J Cell Biol 113:147–154.

Nehls, V., Drenckhahn, D. (1993). The versatility of microvascular pericytes: from mesenchyme to smooth muscle? Histochemistry 99:1–12.

Nehls, V., Herrmann, R., Huhnken, M. (1998). Guided migration as a novel mechanism of capillary network remodeling is regulated by basic fibroblast growth factor. Histochem Cell Biol 109:319–329.

Neufeld, G., Cohen, T., Gengrinovitch, S., Poltorak, Z. (1999). Vascular endothelial growth factor (VEGF) and its receptors. FASEB J 13:9–22.

Neuhaus, J., Risau, W., Wolburg, H. (1991). Induction of blood-brain barrier characteristics in bovine brain endothelial cells by rat astroglial cells in transfilter coculture. Ann NY Acad Sci 633:578–580.

Nguyen, M., Arkell, J., Jackson, C.J. (1999). Thrombin rapidly and efficiently activates gelatinase A in human microvascular endothelial cells via a mechanism independent of active MT1 matrix metalloproteinase. Lab Invest 79:467–475.

Nico, B., Quondamatteo, F., Herken, R., et al. (1999). Developmental expression of ZO-1 antigen in the mouse blood-brain barrier. Brain Res Dev Brain Res 114:161–169.

Noden, D.M. (1988). Interactions and fates of avian craniofacial mesenchyme. Development 103(suppl):121–140.

Noden, D.M. (1990). Origins and assembly of avian embryonic blood vessels. Ann NY Acad Sci 588:236–249.

Oh, S.J., Kurz, H., Christ, B., Wilting, J. (1998). Platelet-derived growth factor-B induces transformation of fibrocytes into spindle-shaped myofibroblasts in vivo. Histochem Cell Biol 109:349–357.

Olofsson, B., Korpelainen, E., Pepper, M.S., et al. (1998). Vascular endothelial growth factor B (VEGF-B) binds to VEGF receptor-1 and regulates plasminogen activator activity in endothelial cells. Proc Natl Acad Sci USA 95:11709–11714.

Orte, C., Lawrenson, J.G., Finn, T.M., Reid, A.R., Allt, G. (1999). A comparison of blood-brain barrier and blood-nerve barrier endothelial cell markers. Anat Embryol 199:509–517.

Ortega, N., Hutchings, H., Plouet, J. (1999). Signal relays in the VEGF system. Front Biosci 4:D141–D152.

Papoutsi, M., Kurz, H., Schächtele, C., et al. (2000). Induction of the blood-brain barrier marker neurothelin/HT7 in endothelial cells by a variety of tumors in chick embryos. Histochem Cell Biol 113:105–113.

Pardanaud, L., Dieterlen-Liévre, F. (1999). Manipulation of the angiopoietic/hemangiopoietic commitment in the avian embryo. Development 126:617–627.

Patt, S., Danner, S., Theallier-Janko, A., et al. (1998). Upregulation of vascular endothelial growth factor in severe chronic brain hypoxia of the rat. Neurosci Lett 252:199–202.

Peault, B.M., Thiery, J.P., Le Douarin, N.M. (1983). Surface marker for hemopoietic and endothelial cell lineages in quail that is defined by a monoclonal antibody. Proc Natl Acad Sci USA 80:2976–2980.

Pennell, N.A., Streit, W.J. (1997). Colonization of neural allografts by host microglial cells: relationship to graft neovascularization. Cell Transplant 6:221–230.

Pepper, M.S. (1997). Transforming growth factor β: vasculogenesis, angiogenesis, and vessel wall integrity. Cytokine Growth Factor Rev 8:21–43.

Plate, K.H. (1999). Mechanisms of angiogenesis in the brain. J Neuropathol Exp Neurol 58:313–320.

Plate, K.H., Breier, G., Farrell, C.L., Risau, W. (1992). Platelet-derived growth factor receptor-β is induced during tumor development and upregulated during tumor progression in endothelial cells in human gliomas. Lab Invest 67:529–534.

Poelmann, R.E., Gittenberger-de Groot, A.C., Mentink, M.M., Delpech, B., Girard, N., Christ, B. (1990). The extracellular matrix during neural crest formation and migration in rat embryos. Anat Embryol 182:29–39.

Provis, J.M., Leech, J., Diaz, C.M., Penfold, P.L., Stone, J., Keshet, E. (1997). Development of the human retinal vasculature: cellular relations and VEGF expression. Exp Eye Res 65:555–568.

Qin, Y., Sato, T.N. (1995). Mouse multidrug resistance 1a/3 gene is the earliest known endothelial cell differentiation marker during blood-brain barrier development. Dev Dyn 202:172–180.

Ramsauer, M., Kunz, J., Krause, D., Dermietzel, R. (1998). Regulation of a blood-brain barrier-specific enzyme expressed by cerebral pericytes (pericytic aminopeptidase N/pAPN) under cell culture conditions. J Cereb Blood Flow Metab 18:1270–1281.

Ribatti, D., Alessandri, G., Vacca, A., Iurlaro, M., Ponzoni, M. (1998). Human neuroblastoma cells produce extracellular matrix-degrading enzymes, induce endothelial cell proliferation and are angiogenic *in vivo*. Int J Cancer 77:449–454.

Risau, W. (1998). Development and differentiation of endothelium. Kidney Int 54:S3–S6.

Risau, W., Dingler, A., Albrecht, U., Dehouck, M.P., Cecchelli, R. (1992). Blood-brain barrier pericytes are the main source of gamma-glutamyltranspeptidase activity in brain capillaries. J Neurochem 58:667–672.

Risau, W., Esser, S., Engelhardt, B. (1998). Differentiation of blood-brain barrier endothelial cells. Pathol Biol 46:171–175.

Risau, W., Hallmann, R., Albrecht, U. (1986). Differentiation-dependent expression of proteins in brain endothelium during development of the blood-brain barrier. Dev Biol 117:537–545.

Rius, C., Smith, J.D., Almendro, N., et al. (1998). Cloning of the promoter region of human endoglin, the target gene for hereditary hemorrhagic telangiectasia type 1. Blood 92:4677–4690.

Roncali, L., Virgintino, D., Coltey, P., et al. (1996). Morphological aspects of the vascularization in intraventricular neural transplants from embryo to embryo. Anat Embryol 193:191–203.

Rosenstein, J.M., Mani, N., Silverman, W.F., Krum, J.M. (1998). Patterns of brain angiogenesis after vascular endothelial growth factor administration *in vitro* and *in vivo*. Proc Natl Acad Sci USA 95:7086–7091.

Rubin, L.L. (1992). Endothelial cells: adhesion and tight junctions. Curr Opin Cell Biol 4:830–833.

Rubin, L.L., Staddon, J.M. (1999). The cell biology of the blood-brain barrier. Ann Rev Neurosci 22:11–28.

Schlatter, P., Konig, M.F., Karlsson, L.M., Burri, P.H. (1997). Quantitative study of intussusceptive capillary growth in the chorioallantoic membrane (CAM) of the chicken embryo. Microvasc Res 54:65–73.

Schlosshauer, B., Bauch, H., Frank, R. (1995). Neurothelin: amino acid sequence, cell surface dynamics and actin colocalization. Eur J Cell Biol 68:159–166.

Schumacher, U., Møllgard, K. (1997). The multidrug-resistance P-glycoprotein (Pgp, MDR1) is an early marker of blood-brain barrier development in the microvessels of the developing human brain. Anat Embryol 108:179–182.

Seiffert, D., Iruela-Arispe, M.L., Sage, E.H., Loskutoff, D.J. (1995). Distribution of vitronectin mRNA during murine development. Dev Dyn 203:71–79.

Seulberger, H., Lottspeich, F., Risau, W. (1990). The inducible blood-brain barrier specific molecule HT7 is a novel immunoglobulin-like cell surface glycoprotein. EMBO J 9:2151–2158.

Seulberger, H., Unger, C.M., Risau, W. (1992). HT7, neurothelin, basigin, gp42 and OX-47—many names for one developmentally regulated immuno-globulin-like surface glycoprotein on blood-brain barrier endothelium, epithelial tissue barriers and neurons. Neurosci Lett 140:93–97.

Shah, N.M., Groves, A.K., Anderson, D.J. (1996). Alternative neural crest cell fates are instructively promoted by TGFβ superfamily members. Cell 85:331–343.

Soker, S., Takashima, S., Miao, H.Q., Neufeld, G., Klagsbrun, M. (1998). Neuropilin-1 is expressed by endothelial and tumor cells as an isoform-specific receptor for vascular endothelial growth factor. Cell 92:735–745.

Soldi, R., Mitola, S., Strasly, M., Defilippi, P., Tarone, G., Bussolino, F. (1999). Role of $\alpha_v\beta_3$ integrin in the activation of vascular endothelial growth factor receptor-2. EMBO J 18: 882–892.

Sondell, M., Kanje, M. (2001). Postnatal expression of VEGF and its receptor flk-1 in peripheral ganglia. Neuroreport 22:105–108.

Sondell, M., Lundborg, G., Kanje, M. (1999). Vascular endothelial growth factor has neurotrophic activity and stimulates axonal outgrowth, enhancing cell survival and Schwann cell proliferation in the peripheral nervous system. J Neurosci 19:5731–5740.

Sondell, M., Sundler, F., Kanje, M. (2000). Vascular endothelial growth factor is a neurotrophic factor which stimulates axonal outgrowth through the flk-1 receptor. Eur J Neurosci 12:4243–4254.

Sosic, D., Brand-Saberi, B., Schmidt, C., Christ, B., Olson, E.N. (1997). Regulation of paraxis expression and somite formation by ectoderm- and neural tube-derived signals. Dev Biol 185:229–243.

Staddon, J.M., Rubin, L.L. (1996). Cell adhesion, cell junctions and the blood-brain barrier. Curr Opin Neurobiol 6:622–627.

Stein, I., Neeman, M., Shweiki, D., Itin, A., Keshet, E. (1995). Stabilization of vascular endothelial growth factor mRNA by hypoxia and hypoglycemia and coregulation with other ischemia-induced genes. Mol Cell Biol 15:5363–5368.

Steinbrech, D.S., Longaker, M.T., Mehrara, B.J., et al. (1999). Fibroblast response to hypoxia: the relationship between angiogenesis and matrix regulation. J Surg Res 84: 127–133.

Stewart, P.A., Wiley, M.J. (1981). Developing nervous tissue induces formation of blood-brain barrier characteristics in invading endothelial cells: a study using quail-chick transplantation chimeras. Dev Biol 84:183–192.

Stone, J., Itin, A., Alon, T., et al. (1995). Development of retinal vasculature is mediated by hypoxia-induced vascular endothelial growth factor (VEGF) expression by neuroglia. J Neurosci 15:4738–4747.

Streit, W.J. (1993). Microglial-neuronal interactions. J Chem Neuroanat 6:261–266.

Streit, W.J. (1996). The role of microglia in brain injury. Neurotoxicology 17:671–678.

Streit, W.J., Graeber, M.B. (1993). Heterogeneity of microglial and perivascular cell populations: insights gained from the facial nucleus paradigm. Glia 7:68–74.

Strong, L.H. (1961). The first appearance of vessels within the spinal cord of the mammal: their developing patterns as far as partial formation of the dorsal septum. Acta Anat 44:80–108.

Strong, L.H. (1964). The early embryonic pattern of internal vascularization of the mammalian cerebral cortex. J Comp Neurol 123:121–138.

Sunderkötter, C., Steinbrink, K., Goebeler, M., Bhardwaj, R., Sorg, C. (1994). Macrophages and angiogenesis. J Leukoc Biol 55:410–422.

Suri, C., Jones, P.F., Patan, S., et al. (1996). Requisite role of angiopoietin-1, a ligand for the TIE2 receptor, during embryonic angiogenesis. Cell 87:1171–1180.

Suri, C., McClain, J., Thurston, G., et al. (1998). Increased vascularization in mice over-expressing angiopoietin-1. Science 282:468–471.

Suzuki, T., Ogata, A., Tashiro, K., Nagashima, K., Tamura, M., Nishihira, J. (1999). Augmented expression of macrophage inhibitory factor MIF in the telencephalon of the developing rat brain. Brain Res 816:457–462.

Teillet, M.A., Kalcheim, C., Le Douarin, N.M. (1987). Formation of the dorsal root ganglia in the avian embryo: segmental origin and migratory behavior of neural crest progenitor cells. Dev Biol 120:329–347.

Topel, I., Stanarius, A., Wolf, G. (1998). Distribution of the endothelial constitutive nitric oxide synthase in the developing rat brain: an immunohistochemical study. Brain Res 788:43–48.

Tsurumi, Y., Murohara, T., Krasinski, K., et al. (1997). Reciprocal relation between VEGF and NO in the regulation of endothelial integrity. Nature Med 3:879–886.

Turley, E.A., Hossain, M.Z., Sorokan, T., Jordan, L.M., Nagy, J.I. (1994). Astrocyte and microglial motility *in vitro* is functionally dependent on the hyaluronan receptor RHAMM. Glia 12:68–80.

van der Zee, R., Murohara, T., Luo, Z., et al. (1997). Vascular endothelial growth factor/vascular permeability factor augments nitric oxide release from quiescent rabbit and human vascular endothelium. Circulation 95:1030–1037.

Vaquero, J., Zurita, M., de Oya, S., Coca, S. (1999). Vascular endothelial growth permeability factor in spinal cord injury. J Neurosurg 90:220–223.

Virgintino, D., Maiorano, E., Bertossi, M., Pollice, L., Ambrosi, G., Roncali, L. (1993). Vimentin- and GFAP-immunoreactivity in developing and mature neural microvessels. Study in the chicken tectum and cerebellum. Eur J Histochem 37:353–362.

Virgintino, D., Robertson, D., Monaghan, P., et al. (1998). Glucose transporter GLUT1 localization in human foetus telencephalon. Neurosci Lett 256:147–150.

Wamil, A.W., Wamil, B.D., Hellerqvist, C.G. (1998). CM101-mediated recovery of walking ability in adult mice paralyzed by spinal cord injury. Proc Natl Acad Sci USA 95: 13188–13193.

Wang, H., Keiser, J.A. (1998). Vascular endothelial growth factor upregulates the expression of matrix metalloproteinases in vascular smooth muscle cells—role of flt-1. Circ Res 83:832–840.

Weinstein, B.M. (1999). What guides early embryonic blood vessel formation? Dev Dyn 215:2–11.

Wekerle, H., Engelhardt, B., Risau, W., Meyermann, R. (1991). Interaction of T lymphocytes with cerebral endothelial cells in vitro. Brain Pathol 1:107–114.

Wilms, P., Christ, B., Wilting, J., Wachtler, F. (1991). Distribution and migration of angiogenic cells from grafted avascular intraembryonic mesoderm. Anat Embryol 183:371–377.

Wilting, J., Brand, S., Huang, R., et al. (1995a). Angiogenic potential of the avian somite. Dev Dyn 202:165–171.

Wilting, J., Brand, S., Kurz, H., Christ, B. (1995b). Development of the embryonic vascular system. Cell Mol Biol Res 41:219–232.

Wilting, J., Christ, B. (1989). An experimental and ultrastructural study on the development of the avian choroid plexus. Cell Tissue Res 255:487–494.

Wilting, J., Kurz, H., Oh, S.J., Christ, B. (1998). Angiogenesis and lymphangiogenesis: analogous mechanisms and homologous growth factors. In: Little, C.D., Mironov, V., Sage, E.H., eds. Vascular morphogenesis: in vivo, in vitro, in mente, pp. 21–34. Boston: Birkhäuser.

Wizigmann-Voos, S., Plate, K.H. (1996). Pathology, genetics and cell biology of hemangioblastomas. Histol Histopathol 11:1049–1061.

Yablonka-Reuveni, Z., Christ, B., Benson, J.M. (1998). Transitions in cell organization and in expression of contractile and extracellular matrix proteins during development of chicken aortic smooth muscle: evidence for a complex spatial and temporal differentiation program. Anat Embryol 197:421–437.

Yablonka-Reuveni, Z., Schwartz, S.M., Christ, B. (1995). Development of chicken aortic smooth muscle: expression of cytoskeletal and basement membrane proteins defines two distinct cell phenotypes emerging from a common lineage. Cell Mol Biol Res 41:241–249.

Yamada, N., Kato, M., Yamashita, H., et al. (1995). Enhanced expression of transforming growth factor-β and its type-I and type-II receptors in human glioblastoma. Int J Cancer 62:386–392.

Yamagata, M., Herman, J.P., Sanes, J.R. (1995). Lamina-specific expression of adhesion molecules in developing chick optic tectum. J Neurosci 15:4556–4571.

Yamagishi, H., Olson, E.N., Srivastava, D. (2000). The basic helix-loop-helix transcription factor, dHAND, is required for vascular development. J Clin Invest 105:261–270.

Yamashita, J., Itoh, H., Hirashima, M., Minetaro, O., Nishikawa, S., Yurugi, T., Naito, M., Nakao, K., Nishikawa, S.I. (2000). Flkl-positive cells derived from embryonic stem cells serve as vascular progenitors. Nature 408:92–96.

Young, H.E., Mancini, R.P., Wright, J.C., Black, A.C., Reagan, C.R., Lucas, P.A. (1995). Mesenchymal stem cells reside within the connective tissues of many organs. Dev Dyn 202:137–144.

Zerlin, M., Goldman, J.E. (1997). Interactions between glial progenitors and blood vessels during early postnatal corticogenesis: blood vessel contact represents an early stage of astrocyte differentiation. J Comp Neurol 387:537–546.

Zhang, Z., Galileo, D.S. (1998). Widespread programmed cell death in early developing chick optic tectum. Neuroreport 9:2797–2801.

Zhao, L.M., Zhang, M.M., Ng, K.Y. (1998). Effects of vascular permeability-factor on the permeability of cultured endothelial cells from brain capillaries. J Cardiovasc Pharmacol 32:1–4.

Zucker, S., Mirza, H., Conner, C.E., et al. (1998). Vascular endothelial growth factor induces tissue factor and matrix metalloproteinase production in endothelial cells: conversion of prothrombin to thrombin results in progelatinase A activation and cell proliferation. Int J Cancer 75:780–786.

Zuo, J., Neubauer, D., Dyess, K., Ferguson, T.A., Muir, D. (1998). Degradation of chondroitin sulfate proteoglycan enhances the neurite-promoting potential of spinal cord tissue. Exp Neurol 154:654–662.

CHAPTER 9

Vascular Development of the Kidney

R. Ariel Gómez and María Luisa S. Sequeira López

The kidney is a highly vascularized organ that in the normal adult human receives 20% of the cardiac output. Appropriate spatial arrangement and interaction of the kidney microvasculature with the various parts of the nephron is crucial for the regulation of specialized regional functions such as glomerular filtration rate, medullary blood flow, and maintenance of a medullar-concentrating gradient. Thus, the proper and timely assembly of vascular units with their respective nephrons is a crucial morphogenetic event leading to the formation of a kidney necessary for independent extrauterine life. This chapter reviews the anatomic development of the kidney vasculature and the mechanisms involved in its differentiation.

ANATOMIC DEVELOPMENT OF THE KIDNEY

Morphogenesis of the kidney is accomplished by the embryonic phylogenetic recapitulation of three successive and overlapping renal structures: the pronephros, the mesonephros, and the metanephros (Fig. 9.1). The *pronephros* is the definitive kidney of amphioxus and cyclostomes. In humans and others mammals, the pronephros is a transient structure without excretory capacity. It appears in the third week of gestation, in the cervical region, between the second and sixth somite and consists of approximately five to seven hollow nephric vesicles, each connected to a respective tubule that eventually drain into a pronephric duct. If the pronephros fails to develop, renal agenesis will result, sometimes accompanied by ipsilateral adrenal, gonad, and lung agenesis. The pronephric duct becomes continuous with the mesonephric duct, also called the wolffian duct.

The *mesonephros* is the definitive kidney of fish and amphibians. In humans its excretory function is transient. Approximately 40 to 42 (meso)nephrons drain into (and are probably induced by) the mesonephric duct. However, at any given time only 30 to 32 mesonephroi are observed due to progressive cranial degeneration as the caudal mesonephros develops. The mesonephros regresses in humans except for a few cranial tubules, the ductuli efferentia that drain the testis into the mesonephric duct. The mesonephric duct becomes the vas deferens, the seminal vesicle, and portions of the epididymis. In females, the mesonephric duct regresses.

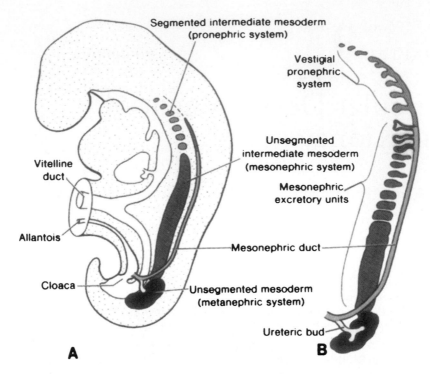

FIGURE 9.1.   A: The relationship of the intermediate mesoderm of the pronephric, mesonephric, and metanephric systems. In cervical and upper thoracic regions, the intermediate mesoderm is segmented; in lower thoracic, lumbar, and sacral regions, it forms a solid, unsegmented mass of tissue, the nephrogenic cord. Note the longitudinal collecting duct, formed initially by the pronephros but later by the mesonephros. B: Schematic representation of excretory tubules of the pronephric and mesonephric systems in a 5-week-old embryo. (From Sadler,[38] with permission.)

Occasional vestigial remnants of it constitute the epoophoron, the oophoron, and Gartner's duct. In females, a normal mesonephric duct is necessary for the development of the müllerian ducts; inadequate mesonephric development results in ipsilateral ureteral-renal agenesis, agenesis of the fallopian tube and contralateral uterus unicornis, and vaginal atresia.

The mesonephric duct also gives origin to the ureteric bud, which appears at 28 days of gestation in a position close to the place where the duct enters the cloaca. Interaction of the ureteric bud with the metanephric blastema around 32 days of gestation initiates a cascade of events resulting in the formation of the *metanephros*, the definitive kidney of mammals, reptiles, and birds.

The ureteric bud induces the mesenchyme to differentiate into tubular and glomerular epithelium (Fig. 9.2). In turn, the induced mesenchyme induces the ureter to grow and branch into the renal mesenchyme. Around the tip of each ureteric branch, the mesenchymal cells group tightly together, a process called condensation; this is followed by the formation of a vesicle that evolves into a comma and then an S-shaped glomerulus. The lower cleft of the S-shaped body is penetrated by mesangial and endothelial cell precursors, which eventually differentiate to give origin to the glomerular capillaries and the glomerular mesangium. In humans, glomerulogenesis is complete by 34 to 35 weeks of gestation. In rats and

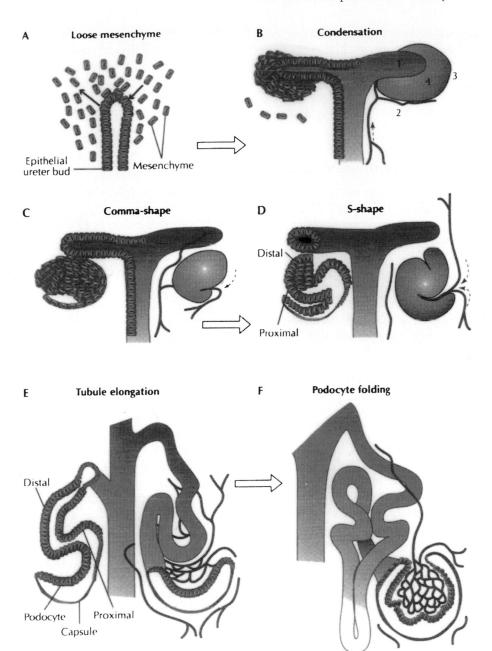

FIGURE 9.2.   Nephrogenesis. A: The branching ureteric epithelium interaction with loose metanephric mesenchyme results in condensation of the mesenchyme (B). Cell lineages shown include (1) ureteric epithelium, (2) vasculature, (3) undifferentiated mesenchyme, and (4) condensed mesenchyme differentiating into epithelia. C and D: These stages are followed by infolding of the primitive glomerular epithelium to form comma- and S-shaped bodies. E: Elongation of the proximal and distal tubular elements subsequently occurs along with further infolding of the glomerular epithelium and vascular structures to form the mature glomerular capillary network (F). The initial phases of glomerular vascularization are believed to occur during the early stages of glomerular differentiation in C and D. (From Ekblom,[39] with permission.)

mice, glomerulogenesis continues for at least 7 to 10 days of postnatal life. Because of this and because the kidney matures in a centrifugal fashion (i.e., juxtamedullary glomeruli develop and mature earlier than outer cortical glomeruli), a section through the kidney of a newborn rat shows the nephron at various stages of development. This experiment of nature has facilitated the use of perinatal rats and mice as models to study the effects of experimental manipulations on nephrovascular development.

Molecular Events in Nephrogenesis

More than 300 genes that are potentially involved in nephrogenesis have been identified. Rather than a single "master" gene, it is likely that the timely combination of multiple genes acting in concert is responsible for most events in kidney development. The responsible cascades of gene activation have yet to be identified. Because of space limitations, only a few genes known to be crucial for kidney morphogenesis will be described here. For more details, the reader may consult the kidney development Web site http://golgi.ana.ed.ac.UK/kidhome.html developed by Davies and Brandli.[1,2]

A crucial event in kidney morphogenesis is the budding of the ureter and its co-inductive interactions with the metanephric mesenchyme.

Budding of the Ureter

This process is regulated by several key molecules including glial cell line–derived neurotropic factor (GDNF) and its receptor, the c-Ret receptor/GDNFR-α receptor molecule, WT-1 and Pax-2. The uninduced mesenchyme secretes the ligand GDNF, a secreted glycoprotein that phosphorylates the c-Ret receptor (a tyrosine kinase receptor) located on the surface of the mesonephric duct, resulting in budding of the ureter[1,2] (Fig. 9.3). It seems that high-affinity binding to the receptor is facilitated by GDNF first binding another receptor molecule called GDNFR-α. In addition to elongation of the ureter, branching within the metanephric mesenchyme is also regulated by the Ret/GDNF system. As development proceeds, GDNF expression is limited to regions of uninduced mesenchyme[3-5] (Fig. 9.3). In correspondence, the receptor is localized to the tips of the branching ureteric tips. Thus, the accurate spatial distribution of these two molecules determines appropriate branching of the ureteric bud and (as described below) appropriate differentiation of the metanephric mesenchyme. Inactivation of c-Ret by gene targeting results in renal agenesis or hypodysplasia. Interestingly, *in vitro* experiments with cultured metanephric mesenchyme from homozygous null mutant mice undergo proper nephrogenesis when induced with wild-type ureter. However, the ureter from Ret-deficient mice cannot induce nephrogenesis of wild-type metanephric mesenchyme.[3] GDNF expression is repressed after conversion of the metanephric mesenchyme to epithelium, limiting excessive, aberrant branching morphogenesis. GDNF knockout mice are incapable of forming ureteric buds; the metanephric mesenchyme is not induced to differentiate and it undergoes apoptosis, and the kidneys do not develop.

Thus, the GDNF/c-Ret system plays a major role in the budding of the primitive ureter from the mesonephric duct and its further branching within the

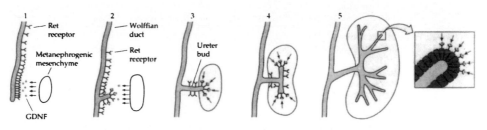

FIGURE 9.3. Ureteric bud growth is dependent on glial cell line–derived neurotropic factor (GDNF) and its receptor. GDNF is secreted by the uninduced metanephric mesenchyme and stimulates the Ret receptor on the mesonephric duct. In response, a ureteric bud is formed. As development proceeds, GDNF expression is limited to regions of uninduced mesenchyme. In parallel, the Ret receptor localizes to the tips of the ureteric buds. (From Gilbert,[40] based on Schuchardt,[41] with permission.)

metanephric mesenchyme. The factors that regulate the spatial restriction of expression of this system's components need to be determined, in particular the restricted and unique expression of GDNF to the metanephric mesenchyme. Elucidation of the molecular mechanisms governing GDNF expression may help understand congenital renal anomalies such as renal ectopy, horseshoe kidney, and supernumerary kidneys.

The WT-1 transcription factor is the product of the Wilms' tumor suppressor gene and is involved in the regulation of signals that control ureteric budding from the mesonephric duct. Although it is expressed by the undifferentiated-uninduced mesenchyme, it is upregulated in the induced mesenchyme.[4] WT-1 is necessary for ureteric bud formation; WT-1 −/− mice do develop ureters.[5] Mutant metanephric mesenchyme cannot be induced by the spinal cord, a strong inducer of mesenchymal differentiation. The latter experiment suggests that WT-1 is involved in the ability of the mesenchyme to maintain its capability to differentiate in response to known inducers of nephrogenesis. WT-1 interacts with many genes involved in cell growth including *Pax-2*.

The *Pax* genes encode for transcription factors having a DNA-binding domain called the paired box. *Pax-2* is expressed in the mouse mesonephric duct and ureteric epithelium at 12 days of embryonic life (E12). It also becomes expressed in the mesenchyme at the beginning of induction. However, it is downregulated as the nephron differentiate. *Pax-2* is necessary for mesonephric duct survival. In *Pax-2* null mutant mice, the mesonephros degenerates between E10 and E12. As a consequence, the mesonephros, the müllerian ducts, the ureters, and the kidneys do not form. In humans, a mutation of the *Pax-2* gene leads to a syndrome encompassing colobomas, vesicoureteral reflux, and renal dysplasia.[6]

Emx 2, a homeobox-containing gene, is the mouse homolog of *empty spiracles*, which in *Drosophila* regulates morphogenesis of the central nervous system. In mice, *Emx 2* seems to be crucial for the development of the urogenital tract. In *Emx 2* mutant mice, the ureteric bud forms and contacts the metanephric mesenchyme. However, further induction and branching morphogenesis does not occur. As a consequence, the kidneys, the urinary tract, and the gonads do not develop.[7] Another interesting transcription factor is BF2, a winged helix protein,

TABLE 9.1. Factors that contribute to nephrogenesis.

GDNF/c-Ret system
  Transcription factors
    WT-1
    Pax-2
    Emx 2
    BF2
    HNF-3
  Growth factors
    EGF
    HGF
    HDGF
    bFGF
    TGF-β
    BMP-2, BMP-7
    PDGF-B
    IGF-1, IGF-2
  Proteases
    Matrix metalloproteinases (MMP)
    MMP-2 (gelatinase A)
    MMP-9
  Extracellular matrix components
    Collagens
    Fibronectin
    Sulfated proteoglycans (syndecan 1)
    Glycoproteins (Wnts, laminins)
    Integrins ($\alpha_8\beta_1$, $\alpha_3\alpha_1$)

GDNF, glial cell line–derived neurotrophic factor; WT-1, Wilms' tumor suppressor gene type 1; Pax-2, paired box 2; Emx 2, homologues of *Drosophila* empty spiracles 2; BF2, brain factor 2; HNF-3, hepatocyte nuclear factor 3; EGF, epidermal growth factor; HGF, hepatocyte growth factor; HDGF, hepatoma-derived growth factor; bFGF, basic fibroblast growth factor; TGF-β, transforming growth factor-β; BMP, bone morphogenetic protein; PDGF-B, platelet derived growth factor B; IGF, insulin growth factor; Wnts, homologues of *Drosophila* morphogen wingles.

present in embryonic stromal cells where it is involved in mesenchymal to epithelial conversion and branching of the ureter within the metanephric mesenchyme.[8]

A variety of factors contribute to the process of nephrogenesis (Table 9.1). These include numerous growth factors, proteases, and components of the extracellular matrix.[9–11] Due to space limitations, these factors are described only briefly here. Outstanding reviews on the role of these factors on kidney morphogenesis have been published elsewhere.[9,10] Deficiency of retinoic acid or retinoic acid receptors result in renal hypoplasia or renal agenesis.[12] Epidermal growth factor (EGF), hepatocyte growth factor (HGF), fibroblast growth factor (FGF), and transforming growth factor-β (TGF-β) are involved in the regulation of ureteric branching and tubulogenesis. According to Sakurai and Nigam,[10] a balance between facilitating growth factors (EGF, HGF insulin-like growth Factor (IGF)) and inhibitory factors (TGF-β) is needed to achieve appropriate branching morphogenesis.

It is likely that the extracellular matrix (ECM) is an important determinant of the regional growth of specific parts of the nephron and therefore of the final architecture and shape of the developing kidney. Unique combinations of ECM elements are likely to be more important than single molecules. Collagens, fibronectin, sulfated proteoglycans such as syndecan 1, and matrix glycoproteins such as Wnts and laminins all contribute to create the appropriate environment conducive to branching morphogenesis.

Remodeling of the ECM is probably fundamental for the conversion of mesenchyme to epithelium and other aspects of nephrogenesis. Remodeling is mediated by a group of matrix metalloproteinases (MMPs), which help promote branching by degrading ECM at the appropriate sites. Tissue inhibitors of MMPs (TIMPS) exert presumably a balancing role by counteracting tissue degradation, also at the appropriate sites during nephrogenesis. At 11 days of gestation, the mouse kidney anlagen express MMP-2 (gelatinase A) and MMP-9. Antibodies against MMP-9 or application of TIMP-1 inhibited *in vitro* branching morphogenesis. Thus, it seems likely that components of the ECM and appropriate remodeling of the ECM contribute to the shape of the organ. Further work is necessary in this important area of research.

DEVELOPMENT OF THE KIDNEY VASCULATURE

Nephrogenesis is linked and occurs simultaneously with the development of the kidney vasculature. Until recently, however, the information available was limited to the work of a few pioneer laboratories. The discovery of key molecules [e.g., vascular endothelial growth factor (VEGF), translocation-Ets-leukemia (TEL), etc.] involved in vascular development in other systems coupled with the availability of new reagents (antibodies, probes) and models (knockout mice) made possible in the last few years an initiation of a more thorough analysis of the events involved in the development of blood vessels of the kidney. Most of our recent knowledge of kidney vascular development comes from studies performed in mice or rats. When known or relevant, information regarding the human species is also discussed here.

Anatomic Development of Kidney Arterioles

In the rat kidney, the first identified arterioles appear around 15 to 16 days of gestation. In the mouse kidney, the first arterioles are seen at 14 to 15 days of gestation. By 20 days of gestation in the rat (1 to 2 days before birth) the basic blueprint of arteriolar development has been established: usually, a single renal artery divides at the level of the kidney hilum into three major lobar arteries. Each lobar artery divides into dorsal and ventral branches, running through the corticomedullary junction paralleling the surface of the kidney. These arteries course the corticomedullary region in an arc-like pattern, receiving the name of arcuate arteries. Each arcuate artery gives numerous branches called interlobular or corticoradial arterioles. The interlobular arterioles divide numerous times into the afferent arterioles of the glomeruli. The afferent arterioles are continued by the glomerular capillaries. Blood circulating through these capillaries leaves the glomeruli through the efferent arterioles, which eventually branch out into the postglomerular capillaries and vasa recta (Fig. 9.4). As mentioned above, the basic organization and major branches of the preglomerular tree are established a few days before birth. However, most of the growth of the arteriolar tree occurs during postnatal life.

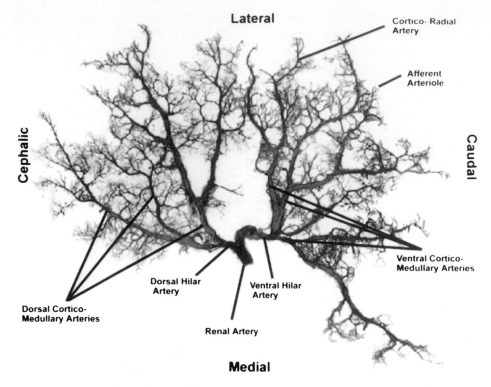

FIGURE 9.4.   Preglomerular arterial tree of the adult rat kidney. The specimen was obtained from a 90-day-old rat after exposure to hydrochloric acid and subsequent microdissection. (Method is described elsewhere.[36])

In correspondence with postnatal glomerulogenesis (more than two-thirds of glomeruli form during the newborn period), there is a marked increase in complexity and surface area of the preglomerular vasculature. This increase in surface area is achieved mainly by an increase in branching and secondarily by an increase in the thickness and elongation of individual vascular segments. Thus, during the first 10 days of postnatal life, the afferent and interlobular arteries increase in number, likely to accommodate the demand for vascularization of new nephrons. The mechanisms that control branching of the kidney vasculature are beginning to be recognized and are described below. In addition to branching, the vasculature must assemble at the right time and supply the appropriate nephrons for a successful organ to be formed. How the vasculature achieves its proper directionality is beginning to be understood and is discussed below.

Origin of the Kidney Vasculature: Angiogenesis versus Vasculogenesis

Angiogenesis is the sprouting of new capillaries from preexisting blood vessels.[13] The process involves dissolution of the basement membrane, remodeling of the extracellular matrix, endothelial cell migration and proliferation, endothelial tube formation, and recruitment of perivascular cells. Angiogenesis is observed during organogenesis and accompanying normal and abnormal tissue growth.[13] In normal adults, vascular turnover is slow, and therefore angiogenesis occurs in a few instances, such as in the uterus during the menstrual period, and during wound healing and exercise-induced muscle growth.[13] In these cases, angiogenesis is activated for brief periods. On the other hand, in pathologic situations, uncontrolled angiogenesis ensues. Examples include tumor growth, rheumatoid arthritis, psoriasis, and retinopathy associated with diabetes or prematurity.

Vasculogenesis is the formation of blood vessels from endothelial cell precursors such as angioblasts. These latter cells differentiate *in situ* and form vascular channels, which after remodeling result in the formation of the appropriate blood vessel(s). Vasculogenesis is common during embryogenesis and in organs of endodermic origin such as the lung. The relative contribution of vasculogenesis versus angiogenesis to the formation of specific blood vessels within a given organ is unknown. According to Noden,[14] regardless of their origin within the embryo, angioblasts will form vascular channels that are appropriate for the site where they have been transplanted. These experiments indicate that the control of vessel differentiation and assembly is determined in great part by the tissue mesenchyme rather than by the endothelial cell per se. As described below, this seems to be the case with the kidney as well.

Both processes, angiogenesis and vasculogenesis, have been postulated as responsible for the formation of the renal vessels. Until recently, it was accepted that the renal vessels originated from branching of preexisting extrarenal vessels (angiogenesis).[15] This hypothesis was based in interspecies grafting experiments whereby undifferentiated mouse embryonic kidneys were grafted onto quail or chicken chorioallantoic membrane. By 7 days, the grafted kidneys developed glomerular structures containing quail endothelial cells, recognizable by their characteristically dark nuclei and prominent nucleoli. Based on these experiments, it was concluded that the chorioallantoid membrane had "vascularized" the mouse glomeruli. It was then interpreted that vascularization of the kidney occurred via sprouting of capillaries, that is angiogenesis, rather than *in situ* vessel formation, vasculogenesis. In this paradigm, the vasculature of the kidney is thought to derive from endothelial cells that invade the organ early during nephrogenesis. From this concept, it was extrapolated that vascularization of the kidney was initiated by a pair of vessels originated from the embryonic aorta.[16] Angiogenic sprouting of those branches, presumably following the ureter, at least initially, resulted in vascularization of the kidney. As the ureter induced differentiation of the glomeruli, sprouting endothelial cells from the invading vessels penetrated the S-shaped bodies and generated the glomerular capillaries. Those interspecies experiments, however, resulted in few glomeruli, and extrarenal vascularization was probably more the exception than the rule. The experiments did not consider the possibility that vascular precursors were present in the metanephric mesenchyme before vessel formation has taken place. In fact, recent experiments challenge the exclusive, angiogenic/extrarenal hypothesis for the origin of the kidney vessels.

Using specific endothelial cells markers, several laboratories have identified vascular precursors in the embryonic kidney before vessels can be detected morphologically.[17,18] Using specific antibodies to Flk-1, a receptor for VEGF present in endothelial cell precursors such as angioblasts and differentiated endothelial cells, several investigators demonstrated the presence of endothelial cell precursors in the prevascular kidney of rats.[17,19] Similar observations have been made in other species, including mice.[20] Others have found capillary precursors in areas of the developing rabbit kidney previously thought to be avascular.[21] Those investigators suggested that the differentiating glomerulus is surrounded by a network of vessel-like structures that eventually connect with differentiated vessels. All of these findings support the pioneering work of Kazimierczak,[22] who described the presence of vascular elements in early stages of kidney development.

In addition to angioblasts, smooth muscle cell progenitors and renin-expressing cell precursors are found in the prevascular metanephric blastema of mice and rats.[18] As discussed below, these progenitors have the potential to, and do, contribute to the formation of kidney arterioles.[18]

These findings suggest that the aforementioned precursor cells have the capability to form the kidney vasculature *in situ* by the process of vasculogenesis. Support for this hypothesis was obtained from several independent laboratories. Abrahamson and colleagues[23] implanted wild-type mouse embryonic kidneys at 11 to 12 days of gestation into the anterior eye chamber of ROSA 26 mice which express β-galactosidase in every cell type. Upon the X gal reaction, cells harboring β-galactosidase turn blue. Thus, after 5 to 7 days *in oculo,* the grafted kidneys underwent nephrogenesis and became vascularized. Using this strategy, the investigators demonstrated that the majority of vascular cells originated from the donor embryonic kidney.

Similar results were obtained when the grafts were placed under the kidney capsule of adult ROSA 26 mice. Transplantation of embryonic kidneys expressing β-galactosidase in endothelial cells (under the control of the Flk-1 receptor locus) yielded similar conclusions. Those experiments as well as recent experiments from our laboratory support the conclusion that not only endothelial but also smooth muscle and juxtaglomerular cells originate *in situ* by a vasculogenetic process.[18] Interestingly, if the host kidney is from a newborn mouse, a time when vascular and nephron development is still ongoing, chimerical vessels are formed containing both donor and host endothelial cells. These results suggest that vascular precursors still present in the neonatal cortex of recipient mice have the potential to differentiate and assemble into the microvasculature of the developing kidney.[23] Recently, Tufro and Gomez[24] cultured avascular metanephric kidneys from rat embryos (E14) on top of blue murine glomerular endothelial cells (MGECs) from Flk-1 +/− mice, and demonstrated that MGECs invaded the metanephric organ and formed capillary-like structures surrounding and within forming nephrons. This process was amplified by low oxygen (3% O_2) and prevented by anti-VEGF neutralizing antibodies, suggesting that VEGF produced by the differentiating nephrons acts as a chemoattractant spatially directing developing capillaries during metanephric development *in vitro*. Although these experiments were performed *in vitro*, they suggest that angiogenesis is also possible under certain experimental conditions. Furthermore, it seems that, depending on the developmental potential of the cells involved, both angiogenesis and vasculogenesis are possible in the development of the kidney vasculature. The contribution of each process in the undisturbed embryonic kidney remains to be determined. It is likely, however, that smaller arterioles and capillaries are formed by vasculogenesis, whereas larger arteries may develop by angiogenesis. This hypothesis needs to be tested.

A

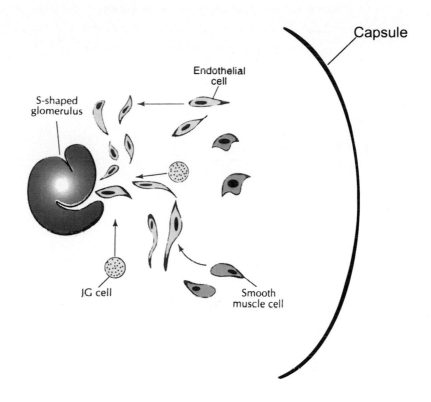

B

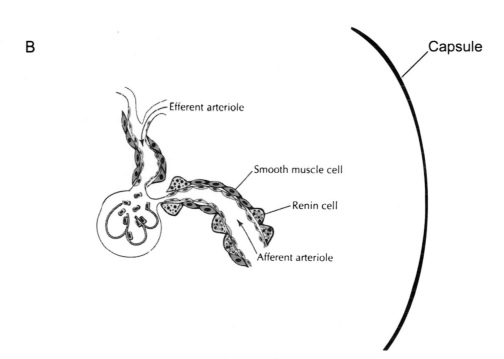

TABLE 9.2. Major factors implicated in kidney
vascularization.

Factors	Receptors
VEGF	Flk-1
	Flt-1
Ephrins A 1–2	Eph
B	
PDGF-B	PDGFR-β
Ets family (Ets-1, TEL)	
Angiopoietins 1–2	Tie-2 (Tek)
Renin-angiotensin system	

PDGF, platelet-derived growth factor; TEL, translocation-Ets-
leukemia; VEGF, vascular endothelial growth factor.

Our conceptualization of both vasculogenesis and angiogenesis in the developing kidney is shown in Figs. 9.5 and 9.6.

Regulation of Vascular Development

The mechanisms involved in the differentiation of the kidney vasculature are becoming clearer. Selected molecules thought to participate in vascular growth are listed in Table 9.2. VEGF and its receptors (Flk-1, Flt-1) have been shown to participate in the development of the glomerular capillaries.[19,25,26]

Flt-1 and Flk-1 are present in angioblast of the embryonic rat kidney (E14), before vessel formation has taken place. Angioblasts seem to form cords prior to their differentiation into endothelial cells. At 19 days of gestation, VEGF is detected in glomerular epithelial and tubular cells, and both receptors are expressed in contiguous endothelial cells. If embryonic kidneys at the prevascular stage are cultured for 5 to 6 days on a chemically defined medium without serum, the metanephric explants undergo nephrogenesis but do not develop their vasculature. However, if the same explants are treated with recombinant human VEGF, the cultured kidneys expand their endothelial cell mass, probably as a result of vasculogenesis.[25] These data suggest that VEGF plays a crucial role in kidney vascularization by promoting endothelial cell differentiation, capillary formation, and maintenance of fenestrated endothelium phenotype. This is accompanied by a corresponding increase in tubular cell mass.[25] Interestingly, the growth of kidney vessels seems to depend on the regulation of VEGF by the tissue oxygen concentration.[25] As mentioned above, when embryonic kidneys are cultured at the usual atmospheric oxygen concentration, vessels do not develop or do so very poorly. However, if the embryonic, prevascular kidneys are cultured in an atmosphere containing 5% oxygen, capillaries develop within and outside the glomerulus. This effect is inhibited by anti-VEGF antibodies. The signal mediating this response seems to involve an increase in VEGF synthesis, which in turn induces angioblasts to differentiate, proliferate, and assemble into endothelial tubes.

Ephrins are glycosylphosphatidylinositol-anchored proteins located either on the cell surfaces (Ephrin-A) or across the cell membrane (Ephrin-B).[27] These membrane-bound ligands bind receptors of the tyrosine kinase family called Eph (erythropoietin-producing hepatocellular receptor), which have an extremely restricted distribution and specificity. In the developing neonatal kidney, EphB1 colocalizes with the Flk-1 receptor to vascular glomerular clefts, forming glomeruli, vessels, and mesenchymal cells. In the central nervous system, the ephrin system regulates axonal targeting. Ephrin-B1 induces renal microvascular cells to assemble into capillaries *in vitro*. Ephrin-A1 is angiogenic in the rabbit cornea and induces chemotactic responses in endothelial cells expressing the EphA2 receptors. This information suggests that ephrins may mediate targeting of endothelial cells after they have acquired their differentiated phenotype.

Another system involved in vascular development is the angiopoietin system, composed of the angiopoietin-1 (Ang-1) that signals through a tyrosine kinase receptor (Tie-2/Tek) expressed only on endothelial cells and early hematopoietic cells, and the Ang-2 that also binds Tie-2 antagonizing Ang-1.[28,29] Tie-1 is an orphan receptor tyrosine kinase that may play a role in metanephric vessel formation. Null mutant mice die in late gestation with impaired vessel integrity.[30,31] Ang-1 and Tie-2 are expressed in glomeruli (Ang-1 was localized in the periphery of the tuft, where podocytes reside, and Tie-2 inside glomerular tufts, where endothelial and mesangial cells reside), and they may be important in the formation and/or maintenance of glomerular capillaries. Tie-2 signaling may also be important for the development of the vasa recta capillaries. Ang 2 has been suggested to act as a natural antagonist of Ang-1 to modulate the formation of the microcirculation.[29] This, however, remains to be confirmed. It has been postulated that Ang-2 has a synergistic activity with VEGF, collaborating with the invading vascular sprouts by blocking a stabilizing or maturing function of Ang-1, whereas in the absence of VEGF, the inhibition of a constitutive Ang-1 signal can contribute to vessel regression.[32]

The Ets family of transcription factors regulate a cascade of genes involved in hematopoiesis (CSF-1, CDD18), renin transcription, endothelial cell differentiation (VEGF, Flk-1, Flt-1), and protease activity (MMP-1, urokinase-plasminogen activator (u-PA)). Severe kidney abnormalities have been described in mice with targeted null mutations of the Ets-1 gene. The abnormalities were observed predominantly in the glomerular capillaries.[33]

Platelet-derived growth factor-B (PDGF-B), a member of the platelet-derived growth factor family, can bind to either the PDGF receptor (PDGFR) α or β. PDGFR-β is a protein tyrosine kinase receptor. Both PDGF-B and PDGFR-β are expressed in the developing human kidney.[34] They are probably involved in mesangial cell differentiation.[35] Mice carrying a null mutation of either PDGF-B or PDGFR-β fail to develop mesangial cells and glomerular capillary tufts.[35]

◀

FIGURE 9.5.   Conceptualization of vasculogenesis in the developing kidney. Nephron vascularization. A: Vascular precursors derived from metanephric mesenchymal cells differentiate *in situ* into endothelial, smooth muscle/mesangial, and renin-producing cells. The lower cleft of the S-shaped glomerulus shown here is being penetrated by endothelial cell precursors that align to form the endothelial lining of the arteriole and also attract smooth muscle, mesangial, and renin cell precursors. These precursors form the coating of the arterioles and the mesangial cells of the glomerulus. Green cells are mesenchymal cells. B: The glomerulus has acquired its arterioles containing smooth muscle and renin-synthesizing cells. Early in gestation, renin-containing cells are abundant along the arterioles. Glomerular capillary loops have also formed and mesangial cells have already penetrated the glomerulus. JG, juxtaglomerular.

A

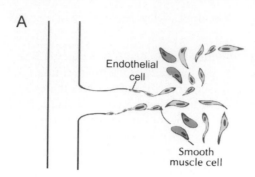

Endothelial
cell

Smooth
muscle cell

FIGURE 9.6. Conceptualization of angiogenesis in the developing kidney. Large kidney arteries. A: Sprouting of endothelial cells from a preexisting vessel leads to a capillary tube formation and then vessel completion by recruitment of smooth muscle cells. Whether smooth muscle cells migrate to the vessel or differentiate from surrounding precursors around the vessel is not clear. B: A new vessel formed by angiogenesis.

B

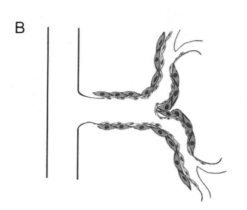

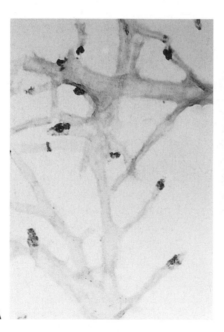

A   B

FIGURE 9.7. Microdissected renal arterial vascular trees from wild-type (A) and angiotensin-converting enzyme (ACE) –/– mice (B). The arterial trees were stained with renin antibody. In ACE –/– mice, the terminal arterioles are sparse, widened, and thickened in contrast with the delicate appearance of afferent arterioles in the wild-type mice. Renin staining is restricted to the distal end of the afferent vessels in ACE +/+ mice. Renin staining is diffuse throughout the vasculature in ACE –/– mice. Similar alterations are found when other genes of the renin-angiotensin system are deleted. (From Hilgers,[42] with permission.)

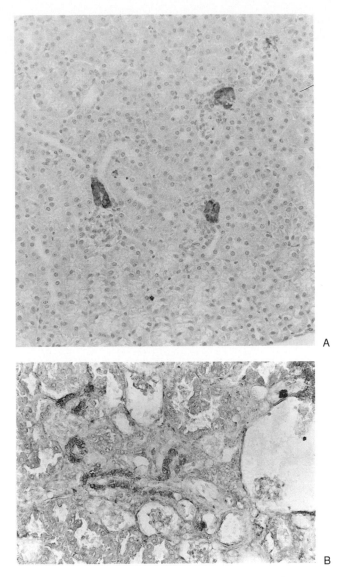

FIGURE 9.8.   Renal morphology and renin distribution in wild-type and ACE –/– mice. A:Wild-type mouse: normal histology and renin distribution. B: ACE –/– mouse: dilated Bowman's spaces, tubular atrophy, interstitial fibrosis, and architectural disorganization are evident. Renin containing cells (brown) surround the arterioles. Similar alterations are found when other genes of the renin-angiotensin system are deleted. (From Hilgers,[43] with permission.)

Although significant advances have been made in our understanding of glo-merular capillary morphogenesis, the formation of the renal arterioles is not as clear. As mentioned above, smooth muscle and renin-containing cells develop *in situ* from mesenchymal progenitors residing within the kidney anlagen. The mechanisms involved in arteriolar assembly are unknown. It is likely, although speculative, that smooth muscle cells and renin cells follow the developing endothelium in response to signals yet to be identified. Numerous studies have shown a temporal and spatial relationship between renin cells and branching of renal arterioles. It is tempting to speculate that these cells either directly or indirectly, through the local generation of angiotensin(s), may regulate this process. We have shown that the renin-angiotensin system participates in the branching of the renal arterioles.[36] Treatment of rats during newborn life (when branching is at its peak) with inhibitors of the renin-angiotensin system leads to arrested branching of the renal arterioles. In addition to their diminished number, the arterioles are shorter and thicker with concentric accumulation of smooth muscle cells. We have speculated that lack of angiotensin leads to immature development of smooth muscle cells, which proliferate excessively in a concentric arrangement as if the immature cells have lost their proper orientation. Similar vascular abnormalities have been observed when angiotensin-converting enzyme, angiotensin receptors, and the angotensinogen and renin genes are deleted by gene targeting. Interestingly, the aberrant vessel development is accompanied by a variety of histologic abnormalities including delayed glomerular development, cyst formation, and generalized architectural disarrangement (Figs. 9.7 and 9.8). The alterations resemble those found in humans treated prenatally with converting-enzyme inhibitors.[37]

CONCLUSION

Great progress has been made in identifying individual genes, molecules, and signals that participate in the vascular development of the kidney. The next challenge is to identify patterns of gene activation and the sequence of events that lead to the precise assembly and spatial arrangement of the kidney vasculature.

REFERENCES

1. Davies, J.A., Bard, J.B.L. (1998). The development of the kidney. Curr Top Dev Biol 39:245–301.
2. Davies, J.A., Brandli, A. (2001). The kidney development database. http://golgi.ana.ed.ac.UK/kidhome.html
3. Schuchardt, A., D'Agati, V., Larsson-Blomberg, L., Costantini, F., Pachnis, V. (1994). Defects in the kidney and enteric nervous system of mice lacking the tyrosine kinase receptor Ret. Nature 367:380–383.
4. Vainio, S., Muller, U. (1997). Inductive tissue interactions, cell signaling, and the control of kidney organogenesis. Cell 90:975–978.
5. Kreidberg, J.A., Sariola, H., Loring, J.M., et al. (1993). WT-1 is required for early kidney development. Cell 74:679–691.
6. Sanyanusin, P., Schimmenti, L.A., McNoe, T.A. (1996). Mutation of the gene in a family with optic nerve colobomas, renal anomalies and vesicoureteral reflux. Nature Genet 13:129.
7. Miyamoto, N., Yoshida, M., Kuratani, S., Matsuo, I., Aizawa, S. (1997). Defects of urogenital development in mice lacking *Emx2*. Development 124:1653–1664.
8. Hatini, V., Huh, S.O., Herzlinger, D., Soares, V.C., Lai, E. (1996). Essential role of stromal mesenchyme in kidney morphogenesis revealed by targeted disruption of winged helix transcription factor BF-2. Genes Dev 10:1467–1478.
9. Kanwar, Y.S., Carone, F.A., Kumar, A., Wada, J., Ota, K. (1997). Role of extracellular matrix, growth factors and proto-oncogenes in metanephric development. Kidney Int 52:589–606(abst).

10. Sakurai, H., Nigam, S.K. (1998). In vitro branching tubulogenesis: implications for developmental and cystic disorders, nephron number, renal repair, and nephron engineering. Kidney Int 54:14–26.

11. Horster, M., Huber, S., Tschöp, J., Dittrich, G., Braun, G. (1997). Epithelial nephrogenesis. Eur J Physiol 434:647–660(abst).

12. Mendelsohn, C., Lohnes, D., Décimo, D. (1994). Function of the retinoic acid receptors (RARs) during development. (II) Multiple abnormalities at various stages of organogenesis in RAR double mutants. Development 120:2749–2771.

13. Risau, W. (1991). Vasculogenesis, angiogenesis and endothelial cell differentiation during embryonic development. In: Feinberg R.N., Sherer, G.K., Auerbach, R., eds. Issues in biomedicine. The development of the vascular system, vol. 14, pp. 58–68. New York: Karger.

14. Noden, D.M. (1989). Embryonic origins and assembly of blood vessels. Am Rev Respir Dis 140:1097–1103.

15. Saxen, L. (1987). Developmental and cell biology series: organogenesis of the kidney, pp. 1–173. New York: Cambridge University Press.

16. Ekblom, P., Sariola, H., Karkinen-Jaaskelainen, M., Saxen, L. (1982). The origin of glomerular endothelium. Cell Differ 11:35–39.

17. Gomez, R.A., Norwood, V.F., Tufro-McReddie, A. (1997). Development of the kidney vasculature. J Microsc Res Tech 39(3):254–260.

18. Sequeira Lopez, M.L.S., Pentz, E.S., Robert, B., Abrahamson, D.R., Gomez, R.A. (2001). Embryonic origin and lineage of juxtaglomerular cells. AJP-Renal 281:F345–F356.

19. Tufro-McReddie, A., Norwood, V.F., Carey R.M., Gomez, R.A. (1999). Vascular endothelial growth factor induces nephrogenesis and vasculogenesis. J Am Soc Nephrol 10:2125–2134.

20. Loughna, S., Landels, E., Woolf, A.S. (1996). Growth factor control of developing kidney endothelial cells. Exp Nephrol 4:112–118.

21. Rakugi, H., Wang, D.S., Dzau, V.J., Pratt, R.E. (1994). Potential importance of tissue angiotensin-converting enzyme inhibition in preventing neointima formation. Circulation 90:449–455.

22. Kazimierczak, J. (1971). Development of the renal corpuscle and the juxtaglomerular apparatus: a light and electron microscopic study. Copenhagen: Munksgaard.

23. Abrahamson, D.R., Robert, B., Hyink, D.P., St. John, P.L., Daniel, T.O. (1998). Origins and formation of microvasculature in the developing kidney. Kidney Int 54:S7–S11.

24. Tufro, A., Gomez, R. (1999). VEGF spatially directs angiogenesis in metanephric development in vitro. J Am Soc Nephrol 9:369A.

25. Tufro-McReddie, A., Norwood, V.F., Aylor, K., Botkin, S., Carey, R.M., Gomez, R.A. (1997). Oxygen regulates vascular endothelial growth factor-mediated vasculogenesis and tubulogenesis. Dev Biol 183:139–149.

26. Kitamoto, Y., Tokunaga, H., Tomita, K. (1997). Vascular endothelial growth factor is an essential molecule for mouse kidney development: glomerulogenesis and nephrogenesis. J Clin Invest 99:2351–2357.

27. Varela-Echavarria, A., Guthrie, S. (1997). Molecules making waves in axon guidance. Genes Dev 11:545–557.

28. Hanahan, D. (1997). Signaling vascular morphogenesis and maintenance. Science 277: 48–50.

29. Yuan, H.T., Suri, C., Yancopoulos, G.D., Woolf, A.S. (1999). Expression of angiopeietin-1, angiopoietin-2, and the Tie-2 receptor tyrosine kinase during mouse kidney maturation. J Am Soc Nephrol 10:1722–1736.

30. Sato, T.N., Tozawa, Y., Deutsch, U., et al. (1995). Distinct roles of the receptor tyrosine kinases tie-1 and tie-2 in blood vessel formation. Nature 376:70–74.

31. Puri, M.C., Rossant, J., Alitalo, K., Bernstein, A., Partanen, J. (1995). The receptor tyrosine kinase TIE is required for the integrity and survival of vascular endothelial cells. EMBO J 14:5884–5891.
32. Maisonpierre, P.C., Suri, C., Jones, P.F., et al. (1997). Angiopoietin-2, a natural antagonist for Tie 2 that disrupts in vivo angiogenesis. Science 277:55–60.
33. Chernavvsky, D., Fernandez, L., Barton, K., Muthusamy, N., Leiden, J., Gomez, R.A. (1999). Expression and function of the ETS-1 gene in the developing kidney. J Am Soc Nephrol 9:360A.
34. Alpers, C.E., Seifert, R.A., Hudkins, K.L., Johnson, R.J., Bowen-Pope, D.F. (1992). Developmental patterns of PDGF B-chain, PDGF-receptor, and a-actin expression in human glomerulogenesis. Kidney Int 42:390–399.
35. Lindahl, P., Hellstrom, M., Kalen, M., et al. (1998). Paracrine PDGF-B/PDGF-RB signaling controls mesangial cell development in kidney glomeruli. Development 125: 3313–3322.
36. Reddi, V., Zaglul, A., Pentz, E.S., Gomez, R.A. (1998). Renin-expressing cells are associated with branching of the developing kidney vasculature. J Am Soc Nephrol 9:63–71.
37. Shotan, A., Widerhorn, J., Hurst, A., Elkayam. U. (1994). Risks of angiotensin-converting enzyme inhibition during pregnancy: experimental and clinical evidence, potential mechanisms, and recommendations for use. Am J Med 96:451–456.
38. Sadler, T.W. (1995). Langman's medical embryology. Lippincott Williams and Wilkins Philadelphia.
39. Ekblom, P. (1984). Basement membrane proteins and gowth factors in kidney differentiation. In: Trelstad, R.L., ed. Role of extracellular matrix in development, pp. 173–206. New York: Alan R. Liss.
40. Gilbert, S.F. (1997). Developmental biology, 5th ed. Sunderland, MA; Sinauer. Associates.
41. Schuchardt, A., et al. (1996). Renal agenesis and hypodisplasia in ret-k-mutant mice result from defects in ureteric bud development. Development 122:1919–1929.
42. Hilgers, K.F. (1997). Angiotensin's role in renal development. Semin Nephrol 17(5): 492–501.
43. Hilgers, K.F. (1997). Aberrant renal vascular morphology and renin expression in mutant mice lacking angiotensin converting enzyme. Hypertension 29:216–121.

CHAPTER **10**

Vascular Development of the Lung

Daphne E. deMello and Lynne M. Reid

The lung has a complex vascular system consisting of a double arterial supply and a double venous drainage. In addition, there is a lymphatic circulation. The development of the lung's vascular system is related temporally to the development of other lung structures such as the airways and alveoli and cannot be considered apart from these structures. Four main cell types make up the wall structure of the lung's vessels: endothelial cells, smooth muscle cells, fibroblasts, and pericytes. The metabolic status of these cells is under the control of a complex array of genes and factors that dictate the expression of a variety of cellular proteins. These proteins determine the normal vessel wall structure at different levels in the pulmonary vascular tree, and its response to injury or other stimuli.

Angioblast precursors are derived from the yolk sac, and contribute to the formation of the pulmonary vasculature by two processes: angiogenesis, the sprouting of new vessels from preexisting ones, contributes to the formation of the proximal lung vessels; and vasculogenesis, the formation of vessels by precursor cells (angioblasts) that form "lakes" or spaces within the mesenchyme, contributes the peripheral vascular component. A third process of "fusion" is necessary to complete the circuit between the proximal and peripheral vasculature.

This chapter discusses these aspects of embryonic lung vascular development, and the changing patterns during fetal and postnatal lung growth until the adult pattern is established. The effect on the developing vasculature of a variety of influences including altered hemodynamics, changes in pO_2, restricted space for growth, and disordered metabolism will also be discussed, because of their potential for producing a drastically remodeled vascular tree with deranged function.

EMBRYONIC ORIGIN OF VESSELS

In the yolk sac of the chick embryo, the assembly of blood vessels involves two processes: angiogenesis, the branching of new vessels from preexisting ones; and vasculogenesis, the development of vessels from blood lakes (Noden, 1989). We have demonstrated that the same processes participate in the formation of the early lung's vasculature (see establishment of the pulmonary vascular circuit, deMello et al., 1997).

Controlling Mechanisms—Genes and Factors

Experiments involving overexpression or knocked-out growth factors or genes indicate that complex regulatory mechanisms orchestrate the assembly of blood vessels within organs such as the lung (Drake et al., 1998; deMello, 1999). For illustration, a few examples are mentioned here. Vascular endothelial growth factor (VEGF) and its receptor proteins [vascular receptor protein (VRP)] are expressed in developing lung epithelial and endothelial cells, respectively, suggesting a role in the processes of assembly of blood vessel wall and in the overall pattern of vessel branching within an organ (Kaipainen et al., 1993; Shifren et al., 1994; Breier et al., 1995; Beck and D'Amore, 1997). Knockout of the VEGF gene or even its reduced expression in heterozygosity results in lethal defects in vessel formation in the mouse embryo (Drake and Little, 1995; Carmeliet et al., 1996; Ferrara et al., 1996). Knockout of the *Flt-1* (VEGF receptor) gene produces a lethal defect in angiogenesis (Fong et al., 1995) and knockout of the *Flk-1* (VEGF receptor) gene, a lethal failure of vasculogenesis (Shalaby et al., 1995). Transforming growth factor-β (TGF-β), is important to the eventual wall structure of the developing vasculature through its influence on the growth, migration, and differentiation of endothelial, smooth muscle, and mesenchymal cells and pericytes (Heimark et al., 1986; Antonelli-Orlidge et al., 1989; Battegay et al., 1991; Dickson et al., 1995; Sato and Rifkin, 1989). Early in mouse gestation, the *Tie-1* (receptor tyrosine kinase) gene expression occurs in angioblasts and continues throughout gestation (Korhonen et al., 1994). In the adult mouse, it is expressed in lung alveolar septal capillaries. *Tie-1* null mice have normal lung vascular branching but ultrastructural defects in the endothelium (Partanen et al., 1996). Thus it is clear that an intricate interplay between a variety of growth factors, receptors, and genes exists to bring about the metamorphosis from the primitive angioblasts in the yolk sac to the mature lung vasculature (see also Chapters 1 to 6).

VASCULAR TOPOGRAPHY IN THE LUNG—ITS RELATION TO AIR SPACES

Double Circulation—Pulmonary and Bronchial

The lung has a double arterial supply, the pulmonary artery and bronchial artery system, and a double venous drainage, the pulmonary and the so-called true bronchial veins (Robertson, 1967; Boyden, 1970; Hislop and Reid, 1972, 1973a,b, 1977, 1981; deMello and Reid, 1991, 1997a). The names do not correspond to each other (Fig. 10.1). Notably, the bronchial arterial supply to the intrapulmonary airways does not drain to the bronchial veins but to the pulmonary veins. This produces a degree of venous admixture in systemic arterial blood, and represents a level where the right and left heart interact. Because of the hemodynamic arrangement of the airway wall, a pressure rise in the left atrium produces congestion and edema of the airway wall as well as of the alveolar.

The pulmonary artery transports blood for oxygenation to the respiratory units and also supplies the pleura. It has two types of branches. The long or conventional arteries arise at an acute angle from the axial vessel and supply the respiratory units at the end of the axial pathway, while the short or supernumerary

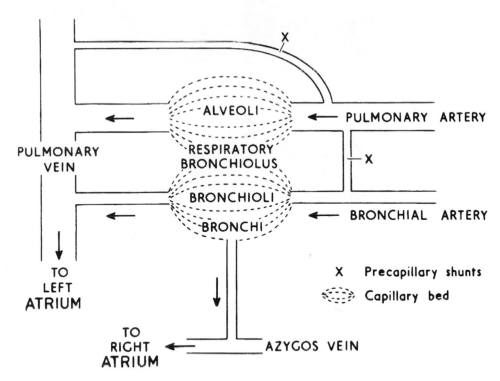

FIGURE 10.1.   The double arterial and venous systems of the lung. The bronchial artery supplies the bronchial capillary bed, the pulmonary artery, and the respiratory region; both the bronchial and pulmonary systems drain to the pulmonary veins. Vascular drainage from the bronchi at the hilum is to the true bronchial veins and to the azygos system. (Reproduced with permission from Reid et al., 1986.)

arteries arise at right angles from the main vessel and supply adjacent respiratory units (Fig. 10.2). The number of supernumerary arteries exceeds that of the conventionals by a ratio of 3:1 or 4:1. Lumen size of successive branches varies, and each can be expected to have a different arterial opening pressure and play a signifcant role in recruitment, for example, with exercise (Fig. 10.3).

The bronchial arteries are the nutrient supply for the conducting airways and hilar structures. The pulmonary venous drainage is to the left atrium and the pattern of pulmonary veins is similar to that of the pulmonary arteries, so that there are conventional and supernumerary venous tributaries (Hislop and Reid, 1973a). The bronchial veins also drain to the left atrium except for a small perihilar region that drains to the right side of the heart via the azygous venous system. Then the lymphatic system with its low-pressure irrigation channels drains through a luminal opening to the azygos systemic venous system and so connects between the two. This connection ensures that metabolic products that reach the lung interstitium circulate through the pulmonary artery capillary bed and are metabolically monitored and often modified.

The arrangement of the lung air spaces and connective tissue framework offers a geographic infrastructure to which the topography of vascular structures needs to be related.

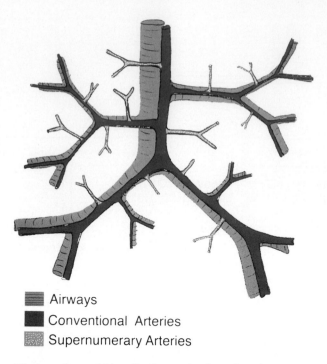

Airways
Conventional Arteries
Supernumerary Arteries

FIGURE 10.2. The bronchoarterial bundle. Conventional arteries arise at acute angles to the main axis and supply the respiratory region at the end of the axial pathway. Supernumerary arteries are short and have varying diameters. They arise at right angles to the main axis and supply the air spaces adjacent to the axis.

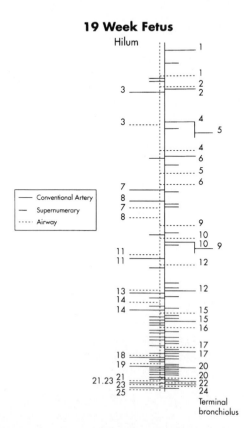

19 Week Fetus

FIGURE 10.3. Reconstruction of the posterior basal artery and airway of the left lower lobe in a 19-week fetus. The airway branches and their accompanying conventional arteries are numbered, and the 58 supernumerary arteries are also shown. Actual length of the artery is 16 mm. (Reproduced with permission from Hislop and Reid, 1972.)

Connective Tissue Septa

The airways with the bronchial artery branches and capillary beds embedded in their walls as well as the pulmonary artery and its branches are enclosed in a connective tissue sheath, the bronchovascular bundle. Proceeding peripherally, the sheath becomes thinner until near the beginning of the acinus it is no longer well developed. Additional connective tissue septa, arising from the pleura, penetrate into the lung. Their density varies widely: some regions, such as the centrolateral regions of the lower lobes are hardly subdivided at all, while other regions such as the edges of the lung (e.g., the anterior sharp edge of the upper lobe, the costodiaphragmatic rim, and the costovertebral border) are relatively heavily subdivided. When present, these septa run at the edge of a unit, be it lobule or acinus; the septa never completely isolate a unit. There is always a bridge of alveolar tissue between a unit and its neighbor so that collateral air drift occurs. The pleura is the only structure that impedes collateral air drift. Since the oblique fissure is incomplete in about 50% of subjects, in such cases collateral ventilation can even occur between two lobes (Reid and Rubino, 1959, Reid, 1959).

VASCULAR STRUCTURE

Cells and Matrix

The cellular components of the vessel wall, whether artery or vein, are essentially the same; the intima is lined by endothelial cells, the smooth muscle cell or its precursors the pericyte or intermediate cell form the media, and the fibroblast occurs in the adventitia. Each coat contributes its own intercellular matrix. Along an arterial pathway, a complete muscular coat becomes partially muscular and then nonmuscular in the precapillary segment (Elliott and Reid, 1965). By electron microscopy, the intermediate cell (within its own basement membrane) and the pericyte (within the basement membrane of the endothelial cell) are identified in the partially and nonmuscular vessel segments (Meyrick and Reid, 1979) (Fig. 10.2). These structural differences confer upon the individual vessel segments specific physiologic responses and reactivity to stimuli. Multiple phenotypically distinct smooth muscle cell populations are present within the arterial media. These individual cell populations have a unique response to different stimuli, such as hypoxia or hyperoxia, and can alter vascular structure by proliferation (see below) (Wohrley et al., 1995; Jones et al., 1999a,b).

Elastic, Muscular, and Nonmuscular Arteries

In fetal life while preacinar vessels are still small, wall structure, i.e., elastic, transitional, or muscular, is determined. The elastic or transitional structure extends until the 9th or 10th airway generation along an axial pathway, distal to which the vessel is muscular, partially muscular, or nonmuscular in the precapillary segment (Fig. 10.4). The vessel size at which muscular structure changes to partially muscular or nonmuscular remains constant throughout fetal life, and therefore progresses distally as gestation proceeds. In some conditions, a well developed muscle coat is present in smaller, more peripheral vessels than normal (see Abnormal Development, below).

MUSCULAR   PARTIALLY   NONMUSCULAR
                  MUSCULAR   AND CAPILLARY

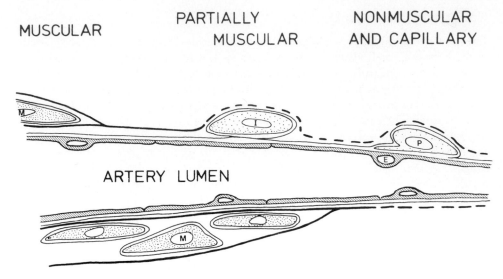

ARTERY LUMEN

FIGURE 10.4.   Diagrammatic representation of arterial structure in vessels larger than a capillary. By electron microscopy, proceeding distally in the nonmuscular segment of the vessel, the typical smooth muscle cell (M) develops features of an intermediate cell (I), and then in the immediate precapillary region has features of a pericyte (P). The intermediate cell and pericyte are precursors of smooth muscle cells. The pericyte lies within the basement membrane of the endothelial cell (E), whereas the intermediate cell has its own. (Reproduced with permission from Meyrick and Reid, 1979.)

In the adult, the elastic structure extends to generation 7 of the airway and to vessels down to a diameter of 3,000 μm, while transitional structure is present in generations 7 to 9 (diameter 3,000 to 2,000 μm), and muscular structure from the 9th generation to alveoli (diameter 2 mm to 30 μm, Fig. 10.3).

ESTABLISHMENT OF THE PULMONARY VASCULAR CIRCUIT

Embryonic Stage—Mouse and Human

Mouse

To determine the contribution and temporal sequence of the processes involved in vascular assembly, the development of the pulmonary vascular circulation was studied in mouse embryos from 9 to 20 days' gestation. Correlation between light and transmission electron microscopy of the developing lungs and scanning electron microscopy of Mercox (methyl methacrylate) lung vascular casts enabled mapping the process of early lung vascular development (deMello et al., 1997).

The first step is the appearance at 9 days of lakes within the primitive mesenchyme surrounding the epithelial lined lung bud. Presumably, angioblast precusors in the mesenchyme initiate this process. Fluid-filled vesicles rupture and their contents coalesce to form intercellular spaces, the ragged membranes reconstruct themselves, and the adjacent cells align to form an endothelial lining.

9 Days

10 Days

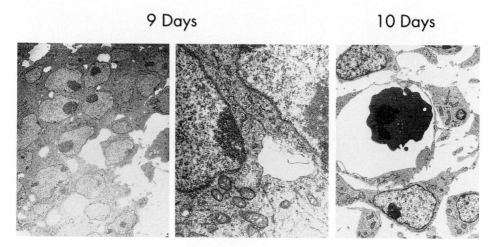

FIGURE 10.5. Transmission electron micrographs of fetal mouse thorax. At 9 days intercellular spaces are seen between densely packed mesenchymal cells around the developing lung bud. Membrane-bound vesicles and membranous fragments within the spaces suggest that the spaces result from discharge of intracytoplasmic vesicles (magnification ×707). At 10 days, hematopoietic precursors are seen within the "lakes," and the lining cells appear thin like endothelial cells (×1,458). (Reproduced with permission from deMello et al., 1997.)

Hematopoietic cells appear within the lakes or spaces (Fig 10.5). This close association as well as presence of similar cell surface antigens on hemangioblasts and hematopoietic cells points to a common cell lineage (Ogawa et al., 1999).

Between 10 and 12 days, the density of these lakes within the peripheral lung mesenchyme increases, but the pulmonary artery is represented by about only four generations of central branches that lie near the hilum. The early peripheral pulmonary vessels develop by the process of vasculogenesis independent of the central vessels, which develop by angiogenesis or sprouting from the main pulmonary vessels. Between 12 and 14 days, generations of central vessels come to number five to seven. Conventional and supernumerary arteries are present. By this time, a connection is established between the central and peripheral systems that permit filling by Mercox of the peripheral vessels and visualization of their casts. The complexity of the peripheral casts increases progressively to term, reflecting both increased peripheral connections and growth of additional peripheral vessels, which now occurs by angiogenesis (Figs. 10.6 to 10.9).

In summary, early lung vascular development in the mouse occurs by two concurrent but separate systems: vasculogenesis in the periphery and angiogenesis centrally. A third process of fusion between the two systems establishes the circulation. Further growth of peripheral vessels then proceeds by angiogenesis.

Perhaps the reason for the two processes is that the hemangioblast precursors that migrate with the primitive lung mesenchyme carry the information that dictates the eventual overall topography of the vascular tree. The developing airway and its accompanying artery is then "drawn" to the sites of vasculogenesis by a chemoattractant mechanism to permit completion of the vascular circuit and formation of the air-blood barrier. This important structure, essential for all independent air breathing and gas exchange, results from fusion of the epithelial

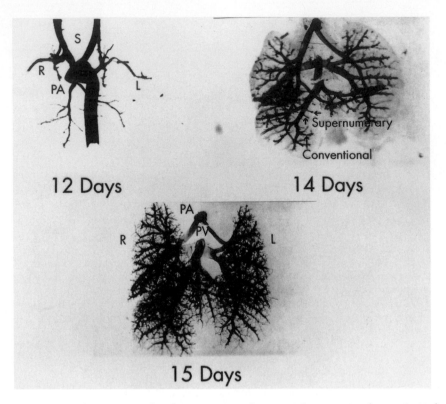

FIGURE 10.6.   Photomicrographs of Mercox casts of mouse pulmonary vasculature. At 12 days, there is no vascular connection to the peripheral lung and only four generations of arterial central branches are present. At 14 days, peripheral vessels of both the arterial (upper) and venous (lower) systems are seen. At 15 days, the density of small peripheral vessels is increased. S, systemic; R, right; L, left; PA, pulmonary artery; PV, pulmonary vein. (Reproduced with permission from deMello et al., 1997.)

and capillary endothelial basement membranes, and is immediately proximal to the pulmonary venous segment. It is possible that the process of vasculogenesis plays a role in determining the site for the formation of the air-blood barrier. A lethal disorder, alveolar capillary dysplasia (see below), results in human babies in whom this process is disturbed. The genes, factors, receptors, and mechanisms that orchestrate this seminal event in lung vascular development remain to be identified.

Human

Using the Carnegie collection of human embryos (O'Rahilly and Muller, 1987), we studied the development of the human lung vasculature using serial histologic sections of embryos from Streeter stages 10 to 23 (deMello and Reid, 1999). At stage 14 (32 days), lakes appeared within the primitive mesenchyme around the lung bud, which at this time was in the neck (Fig. 10.10). By stage 20 (50.5 days), there is a profusion of lakes in the subpleural mesenchyme, but the pulmonary artery extends only as far as the third or fourth airway branch (Fig. 10.11). There are about six airway generations at this time and an expanse of mesenchyme inter-

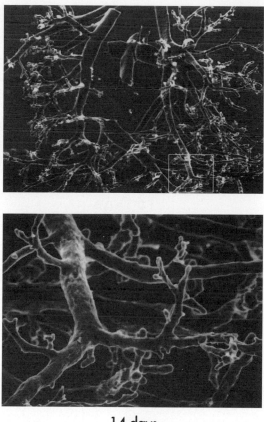

FIGURE 10.7.   Scanning electron micrographs of 14-day fetal mouse lung vascular casts. Filling of some terminal buds signals fusion between the central and peripheral systems. The bottom micrograph is an enlargement of the area in the rectangle in the upper figure (magnification: upper ×69, lower ×320). (Reproduced with permission from deMello et al., 1997.)

venes between the proximal hilar structures and the distal pleural surface. The first vascular connection between central and peripheral systems is a venous one apparent at stage 22 (54 days, Fig. 10.12). The pulmonary artery branches at this time are thick-walled, their distal tips lack a lumen, and they end two to three generations proximal to the distal tip of the airway. Between 12 and 16 weeks, a rich capillary network develops in the subpleural mesenchyme, and the pulmonary artery ends two to three generations proximal to the last airway branches (Fig. 10.13). By 22 to 23 weeks, the capillary network is closely apposed to the air-space epithelium and the capillary lumina bulge into the air spaces, an appearance heralding the formation of the air-blood barrier. In the human, the earliest air-blood barriers have been identified by ultrastructure at 19 weeks of gestation (Dimaio et al., 1989). The failure of air-blood barriers and alveolar capillaries to form leads to a rare disorder, alveolar capillary dysplasia (ACD) (Janney et al., 1981; Cullinane et al., 1992). The condition may be focal and affect a lobe or part of lobe. When widespread throughout the lung, it is uniformly fatal. ACD may be accompanied by misalignment of central vessels (Wagenvoort, 1986; Cater et al., 1989; Langston, 1991; Boggs et al., 1994).

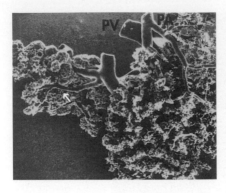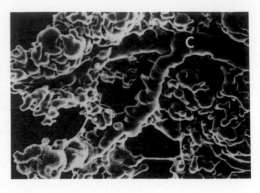

15 Days

FIGURE 10.8.   Scanning electron micrographs of 15-day fetal mouse lung vascular casts. Extensive connections between the central and peripheral systems is reflected by a complex peripheral network. The blind-ending precapillary branches approach the future capillary bed at right angles. The arrow points to the most likely junction between the peripheral and central (C) systems. Note that the central vessels have a smooth profile, whereas the peripheral vessels are irregular, perhaps reflecting their derivation by angiogenesis or vasculogenesis, respectively (magnification: left ×50, right ×320). (Reproduced with permission from deMello et al., 1997.)

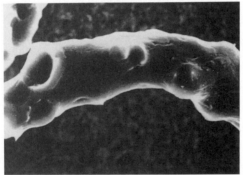

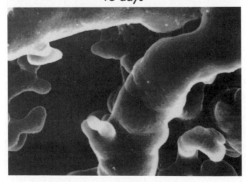

FIGURE 10.9.   At high magnification, scanning electron micrographs of fetal mouse lung vascular casts reveal surface pits at 14 days, probably indicating sites at which filamentous connections have broken. At 15 days, the surface is smoother and several filamentous structures are seen, probably identifying sites of future vascular connections (magnification ×1,200). (Reproduced with permission from deMello et al., 1997.)

32 days

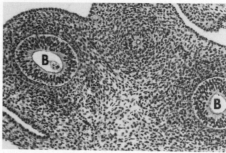

Stage 14

FIGURE 10.10.   Low- (left) and high- (right) magnification views of a transverse section of a human embryo at the level of the tracheal bifurcation reveals dense mesenchyme enveloping the right and left bronchi (B) (magnification ×40). The esophagus is the thick-walled structure with a small lumen seen at the top of the right figure. Blood lakes containing hematopoietic precursors (center of the right figure) are already present in the mesenchyme, but no pulmonary artery accompanies the bronchus as yet (×200). (Reproduced with permission from deMello, 1999.)

50½ days

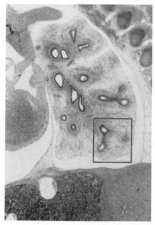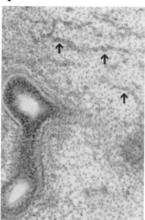

Stage 20

FIGURE 10.11.   Low- (left) and high- (right) magnification views of a sagittally sectioned human embryo (magnification: left ×40, right ×200). At low power, the heart is seen at left and the bronchial generations at right. The bronchial artery accompanies only the first few generations of bronchi, while in the peripheral subpleural region (box) an extensive network of sinusoids resulting from coalescence of blood lakes has formed. No connection is seen between the central and peripheral systems. (Reproduced with permission from deMello, 1999.)

54 days

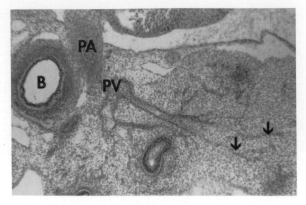

Stage 22

FIGURE 10.12.   Photomicrograph of the human embryonic lung. A pulmonary vein is shown by serial sections to extend from the peripheral network into the hilar pulmonary vein (PV). The lobar pulmonary artery (PA) accompanies the bronchus (B) at the hilum (magnification ×100). (Reproduced with permission from deMello, 1999.)

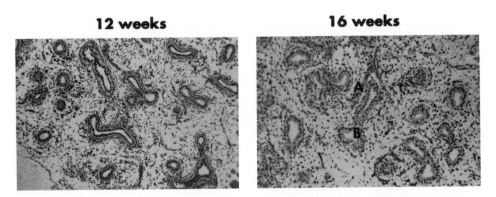

FIGURE 10.13.   Photomicrographs of human fetal lung. At 12 weeks, the pleura is seen at the lower left corner. The subpleural airway buds are not accompanied by a pulmonary artery and are widely separated by intervening mesenchyme containing a vast network of vessels (magnification ×100). By 16 weeks, the pulmonary artery (A) accompanies the airway but has still not reached the distal end. The mesenchyme between airways is significantly reduced compared with that at 12 weeks (×200). (Reproduced with permission from deMello, 1999.)

By 23 weeks, the pulmonary artery has extended as far as the distal tip of the airway to just beneath the distal pleural surface (Fig. 10.14).

As in the mouse, the three processes of angiogenesis, vasculogenesis, and fusion are also involved in the early formation of the human lung vasculature. The venous segment is completed first.

22-23 weeks

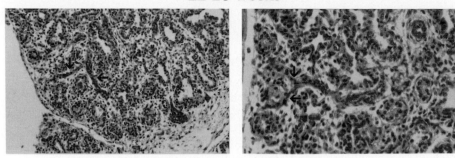

FIGURE 10.14.   Photomicrographs of human fetal lung. At 22 weeks, the pleura is seen at the lower left corner, and the pulmonary artery (arrows) accompanies the airways to the distal subpleural tip (magnification ×200). At 23 weeks the distal end of the pulmonary artery (arrows) encircles the end of the airway (×400). (Reproduced with permission from deMello, 1999.)

Fetal Stage—Laws of Human Lung Development

The stage of fetal development establishes the lung architecture associated with viability and that is present at term. There is a close association between the developing vessels and airways. The following three laws of lung development summarize fetal and postnatal events linked to growth (Reid, 1994):

Law 1—Airways. The airways, i.e., bronchi and bronchioli, develop within the first 16 weeks of intrauterine life.

Law 2—Alveoli. At birth, the air spaces are primitive saccules numbering about 20 million. The alveoli multiply postnatally particularly rapidly in the first 4 years. By the age of 8 about 350×10^6 are present, close to the adult number.

Law 3—Vessels. Preacinar vessels develop at the same time as the airways, and intra-acinar vessels appear as alveoli grow.

The pulmonary arteries and veins have many more branches than do the airways. Whereas the pulmonary arteries travel with the airways in a broncho-arterial sheath, the veins are present at the periphery of an acinus or lobule within their own connective tissue sheath.

Bronchial Arteries

The primitive bronchial arteries arise from the dorsal aorta in the neck close to the celiac arteries. They regress by the 6th week of gestation (Boyden, 1970). If this primitive artery does not regress with growth, it will migrate with the celiacs and come to lie below the diaphragm. Such an anomalous arrangement is often seen with an extralobar sequestrated segment. Between the 9th and 12th weeks, the definitive bronchial arteries arise from the aorta and induce connections with the precapillary plexus within the airway walls.

The injection studies of the pulmonary arterial bed in normal fetal lungs did not show pulmonary to bronchial artery anastomoses larger than 15 μm in diameter. We did show that by 3 days after birth, cross-filling between the two arterial systems occurs.

Bronchopulmonary Anastomoses

Robertson (1967), using a microangiographic technique combined with serial reconstruction, demonstrated a small number of bronchopulmonary anastomoses ranging in size from 35 to 100 μm in 16% of fetuses. He also identified bronchial arteries supplying pulmonary parenchyma and pulmonary arteries supplying bronchial walls. These vessels were all obliterated by 10 weeks postnatally. The anastomoses increased during the first months of life. Wagenvoort and Wagenvoort (1967) found similar results and concluded that the anastomoses were more important in the perinatal period than in the adult.

Conventional and Supernumerary Pulmonary Arteries

As additional airway branches form during fetal life, both conventional and supernumerary artery branches appear. In conditions such as diaphragmatic hernia in which the number of airway branches is reduced, conventional and supernumerary artery and vein numbers are also proportionately reduced (Kitagawa et al., 1971).

By the 16th week of intrauterine life, the full complement of preacinar vessels is present and these increase in size until birth. Intra-acinar vessels develop as respiratory bronchioli and saccules form, and both conventional and supernumerary vessels are identified. A "baffle valve" is present at the origin of the supernumerary arteries and its rich innervation suggests a control mechanism for regulating blood flow. The conventional and supernumerary arteries also have a different response to pharmacologic agents (Bunton et al., 1995, 1996). There is variation in the diameter of successive branches of both conventional or supernumerary types. This indicates that adjacent branches have different critical opening pressures. All of these features point to a role for the supernumerary arteries in vessel recruitment, for example as flow increases with exercise. The supernumeraries also function as a collateral channel when an axial artery is blocked. Flow through a supernumerary proximal to the block can link to other supernumarary channels and supply the axial artery distal to the block. Plexiform lesions, seen in some patients with pulmonary vascular occlusive disease, develop at the origin of the supernumerary branches.

Intra-Acinar Vessels

As gestation progresses, the increasing density of intra-acinar vessels reflects the increasing growth and maturity of the peripheral or acinar region of the lung. As law 3 indicates, above intra-acinar vessels are formed as the alveoli multiply. Therefore, in a unit area of lung, vessel density can be related to alveolar number as in an alveolar-to-arterial ratio. In certain conditions where intrauterine pulmonary blood flow is altered, as in congenital heart disease (see below), arterial development may not keep pace with alveolar multiplication. This results in an increased

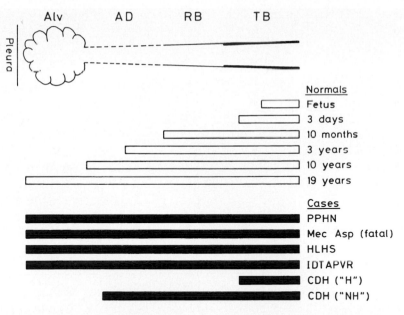

FIGURE 10.15.   Diagrammatic representation of the acinus. The bars indicate the size of the airway landmark that has the highest percentage of partially muscular arteries; i.e., distal to this level, most arteries are nonmuscular. At birth, muscular arteries are not present beyond the level of the terminal bronchiolus. With age, more peripheral vessels become muscularized, so that in the adult (19 years) even some vessels within the alveolar wall are muscularized. In disease, "precocious muscularization" accelerates the process of muscularization. PPHN, persistent pulmonary hypertension of the newborn; Mec Asp (fatal), fatal cases of meconium aspiration; HLHS, hypoplastic left heart syndrome; IDTAPVR, infradiaphragmatic total anomalous pulmonary venous return; CDH, congenital diaphragmatic hernia; H, honeymoon period; NH, no honeymoon period. (Reproduced with permission from Reid et al., 1986.)

alveolar-to-arterial ratio and in a reduction in total arterial number or vascular volume.

Vessel size reflects the patient's age and the level of the vessel in a pathway. Before birth, the diameters of the main vessels are related to the volume of the lung supplied. Preacinar structure is set early as seen in embryonic development. As gestation progresses, the length of the vascular pathway and the diameters of the conventional and supernumerary vessels increase. The structure of a vessel wall varies with its size and location in the vascular tree (see Vascular Structure, above) (Fig. 10.15).

Vascular Morphometric Techniques

Intra-acinar vascular density, size, and structure may all be altered by disease states that affect blood flow or environmental conditions during critical stages of vessel development (see Abnormal Development, below). A variety of techniques have been used to study the pulmonary vasculature including light and electron microscopy and immunohistochemistry (deMello and Reid, 1997a; Ohar et al., 1998). For a reliable quantitative assessment of vessel number, diameter, and wall

FIGURE 10.16.   Arteriogram of the human lung at different ages. Top left: newborn; bottom left, 18 months; right, adult. At birth, all preacinar arteries are present, but the background haze contributed by filling of intra-acinar arteries does not appear until later. (Reproduced with permission from Reid, 1978.)

thickness, the vessels and lung volume must be prepared in a standardized way. It is satisfactory to first distend the vessels to a supraphysiologic pressure to eliminate native vascular tone, and then to fix the lungs in maximal distention. Vascular morphometry may then be done on postfixation angiograms (if a radiopaque material was used for vascular perfusion), or histologic sections of the lung. Using such techniques, normal values have been established for vessels at different ages. The increasing complexity of the pulmonary vasculature from the fetus, to the newborn and adult, is reflected in the series of barium angiograms at the different ages (Fig. 10.16).

Perinatal Adaptation

With air expansion of the lung at birth, a series of structural changes occurs in the arterial wall, especially in the precapillary resistance segment. This results in an increase in vessel diameter and a drop in wall thickness, both of which contribute to the sharp reduction in postnatal pulmonary arterial pressure (Fig. 10.17) (Haworth and Hislop, 1981, 1982; Haworth et al., 1987; deMello et al., 1988, 1992). The cellular mechanisms that trigger these changes are complex and not yet entirely clear. However, failure of normal postnatal adaptation has a disastrous effect on the newborn and is sometimes fatal (see Abnormal Development, below).

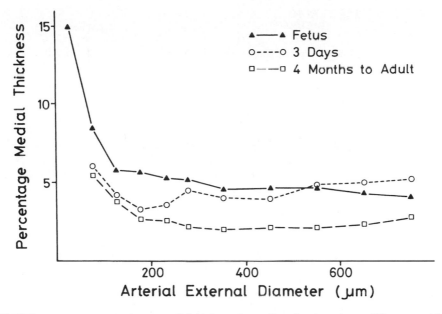

FIGURE 10.17.   Mean percentage medial thickness for arteries of various sizes at differnt ages. There is a drop from the fetal level with age. (Reproduced with permission from Reid et al., 1986.)

Childhood

After birth, preacinar arteries increase in size, but the nature of their structure is virtually settled. The intra-acinar vasculature shows major change and development. The acinus at birth consists of three to four generations of respiratory bronchioli, one or two of alveolar ducts and a few saccules in the alveolar region (Fig. 10.18). It is here that alveolar multiplication occurs, increasing the alveolar number from ~20 million at birth to ~350 million by 8 years (Dunnill, 1962; Davies and Reid, 1970). Alveolar multiplication slows, although the thoracic cage continues to increase in size, so that with age alveolar size increases. As law 3 indicates, intra-acinar vessels develop as new alveoli are formed. These are conventional arteries until 18 months of age and supernumeraries until 8 years. Up to 4 years of age, there is a marked increase in the number of intra-acinar arteries and this corresponds with the rapid increase in the number of alveoli. After 5 years, the concentration of arteries reduces, probably because of the increase in alveolar size, as the ratio of alveoli to arteries remains fairly constant throughout childhood. In a unit area of lung, there are more veins than arteries, presumably to facilitate blood flow from arteries to veins.

ABNORMAL DEVELOPMENT

Normal growth and development depend on a normal genetic endowment and environment. Growth represents adaptation to the normal and timely operation of range of stimuli. The following examples are a small selection of the range of abnormal stimuli that operate to modify growth.

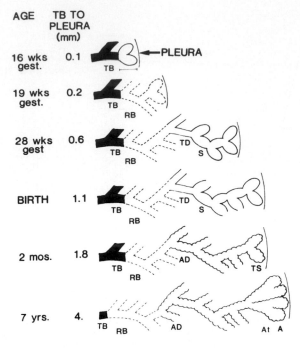

FIGURE 10.18.   Diagrammatic representation of acinar length in the fetus and child. AD, alveolar duct; AS, alveolar sac; At, atrium; RB, respiratory bronchiolus; S, saccule; TB, terminal bronchiolus; TD, terminal duct; TS, terminal sac. (Reproduced with permission from Hislop and Reid, 1981.)

High Flow

The pulmonary circulation is modified by changes in hemodynamic conditions. High flow and pressure cause major changes whether during the growing period or in the adult. Often congenital cardiac lesions do not produce major changes *in utero*. The hemodynamic effect of many congenital lesions is not apparent *in utero*; it is postnatal adaptations in the circulatory flow patterns that lead to high pressure or flow to the lungs (Hislop et al., 1975; Haworth et al., 1977; Reid, 1978; Rabinovitch et al., 1980) (Fig. 10.15).

An interesting exception to this is seen in large systemic arteriovenous communications. A large shunt through such a lesion leads to an increase in blood volume and an increased cardiac output that affects the systemic and the pulmonary circulation. Pulmonary vascular obstructive disease (POVD) virtually develops *in utero*. The remodeled pulmonary circulation with high resistance leads to reduced pulmonary flow in the later stages of gestation. At birth, the lung fields may even be oligemic. If the systemic malformation is repaired, the large cardiac output is diverted to the lungs often with disastrous and fatal outcome.

A large arteriovenous communication in the lung also represents an increase in cardiac output. Here, the shunt causes a "steal" from the adjacent lung so that its vascular bed is hypoplastic. Correction of the lung shunt then leads to "flooding" of the restricted hypoplastic parts of the pulmonary arterial system with rise in pressure and often serious edema. A similar complication sometimes occurs after correction of the obstructive vascular lesions, as seen in chronic thromboembolism of lung.

Hypoxia

Babies born at altitude are typically small for gestational age with generalized growth retardation. But this retardation does not give rise to persistent pulmonary hypertension of the newborn. Exposure to hypoxia at birth often leads to remodeling of pulmonary arterial structure with development of hypertension, although the veins are relatively spared (Meyrick and Reid, 1981; Rabinovitch et al., 1981). In all the species studied so far, the individual response to acute hypoxia varies. In some species them is no response to acute hypexia. The structural remodeling of pulmonary hypertension is thus a postnatal and acquired lesion.

One model has captured considerable attention. The calf is susceptible to altitude and after birth quickly develops pulmonary hypertension (Wohrley et al., 1995). This is a model of hypoxic pulmonary hypertension in the newborn, but it differs from the idiopathic persistent pulmonary hypertension of the newborn seen in humans. In these infants, structural vascular changes develop before birth (Haworth and Reid, 1976; Fox et al., 1977; Murphy et al., 1981, 1984).

The effect of hypoxia and the sequence of recovery events has been studied in both male and female, young and adult rats. A group of infant rats, 8 days old, and adult rats, 9 weeks old, was exposed to half atmospheric pressure for 1 month and then allowed to recover in room air for up to 3 months. The level of increased pulmonary arterial pressure (Ppa) was similar in male and female infants and in male adult rats but significantly lower in adult female rats. In all groups, only partial regression occurred. In male and female adult rats recovery values were similar but infant rats had more elevated Ppa than adults. The structural changes that developed during hypoxia, especially the abnormal presence of muscle in small and peripheral intra-acinar arteries, were more severe in male adults, compared with female, and in infants of both sexes compared with male adults. Residual structural changes were present in all rat groups but were most severe in the infants (Rabinovitch et al., 1981). In the infants, perhaps the alveoli had not multiplied normally but had increased in size in the posthypoxic period. The reduction in arterial density in the infants reflects greater susceptibility to a process whereby hypoxia causes arteries to close off or be resorbed (Sandison, 1928).

Hypoxia causes an increase in the lung volume due to overinflation of alveoli both in adult and young rats. Hypoxia also causes lumen narrowing of the central and preacinar arteries and reduction in the density of small arteries indicating severe volume restriction of the microcirculation (Meyrick and Reid, 1981).

Hyperoxia

Hyperoxia does considerable damage to arteries as well as veins, causing obliterative damage to both these segments (Jones et al., 1984). The resulting restriction in volume of the vascular bed leads to pulmonary hypertension with increased wall thickness and extension of smooth muscle into the walls of smaller arteries and veins than normal. An intriguing cellular adaptation is seen in the injury of hyperoxia. The newly muscularized small arteries transform their structural phenotype by recruitment of interstitial fibroblasts into their wall. These cells migrate to the endothelial cells, align themselves circumferentially around the lumen, lay down an external elastic lamina, differentiate to a smooth muscle cell morphology, and then add an internal elastic lamina (Jones, 1992; Jones et al., 1984, 1999a,b; Powell et al., 1998). Another model of a contracted pulmonary arterial bed is that produced by chronic infusion of platelet activating factor (PAF) in the rabbit. In this model, the wall structure of the pulmonary vasculature is also significantly altered (Ohar et al., 1991, 1994).

Liver Disease and Cavopulmonary Anastomoses

In cirrhosis of the liver, spider nevi develop deep in the lung and on the pleural surface (Berthelot et al., 1966). These give rise to hypoxemia. As in spider nevi of the skin, these represent dilatation of vessels on the arterial side of the capillary bed. A diffuse dilatation as judged by hypoxemia also develops in acute hepatic necrosis (Williams et al., 1979); if the patient recovers, the hypoxemia corrects. Presumably the blood vessels have recovered their normal level of compliance and no longer create a shunt.

Cavopulmonary anastomoses (CPA) are produced to correct certain congenital heart lesions such as heterotaxy or transposition with pulmonary stenosis. In the classic Glenn anastomosis, the superior vena cava is anastomosed to the right pulmonary artery. In a significant proportion of these cases, "shunts" as in cirrhosis can be identified from the hypoxemia. Rather misleadingly, these lesions have been called pulmonary arteriovenous malformations (PAVMs) (Srivastava et al., 1995). They have been shown to correlate with the diversion of hepatic blood away from lung. Transplantation has led to recovery of the lung after cirrhosis and redirection of hepatic blood back to the pulmonary bed to recovery after CVPA (Laberge et al., 1992).

Cardiolipin

The health of the blood vessels of the embryo and infant can be damaged by the presence in the mother of abnormal immunoglobulins, e.g., lupus anticoagulant or anticardiolipin antibodies. Maternal deficiencies in such clotting factors as protein C, protein S, or antithrombin III are also associated with fetal abnormalities, notably thrombosis of blood vessels in placenta and in lung (Sammaritano et al., 1990; Krous, 1993). Growth retardation is part of this syndrome.

Kidney-Lung Axis

Interference with lung development especially vascular lung development follows disturbance of the kidney-lung axis (deMello and Reid, 1997b). Potter's syndrome describes the constellation of features associated with hypoplastic or absent kidneys. Small lungs and abnormal facies were blamed on reduced urine production and oligohydramnios, causing a failure of lung expansion (Potter 1946a,b, 1965). That oligohydramnios is not a consistent feature (polyhydramnios may be present) calls into doubt this simple interpretation. These small lungs have major reductions in airway number indicating impaired growth well before the 16th week of intrauterine development when urine production is not significant.

The studies of Clemmons (1977) make it clear that it is a metabolic dialogue between kidney and lung that is impaired. In the chick embryo, he demonstrated that the kidney produces much hydroxyproline for export to other organs as a building block for collagen. Nephrotoxins reduced the production of this substance, and radiolabeling showed reduced uptake in the lung. The reduced amniotic fluid is as likely an expression of reduced lung liquid as its cause.

Premature rupture of membranes exerts its effect in the late stages of development. Drainage of lung liquid prepares the lung for delivery. A secreting lung becomes an absorbing one. In the last weeks before normal delivery, the small lungs experience an increase in saccular number. So premature rupture of membranes presumably contributes to small lungs by impairing this final burst of airspace multiplication.

In renal agenesis or dysplasia, the lungs are variously affected. In renal agenesis the total lung volume was reduced in all eight cases studied. The alveoli were smaller than normal. In seven cases, the arteries were reduced in size for age, although in each case the arteries were large for the volume of lung. This was greater in infants with renal agenesis, where the arteries appeared large, tortuous, and crowded. The thickness of the arterial muscular wall was usually less than normal for age. Conventional and supernumerary arteries are both reduced in number presumably following the airway reduction (Hislop et al., 1979).

Congenital Diaphragmatic Hernia (CDH)

Congenital diaphragmatie hernia refers to a lesion present at birth, *not* to an inherited defect, although some few cases of familial CDH are reported. Surgical models of CDH have been produced, notably in sheep (Nobuhara et al., 1998a,b). A lamb is delivered by surgical section, a hole is produced in the diaphragm, and gut is placed in the chest. The lamb is returned to the uterus and gestation proceeds to term. The pattern and degree of disturbed lung growth is modified by the age when the hernia is produced. The earlier the hernia is produced, the greater is the degree of lung growth restriction (DiFiore et al., 1994).

In the vicarious experiments offered by infants delivered with a diaphragmatic hernia (Geggel et al., 1985; Beals et al., 1992), the importance of impaired vascular development is revealed. Animal studies indicate that the cause of the hernia is a failure of the mesenchyme of the pleuroperitoneal fold to differentiate normally, rather than failure of the pleuroperitoneal canal to close as was previously believed (Allan and Greer, 1997a,b; Kluth, 1999). These same studies have confirmed that the hypoplastic lung is secondary to the migration of liver and gut into the peural cavity encroaching on space designed for lung occupancy. Usually both lungs are affected with the ipsilateral more reduced in volume than the contralateral and the lower lobe usually affected more than the upper lobe. Particularly in the more affected regions, airway generation number is greatly reduced, pointing to an early impediment to the lung. Alveoli are too small and too few, although differentiation to alveoli usually occurs. The arterial branches are reduced in number from the hilum to the periphery and are too small for their position in the branching pattern, and too muscular as judged by wall thickness and the peripheral extension of muscularized arteries (Kitagawa et al., 1971, Geggel et al., 1985).

232 D.E. deMello and L.M. Reid

Studies have been done in a nitrofen model in the rat (Allan and Greer, 1997b; Greer et al., 1999). Administration of this pesticide to the dams during mid-gestation results in diaphragmatic hernia in the infants. There is general growth retardation, a widespread toxic effect, not just a localized injury.

SUMMARY

Recent studies in mouse and human lung have revealed that the central vessels develop by angiogenesis, the peripheral ones by vasculogenesis, and that by a third lytic process a luminal connection between the two is established to complete the vascular loop. The venous channels and their central connections are the first to develop.

Early in fetal life, the complete airway number is present. Alveolar and vascular development continues during the fetal period to establish the normal infrastructure present at the age of viability and then at term. Considerable multiplication and remodeling in the alveolar region and the various segments of the vascular loop continue through childhood and into adolescence.

The double arterial supply (bronchial and pulmonary artery) and the double venous drainage ("true" bronchial vein and pulmonary vein) constitute regions of special hemodynamic function, e.g., the pulmonary veins receive all pulmonary artery and bronchial artery drainage from the lungs and pleura; the so-called true bronchial veins drain only a small region at the lung's hilum. The pulmonary artery circulation can also be considered "double." From the axial pulmonary arteries, those that pass to the lung's distal pleural surface, two types of branch arise. The "conventional" or "long" branches run with the airways and supply the capillary bed beyond the terminal while the "supernumerary" or "short" branches, which are three or four times more numerous, typically pass at right angles from the parent artery and quickly divide into a capillary bed to supply the alveolar region abutting the conventional arteries. The supernumerary arteries have a different profile of pharmacologic response from the conventional. A special "baffle" valve, well innervated, is present at the origin of the supernumerary.

With the long evolution of normal lung vascular development, impairment can be caused at any age and by a variety of events, some primarily vascular, some in other systems that modulate growth adaptation by the vascular bed. Examples were given to illustrate cellular and vessel growth adaptation to a variety of conditions, e.g., high pulmonary blood flow, hypoxia, hyperoxia, the cytokine disturbance of cirrhosis of the liver, or the diversion of blood from the lung as in the Glenn shunt.

A metabolic deficiency arising from prenatal absence or injury to the kidney(s) interferes with lung growth due to a selective effect on its arteries. The crowding of the lung that follows a congenital diaphragmatic hernia leads to a hypoplastic lung with serious impairment and remodeling of the pulmonary arterial tree.

The newer techniques serving molecular and genetic biology make possible analysis of the physicochemical changes that control and mediate normal lung development and those that determine and modulate abnormal adaptation. Genomics and proteomics must focus not just on arteries, capillaries, and veins, but on the various anatomic levels identifiable within each segment, and also separately on the various cell types within each segment. New techniques that permit imaging of mesoscale structures, that is, materials just a few atoms or so in diameter (from 1 to 100 nanometers) will produce an overwhelming database. This will challenge the scientist to imagine and create new methods, including new mediators, to manage, treat, and correct disease.

ACKNOWLEDGMENTS

This work is supported in part by National Institutes of Health grant HL 55600. The authors thank Ms. Kim Basler for secretarial assistance.

REFERENCES

Allan, D.W., Greer, J.J. (1997a). Embryogenesis of the phrenic nerve and diaphragm in the fetal rat. J Comp Neurol 382(4):459–468.

Allan, D.W., Greer, J.J. (1997b). Pathogenesis of nitrofen-induced congenital diaphragmatic hernia (CDH) in fetal rats. J Appl Physiol 83(2):338–347.

Antonelli-Orlidge, A., Saunders, K.B., Smith, S.R., et al. (1989). An activated form of transforming growth factor β is produced by cocultures of endothelial cells and pericytes. Proc Natl Acad Sci USA 86:4544–4548.

Battegay, E.J., Raines, E.W., Seifert, R.A., et al. (1991). TGF-β induces bimodal proliferation of connective tissue cells via complex control of an autocrine PDGF loop. Cell 63:515–524.

Beals, D.A., Schloo, B.L., Vacanti, J.P., et al. (1992). Pulmonary growth and remodeling in infants with high-risk congenital diaphragmatic hernia. J Pediatr Surg 27(8):997–1001; discussion, 1001–1002.

Beck, L., Jr., D'Amore, P.A. (1997). Vascular development: cellular and molecular regulation. FASEB J 11(5):365–373.

Berthelot, P., Walker, J.G., Sherlock, S., Reid, L. (1966). Arterial changes in the lungs in cirrhosis of the liver—lung spider nevi. N Engl J Med 274:291–298.

Boggs, S., Harris, M.C., Hoffman, D.J., et al. (1994). Misalignment of pulmonary veins with alveolar capillary dysplasia: affected siblings and variable phenotypic expression. J Pediatr 124:125–128.

Boyden, E.A. (1970). The developing bronchial arteries in a fetus of the 12th week. Am J Anat 129:357–368.

Breier, G., Clauss, M., Risau, W. (1995). Coordinate expression of vascular endothelial growth factor receptor-1 (flt-1) and its ligand suggests a paracrine regulation of murine vascular development. Dev Dyn 204(3):228–239.

Bunton, D., MacDonald, A., Brown, T., Tracey, A., McGrath, J.C., Shaw, A.M. Related Articles 5-hydroxytryptamine- and U46619-mediated vasoconstriction in bovine pulmonary conventional and supernumerary arteries: effect of endogenous nitric oxide. Clin Sci (Lond). 2000 Jan;98(1):81–89.

Bunton, D.C., Shaw, A.M., Fisher, A., McGrath, J.C., Montgomery, I., Daly, C., MacDonald, A. Related Articles V-shaped cushion at the origin of bovine pulmonary supernumerary arteries: structure and putative function. J Appl Physiol. 1999 Dec; 87(6):2348–2356.

Carmeliet, P., Ferreira, V., Breier, G., et al. (1996). Abnormal blood vessel development and lethality in embryos lacking a single VEGF allele. Nature 380:435–439.

Cater, G., Thibeault, D.W., Beatty, E.C., et al. (1989). Misalignment of lung vessels and alveolar capillary dysplasia: a cause of persistent pulmonary hypertension. J Pediatr 114:293–300.

Clemmons, J.J.W. (1977). Embryonic renal injury.: a possible factor in fetal malnutrition. Pediatr Res 11:404.

Cullinane, C., Cox, P.N., Silver, M.M. (1992). Persistent pulmonary hypertension of the newborn due to alveolar capillary dysplasia. Pediatr Pathol 12:499–514.

Davies, G., Reid, L. (1970). Growth of the alveoli and pulmonary arteries in childhood. Thorax 25:669–681.

deMello, D.E. (1999). Structural elements of human fetal and neonatal lung vascular development. In: Weir, K.E., Archer, S.L., Reeves, J.T., eds. The fetal and neonatal pulmonary circulations (AHA monograph series), vol. 4, pp. 37–64. Armonk, NY: Futura.

deMello, D.E., Gashi-Luci, L., Hu, L.M., et al. (1988). Effect of hypoxia on post-natal vascular adaption in the newborn rabbit. FASEB J 2:A1181.

deMello, D.E., Reid, L.M. (1991). Pre and post-natal development of the pulmonary circulation. In: Chernick, V., Mellins, R.B., eds. Basic mechanisms of pediatric respiratory disease: cellular and integrative, pp. 36–54. Philadelphia: BC Decker.

deMello, D.E., Reid, L.M. (1997a). Arteries and veins. In: Crystal, R.G., West, J.B., Barnes, P.J., et al., eds. The lung: scientific foundations, 2nd ed., pp. 1117–1127. Philadelphia: Lippincott-Raven.

deMello, D.E., Reid, L.M. (1997b). The kidney/lung loop. In: Thomas, D.F.M., ed. Urological disease in the fetus and infant, vol. 5, pp. 62–77. Oxford: Butterworth Heinemann.

deMello, D.E., Reid, L.M. (2000). Embryonic and early fetal development of human lung vasculature and its functional implications. Pediatr Dev Pathol 3:439–449.

deMello, D.E., Sawyer, D., Galvin, N., et al. (1997). Early fetal development of the mouse pulmonary vasculature. Am J Respir Cell Mol Biol 16:568–581.

deMello, D.E., Thomas, K., Heyman, S., et al. (1992). Ultrastructural features of resistance arteries of newborn rabbits during perinatal adaptation in normoxia and hypoxia. Lab Invest 66(1):4P.

Dickson, M.C., Martin, J.S., Cousins, F.M., et al. (1995). Defective hematopoiesis and vasculogenesis in transforming growth factor-β 1 knock-out mice. Development 121:1845–1854.

DiFiore, J,W., Fauza, D.O., Slavin, R., et al. (1994). Experimental fetal tracheal ligation reverses the structural and physiological effects of pulmonary hypoplasia in congenital diaphragmatic hernia. J Pediatr Surg 19:248–257.

Dimaio, M., Gil, J., Ciurea, D., et al. (1989). Structural maturation of the human fetal lung: a morphometric study of the development of air blood barriers. Pediatr Res 26:88–93.

Drake, C.J., Little, C.D. (1995). Exogenous vascular endothelial growth factor induces malformed and hyperfused vessels during embryonic neovascularization. Proc Natl Acad Sci USA 92:7657–7661.

Drake, C.J., Hungerford, J.E., Little, C.D. (1998). Morphogenesis of the first blood vessels. Ann NY Acad Sci 857:155–179.

Dunnill, M.S. (1962). Postnatal growth of the lung. Thorax 17:329–333.

Elliott, F.M., Reid, L. (1965). Some new facts about the pulmonary artery and its branching pattern. Clin Radiol 16:193–198.

Ferrara, N., Carver-Moore, K., Chen, H., et al. (1996). Heterozygous embryonic lethality induced by targeted inactivation of the VEGF gene. Nature 380:439–442.

Fong, G.H., Rossant, J., Gertsenstein, M., et al. (1995). Role of the flt-1 receptor tyrosine kinase in regulating the assembly of vascular endothelium. Nature 376:66–70.

Fox, W.W., Gewitz, M.H., Dinwiddie, R., et al. (1977). Pulmonary hypertension in the perinatal aspiration syndromes. J Pediatr 59:205–211.

Geggel, R., Murphy, J., Langleben, D., et al. (1985). Congenital diaphragmatic hernia: arterial structural changes and persistent pulmonary hypertension after surgical repair. J Pediatr 107:457–464.

Greer, J.J., Allan, D.W., Martin-Caraballo, M., Lemke, R.P. (1999). An overview of phrenic nerve and diaphragm muscle development in the perinatal rat. J Appl Physiol (Invited Review) 86(3):779–786.

Haworth, S.G., Hall, S.M., Chew, M., et al. (1987). Thinning of fetal pulmonary arterial wall and postnatal remodeling: ultrastructural studies on the respiratory unit arteries of the pig. Virchows Arch 411:161–171.

Haworth, S.G., Hislop, A.A. (1981). Adaptation of the pulmonary circulation to extra-uterine life in the pig and its relevance to the human infant. Cardiovasc Res 15(2):108–119.

Haworth, S.G., Hislop, A.A. (1982). Effect of hypoxia on adaptation of the pulmonary circulation to extra-uterine life in the pig. Cardiovasc Res 16:293–303.

Haworth, S.G., Reid, L. (1976). Persistent fetal circulation: newly recognized structural features. J Pediatr 88:614–620.

Haworth, S.G., Sauer, U., Buhlmeyer, K., et al. (1977). Development of the pulmonary circulation in ventricular septal defect: a quantitative structural study. Am J Cardiol 40:781–788.

Heimark, R.L., Twardzik, D.R., Schwartz, S.M. (1986). Inhibition of endothelial regeneration by type-β transforming growth factor from platelets. Science 233:1078–1080.

Hislop, A., Haworth, S.G., Shine-Bourne, E.A., et al. (1975). Quantitative structural analysis of pulmonary vessels in isolated ventricular septal defect in infancy. Br Heart J 37:1014–1021.

Hislop, A., Hey, E., Reid, L. (1979). The lungs in congenital bilateral renal agenesis and dysplasia. Arch Dis Child Int 54:32–38.

Hislop, A., Reid, L. (1972). Intrapulmonary arterial development in fetal life-branching pattern and structure. J Anat 113:35–48.

Hislop, A., Reid, L. (1973a). Fetal and childhood development of the intrapulmonary veins in man—branching pattern and structure. Thorax 28:313–319.

Hislop, A., Reid, L. (1973b). Pulmonary arterial development during childhood: branching pattern and structure. Thorax 28:129–135.

Hislop, A., Reid, L. (1977). Formation of the pulmonary vasculature. In: Development of the lung, In: Lung biology in health and disease, vol. 6, pp. 37–86. New York: Marcel Dekker.

Hislop, A., Reid, L. (1981). Growth and development of the respiratory system: anatomical development. In: Davis, J.A., Dobbing, J., eds. Scientific foundations of paediatrics, 2nd ed., pp. 390–431. London: Heinemann.

Janney, C.G., Askin, F.B., Kuhn C. (1981). Congenital alveolar capillary dysplasia—an unusual cause of respiratory distress in the newborn. Am J Clin Pathol 76:722–727.

Jones, R. (1992). Ultrastructural analysis of contractile cell development in lung mirovessels in hyperoxic pulmonary hypertension. Fibroblasts and intermediate cells selectively reorganize non-muscular segments. Am J Pathol 141:1491–1505.

Jones, R., Jacobson, M., Steudel, W. (1999a). Alpha-smooth-muscle actin and microvascular precursor smooth-muscle cells in pulmonary hypertension. Am J Respir Cell Mol Biol 20:582–594.

Jones, R., Steudel, W., White, S., Jacobson, M., Low, R. (1999b). Microvessel precursor smooth muscle cells express head-inserted smooth muscle myosin heavy chain (SM-B) isoform in hyperoxic pulmonary hypertension. Cell Tissue Res 295:453–465.

Jones, R., Zapol, W.M., Reid, L. (1984). Pulmonary artery remodeling and pulmonary hypertension after exposure to hyperoxia for 7 days. A morphometric and hemodynamic study. Am J Pathol 117:273–285.

Kaipainen, A., Korhonen, J., Pajusola, K., et al. (1993). The related flt4, flt1, and kdr receptor tyrosine kinases show distinct expression patterns in human fetal endothelial cells. J Exp Med 178(6):2077–2088.

Kitagawa, N., Hislop, A., Boyden, E.A., et al. (1971). Lung hypoplasia in congenital diaphragmatic hernia. A quantitative study of airway, artery and alveolar development. Br J Surg 58:342–346.

Kluth, D. (1999). Congenital diaphragmatic hernia: the impact of embryological studies. Personal communication.

Korhonen, J., Polvi, A., Partanen, J., et al. (1994). The mouse *tie* receptor tyrosine kinase gene: expression during embryonic angiogenesis. Oncogene 9:395–403.

Krous, F.T. (1993). Placental thrombi and related problems. Semin Diagn Pathol 10:275–283.

Laberge, J.M., Brandt, M.L., Lebecque, P., et al. (1992). Reversal of cirrhosis-related pulmonary shunting in two children by orthotopic liver transplantation. Transplantation 53:1135–1138.

Langston, C. (1991). Misalignment of pulmonary veins and alveolar capillary dysplasia. Pediatr Pathol 11:163–170.

Meyrick, B., Reid, L. (1979). Ultrastructural features of the distended pulmonary arteries of the normal rat. Anat Rec 193:71–97.

Meyrick, B., Reid, L. (1981). The effect of chronic hypoxia on pulmonary arteries in young rats. Exp Lung Res 2:257–271.

Murphy, J.D., Rabinovitch, M., Goldstein, J.D., et al. (1981). The structural basis of persistent pulmonary hypertension of the newborn infant. J Pediatr 98:962–967.

Murphy, J.D., Vawter, G., Reid, L. (1984). Pulmonary vascular disease in meconium aspiration. J Pediatr 104:758–762.

Nobuhara, K.K., DiFiore, J.W., Ibla, J.C., et al. (1998b). Insulin-like growth factor-I gene expression in three models of accelerated lung growth. J Peditar Surg 33:1057–1061.

Nobuhara, K.K., Fauza, D.O., DiFiore, J.W., et al. (1998a). Continuous intrapulmonary distension with perfluorocarbon accelerates neonatal (but not adult) lung growth. J Pediatr Surg 33:292–298.

Noden, D.M. (1989). Embryonic origins and assembly of blood vessels. Am Rev Respir Dis 140:1097–1103.

Ogawa, M., Kisumoto, M., Nishikawa, S., Fujimoto, T., Kodama, H., Nishikawa, S.I. (1999). Expression of α4-integrin defines the earliest precursor of hematopoietic cell lineage diverged from endothelial cells. Blood 93:1168–1177.

Ohar, J.A., Waller, K.S., deMello, D.E., Lagunoff, D. (1991). Administration of chronic intravenous platelet activating factor induces pulmonary arterial contracture and hypertension in rabbits. Lab Invest 65:451–458.

Ohar, J.A., Waller, K.S., deMello, D.E., Williams, T., Lukes, D. (1998). Computerized morphometry of the pulmonary vasculature over a range of intravascular distending pressures. Anat Rec 252(1):92–101.

Ohar, J.A., Waller, K.S., Pantano, J., deMello, D.E., Dahms, T.E. (1994). Chronic platelet-activating factor induces a decease in pulmonary vascular compliance, hydroxyproline and loss of vascular matrix. Am J Respir Crit Care Med 149:1628–1634.

O'Rahilly, R., Muller, F. (1987). Developmental stages in human embryos. Washington, DC: Carnegie Institution of Washington.

Partanen, J., Puri, M.C., Schwartz, L., et al. (1996). Cell autonomous functions of the receptor tyrosine kinase TIE in a late phase of angiogenic capillary growth and endothelial cell survival during murine development. Development 122:3013–3021.

Potter, E.L. (1946a). Bilateral renal agenesis. J Pediatr 29:68–76.

Potter, E.L. (1946b). Facial characteristics of infants with bilateral renal agenesis. Am J Obstet Gynecol 51:885–888.

Potter, E.L. (1965). Bilateral absence of ureter and kidney. Am J Obstet Gynecol 25: 3–12.

Powell, P., Wang, C.C., Horunuchi, H., et al. (1998). Differential expression of fibrolast growth factors 1–4 and ligand genes in late fetal and early post-natal rat lung. Am J Respir Mol Biol 19:563–572.

Rabinovitch, M., Gamble, W.J., Miettinen, O.S., Reid, L. (1981). Age and sex influence on pulmonary hypertension of chronic hypoxia and on recovery. Am J Physiol 240:H62–H72.

Rabinovitch, M.S., Haworth, S., Vance, Z., et al. (1980). Early pulmonary vascular changes in congenital heart disease studied in biopsy tissue. Hum Pathol 11:499–509.

Reid, L. (1959). The connective tissue septa in the adult human lung. Thorax 14:138–145.

Reid, L. (1978). The pulmonary circulation: remodeling in growth and disease. The 1978 J. Burns Amberson Lecture. Am Rev Respir Dis 119:531–546.

Reid, L. (1994). Structural remodeling of the pulmonary vasculature by environmental change and disease. In: Wagner, W.W. Jr., Weir, E.K., eds. The pulmonary circulation and gas exchange, vol. 5, pp. 77–105. Armonk, NY: Futura.

Reid, L., Fried, R., Geggel, R., Langleben, D. (1986). Anatomy of pulmonary hypertensive states. In: Bergofsky, E.H., ed. Abnormal pulmonary circulation, vol. 7, pp. 221–263. New York Churchill Livingstone.

Reid, L., Rubino, M. (1959). The connective tissue septa in the fetal human lung. Thorax 14:3–13.

Robertson, B. (1967). The normal intrapulmonary arterial pattern of the human late fetal and neonatal lung. Acta Paediatr Scand 56:249.

Sammaritano, L.R., Gharavi, A.E., Lockshin, M.D. (1990). Antiphospholipid antibody syndrome: immunologic and clinical aspects. Semin Arthritis Rheum 20:81–96.

Sandison, J.C. (1928). Observations on the growth of blood vessels as seen in the transparent chamber introduced into the rabbits ear. Am J Anat 41:475–496.

Sato, Y., Rifkin, D.B. (1989). Inhibition of endothelial cell movement by pericytes and smooth muscle cell: activation of a latent transforming growth factor-β1-like molecule by plasmin during co-culture. J Cell Biol 109:309–315.

Shalaby, F., Rossant, J., Yamaguchi, T.P., et al. (1995). Failure of blood-island formation and vasculogenesis in Flk-1 deficient mice. Nature 376:62–66.

Shifren, J.L., Doldi, N., Ferrara, N., et al. (1994). In the human fetus, vascular endothelial growth factor is expressed in epithelial cells and myocytes, but not vascular endothelium: implications for mode of action. J Clin Endocrinol Metab 79(1):316–322.

Srivastava, D., Preminger, T., Lock, J.E., et al. (1995). Hepatic venous blood and the development of pulmonary arteriovenous malformations in congenital heart disease. Circulation 92:1217–1222.

Wagenvoort, C.A. (1986). Misalignment of lung vessels: a syndrome causing persistent neonatal pulmonary hypertension. Hum Pathol 17:727–730.

Wagenvoort, C.A., Wagenvoort, N. (1967). Arterial anastomoses, broncho-pulmonary arteries and pulmo-bronchial arteries in perinatal lungs. Lab Invest 16:13.

Williams, A., Trewby, P., Williams, R., Reid, L. (1979). Structural alterations to the pulmonary circulation in fulminant hepatic failure. Thorax 34:447–453.

Wohrley, J.D., Frid, M.G., Moiseeva, E.P., Orton, E.C., Belknap, J.K., Stenmark, K.R. (1995). Hypoxia selectively induces proliferation in a specific subpopulation of smooth muscle cells in the bovine neonatal pulmonary arterial media. J Clin Invest 96:273–281.

CHAPTER **11**

Vascularization of the Placenta

Ronald J. Torry, Joanna Schwartz, and Donald S. Torry

The development of a functional vascular network involves vasculogenesis and angiogenesis. Vasculogenesis is the de novo formation of blood vessels from the differentiation of angioblasts (Flamme et al., 1997; Risau, 1997). These vessels then expand by angiogenesis, the formation of new blood vessels from preexisting vessels, by either sprouting or nonsprouting mechanisms (Risau, 1997). There have been intensive efforts to determine the molecular mechanisms regulating vascular growth and development (Carmeliet and Collen, 1997), and much of the rationale for this has stemmed from the increasing clinical importance (Folkman, 1995) and therapeutic potential of modulating angiogenesis during various disease states (Battegay, 1995; Isner and Asahara, 1999). In normal adult tissue, vascular growth is a relatively rare event, indicated by capillary endothelial cell proliferation rates of only 0.01% to 0.14% in most tissues (Engerman et al., 1967). Significant angiogenesis in the adult is generally limited to pathologic conditions such as arthritis, retinopathy, chronic inflammation, wound healing, and solid tumor growth. However, the female reproductive tract of primates undergoes substantial vascular growth and remodeling associated with the menstrual cycle and pregnancy (Findlay, 1986; Reynolds et al., 1992; Gordon et al., 1995; Torry and Torry, 1997; Rees and Bicknell, 1998).

This chapter focuses on the vascular development of the placenta. Specifically, it briefly reviews changes in the maternal vasculature during implantation and early placentation, expansion of the placental vasculature, remodeling of the maternal arterial vasculature during pregnancy, and conditions that alter placental vascularity. Finally, since relatively little is known regarding key regulators of vascular growth in the placenta, this chapter highlights some of the newer and/or more prominent potential regulators of vascular growth in the placenta. Due to differences in placentation between species (Leiser and Kaufmann, 1994) and the unique characteristics of some types of aberrant placentation in humans, we concentrate on the vascularity of the human placenta. However, since mechanistic studies are often necessarily limited in humans, we include key animal studies when warranted.

ROLE OF THE VASCULATURE IN PLACENTATION

Development of the human placenta is initiated by attachment of the conceptus to the endometrium of the mother during implantation. Although the regulation of this process is likely controlled by the trophoblast, ultimately viability of the conceptus is dependent on at least three temporally different vascular processes: (1) adequate uterine angiogenesis/vascularity at the time of implantation, (2) development and expansion of the villous vasculature, and (3) remodeling of the maternal circulation. Conceptually, these vascular processes are not unlike those associated with the growth of solid tumors, and several characteristics of the resulting placental vasculature are similar to those observed in tumors (Torry and Rongish, 1992).

Implantation and placentation require a complex and well-coordinated series of events (Cross et al., 1994; Jaffe, 1998), the cellular and molecular mechanisms of which are beginning to be elucidated (Rinkenberger et al., 1997; Cross, 1998). Disruption of the early events of placentation is important since many human pregnancies abort before the pregnancy is clinically evident (Wilcox et al., 1988). Inadequacies in later events (i.e., poor remodeling of the maternal vasculature) contributes to the pathophysiology of preeclampsia and/or intrauterine growth retardation (IUGR) of the fetus (Khong et al., 1986; Starzyk et al., 1999). Therefore, vascular insufficiencies during development of the placenta could profoundly affect maternal/fetal outcome (Torry et al., 1998a).

The functional significance of angiogenesis in reproductive physiology has been demonstrated experimentally (Klauber et al., 1997). Inhibition of angiogenesis with a single dose of AGM-1470 (O-chloracetylcarbamoyl fumagillol), an analogue of fumagillin with antiangiogenic effects (Ingber et al., 1990), profoundly interrupted placentation in mice. AGM-1470 treatment on embryonic day 1 (before implantation) or embryonic day 7 (after implantation) resulted in either no placental labyrinth development or decreased labyrinth development, respectively. In both cases, resorption of all embryos occurred. Uterine vascularity was also decreased significantly in both pregnant and nonpregnant mice treated with AGM-1470 (Klauber et al., 1997). These results confirm that angiogenesis is a critical component of normal endometrial maturation and subsequent implantation, and highlights the importance of the vasculature in placentation and reproductive physiology in general (Jaffe, 2000).

VASCULAR DEVELOPMENT AND ARCHITECTURE

Endometrium

Following menstruation, the existing vasculature in the stratum basalis gives rise to a new capillary network that proliferates rapidly to maintain the integrity of the growing endometrium during the next menstrual cycle (Kaiserman-Abramof and Padykula, 1989). These capillaries eventually differentiate into arterioles and arteries as they become enveloped by smooth muscle cells (Ramsey and Donner, 1980). Of particular interest for this review is the subepithelial capillary network since the development of this rich vascular network is critical for successful implantation and pregnancy.

Endometrial angiogenesis occurs during the proliferative and secretory phases of the human menstrual cycle (Goodger and Rogers, 1995b; Rogers et al., 1998). Although human endometrial explants demonstrate peaks of angiogenic-migratory activity *in vitro* (Rogers et al., 1992), there is such variability in capillary endothelial cell mitotic indices throughout the menstrual cycle that discrete periods of endothelial cell proliferation (Goodger and Rogers, 1994; Gargett et al., 1999) or endothelial cell expression of angiogenic integrin $\alpha_v\beta_5$ cannot be determined (Rogers et al., 1998). These findings suggest that endothelial cell migration and proliferation are not necessarily linked processes in the endometrium (Goodger and Rogers, 1995b). Regardless of the exact mechanism involved, the magnitude of vascular growth appears to keep pace with growth of the endometrial stroma since there are no significant changes in vascular density noted throughout the human menstrual cycle (Shaw et al., 1979). Vascular growth is required to support the proliferation and repair of the endometrium during the menstrual cycle, and it provides a richly vascularized, receptive endometrium for implantation (Enders et al., 1983). Indeed, endometrial angiogenesis is a critical modulator of successful implantation (Klauber et al., 1997) and it is also associated with several endometrial disease processes (Abulafia and Sherer, 1999). The exact regulator(s) of endometrial vascular growth is (are) not known; however, the tissue is a rich source of many angiogenic molecules (Giudice, 1994; Smith, 1998).

In addition to vascular growth, there are changes in vascular permeability throughout the menstrual cycle that facilitate progression of the thin, dense menstrual endometrium to the thickened, highly edematous secretory endometrium. The changes in vascular permeability of the uterine capillary network result in edema formation and can be associated with extravascular fibrin deposition (Torry et al., 1996). Increased vascular permeability, fibrin deposition, and blood vessel growth are also associated with corpus luteum formation in humans (Kamat et al., 1995). Finally, the existing capillaries of the endometrium also undergo significant dilation in the postovulatory period (Peek et al., 1992). These anatomic and functional changes in the microvasculature of the endometrium facilitate successful implantation (Enders et al., 1983; Klauber et al., 1997).

Implantation

Ethical concerns have discouraged examination of the vasculature during early blastocyst implantation in humans. Therefore, much of our knowledge of the vascular changes associated with early implantation is based on animal studies, and the sometimes significant differences in placentation between species may preclude extrapolation to human pregnancies. Nevertheless, there are several consistent vascular changes within the maternal uterine vasculature during implantation. One change is a significant increase in vascular permeability at the implantation site (Abrahamsohn et al., 1983; Christofferson and Nilsson, 1988) that occurs even prior to blastocyst invasion (Christofferson and Nilsson, 1988). Ultrastructurally, the endothelial cells of uterine capillaries near the implant site are thickened and demonstrate indices of general metabolic activation. These changes, which occur mostly in venous capillaries and postcapillary venules, are not evident at sites distant from the implantation area (Abrahamsohn et al., 1983). Interestingly, changes in vessel permeability also occur at nonuterine implantation sites, suggesting the changes are mediated by inherent properties of the embryo (Rogers et al., 1988). The exact mechanism promoting vascular permeability at the early implantation site is not known.

Remodeling of endometrial capillaries at the time of implantation also occurs. The implantation site itself is surrounded by capillaries with larger diameters than those capillaries at areas distant from the implantation site in both rats (Welsh and Enders, 1991; Rogers, 1992; Tawia and Rogers, 1992) and rabbits (Leiser and Beier, 1988). In the mink, with an endotheliochorial placenta, the transition to the intermediate placenta is characterized by proliferation of both maternal and fetal blood vessels, and, as the pregnancy progresses, the maternal capillaries are transformed into large sinusoids (Pfarrer et al., 1999b). Early vascular changes in the developing porcine placenta (nonimplanting, epitheliochorial placenta) are essentially similar to those noted above with the subepithelial capillaries demonstrating variable diameters from 3 to 14 μm in the precontact (days 9.5 to 12.5 post coitum) stage (Dantzer and Leiser, 1994). These morphologic changes in capillaries may be mediated by trophoblast interactions. For example, in the early lacunar stage of macaques (1 to 2 days after initiation of implantation), the cytotrophoblast and syncytiotrophoblast are located at the maternal side of the vascular lacunae. Later, syncytiotrophoblasts line the lacunae and are contiguous with the maternal circulation (Enders, 1995). Within 8 days after onset of implantation, cytotrophoblasts are in the lumen and media of spiral arterioles and have displaced maternal endothelial cells.

The permeability and remodeling changes noted above are also accompanied by endothelial cell proliferation, which increases both before and during implantation in the rat (Welsh and Enders, 1991; Goodger and Rogers, 1993). Endometrial endothelial cell proliferation increases significantly from day 3 to day 5 of pregnancy. After implantation, the proliferation index increases dramatically but only at the site of implantation (Goodger and Rogers, 1993). Endothelial cell proliferation appears to keep pace with expansion of the stromal tissue since there are no significant differences in endothelial cell density either before or after implantation (Goodger and Rogers, 1995a). By late day 8 in the rat, migrating maternal endothelial cells are found in the mesometrial chamber lumen, and by day 9 there is prominent endothelial cell mitosis and clear vascular continuity between maternal sinusoids and the mesometrial chamber (Welsh and Enders, 1991). Since these findings were independent of direct conceptus contact at the time, the authors proposed that maternal angiogenesis provides for initial vascular supply before formation of the chorioallantoic placenta (Welsh and Enders, 1991).

First Trimester

The earliest studies performed on human samples typically entail specimens obtained 4 to 5 weeks postconception (p.c.) and usually involve spontaneous abortion or planned termination of apparently normal pregnancies. In humans, vasculogenesis is noted ultrastructurally in early tertiary villi at 21 to 22 days p.c. (Demir et al., 1989). At this stage, clusters of "haemangiogenetic cells" are present and some demonstrate primitive lumen formation. By day 28 p.c., most villi demonstrate vasculogenic activity as well as the presence of true capillaries with larger lumens and flattened endothelial cells. The basal lamina is still not evident, but presumptive pericytes are in close apposition to the basal aspect of the vessels.

Formation of the placental capillary basal laminae begins at about 6 weeks p.c., but complete capillary basal laminae appear only during the last trimester of pregnancy. Vasculogenesis appears sporadically through 14 weeks p.c. Thus, prevasculogenic stages occur prior to 22 days p.c., while early placental vasculogenesis occurs from day 22 to about the 26th week p.c. (Demir et al., 1989).

Histomorphometric analysis of normal pregnancies terminated between 6 and 15 weeks of gestation has demonstrated an increase in the vascularity of human villi (Jauniaux et al., 1991). More recent studies utilizing anti-CD34 immunohistochemistry to identify mature and precursor endothelial cells (te Velde et al., 1997) have generally confirmed these findings. The presence of hemangioblastic cords without obvious lumens is evident in early samples from the first trimester placenta, while none of these cords are found in term placenta. In addition, morphometric techniques demonstrated that vascularity of the villi increase through the first trimester. This increase is due to an increase in the number of peripheral vascular structures (those situated against the trophoblast layer), since central vessels (those with no connection to trophoblast) remained largely unchanged during the first trimester (Fig. 11.1). This, coupled with a selective reduction in villous stromal area throughout the same time period produced a steady increase in the vascular volume density of each villus (te Velde et al., 1997). The authors concluded that the formation of mature vascular structures from primitive hemangioblastic cords is the main mechanism for vascular development within the villi and that a process of margination of the capillaries occurs in the first trimester (te Velde et al., 1997).

The origin of villous hemangioblasts is not known. Based strictly on morphologic criteria, they are thought to differentiate from mesenchymal cells (Demir et al., 1989), and ultrastructual studies have demonstrated a close spatial relationship between villous mesenchymal cells and cytotrophoblasts during endothelial cell differentiation and vasculogenesis in 12 to 14-week human villi (Asan et al., 1999). These results suggest that cytotrophoblast may directly or indirectly influence early vasculogenesis during human placentation, but more sophisticated cell biologic, biochemical, or molecular studies to confirm this hypothesis have not, to our knowledge, been performed. Similarly, the contribution, if any, of circulating endothelial stem cells (Isner and Asahara, 1999) to the vascularization process of the early placenta is not known.

Second and Third Trimesters

The human placenta experiences dramatic growth including elongation of the terminal villi during the second trimester (Jackson et al., 1992). The villi also undergo remodeling as characterized by increased villus capillarization, decreased villus diameters, and decreased thickness of the villus membranes. These adaptive changes help to decrease the diffusion distances between fetal and maternal circulations. The decrease in villus cross-sectional area and circumference as well as the relative increase in capillarity of the existing villi are complete by the 28th week of gestation in humans (Stoz et al., 1988).

At term, the human stem villi contain paired vessels, both surrounded by smooth muscle cells, but the vein typically has a larger diameter than the artery (Tanaka et al., 1999). The terminal villi microvasculature is quite extensive with an average vessel diameter of 12.3 to 14.5 μm and a length of 3 to 5,000 μm. This allows

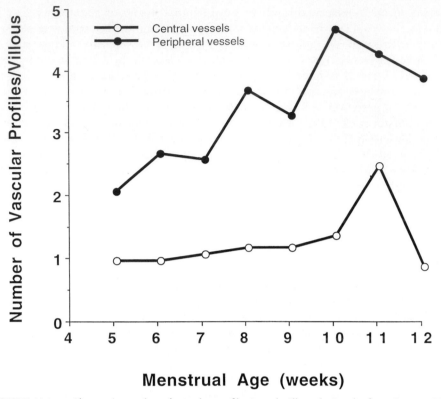

FIGURE 11.1.   Changes in number of vascular profiles in each villous during the first trimester. Vascular structures were identified using anti-CD34 immunocytochemistry. The number of peripherally situated vasculature structures increased throughout the first trimester while the central vessels remained largely unchanged. (Adapted from te Velde et al., 1997.)

one coiled capillary loop to perfuse three to five terminal villi (Kaufmann et al., 1985). More recently, using confocal microscopy, the terminal villi capillaries are shown to exist either in a single loop or anastomotic system (Jirkovska et al., 1998). In the bovine terminal villi, there are substantial capillary interconnections, forming an extensive vascular network (Leiser et al., 1997).

The vasculosyncytial membrane, which encompasses the distance between the intervillous space and a villus capillary, represents an important area for the exchange of materials between mother and fetus. This barrier is only 1 to 2 μm thick and results from the remodeling and obtrusion of segments of the fetal capillaries into the trophoblast layer of the villi (Burton and Tham, 1992). Although some describe sinusoidal-like dilations of parts of the capillary loops (Kaufmann et al., 1985), more recent studies have failed to confirm this finding (Jackson et al., 1992). Nevertheless, the thickness of this barrier decreases with increasing gestational age within the first trimester (Jauniaux et al., 1991). Remodeling of the villous capillaries and the villi themselves are thought to facilitate the exchange of materials between fetal and maternal circulations (Jackson et al., 1992).

One of the greatest vascular changes that occurs in human pregnancy is the substantial arterial remodeling initiated by trophoblast invasion of the maternal circulation (Jaffe, 1998). The cytotrophoblasts colonize the spiral arteries/arterioles where they displace maternal endothelial cells (Enders and Blankenship, 1997). This appears to be an innate feature of trophoblasts since similar processes occur at nonuterine implantation sites (Rogers et al., 1988). In addition to anchoring the fetus to the uterine wall, the invasion of trophoblasts into the maternal circulation converts each vascular segment from a high-resistance/low-capacitance vessel to a low-resistance/high-capacitance vessel (Khong et al., 1986). This "physiological change" within spiral arteries facilitates increased maternal blood flow into the intervillous spaces. In humans, some vascular remodeling precedes direct invasion of the spiral arteries by trophoblasts (Craven et al., 1998). Although the invading cytotrophoblasts demonstrate a preference for the maternal vasculature in the macaque (Enders et al., 1996), this preference in the human is debatable (Pijnenborg, 1998).

The endovascular and stromal invasion of trophoblasts requires cell-cell and cell-matrix interactions through various adhesion molecules and corresponding receptors. The expression of integrins on the cell surface of trophoblasts is thought to play a critical role in determining their gene expression and cellular behavior (Burrows et al., 1996). Indeed, some studies have shown that trophoblasts change their adhesion receptor expression profile to mimic many of those expressed by maternal endothelial cells (Blankenship and Enders, 1997a; Damsky and Fisher, 1998), including some expressed by angiogenic endothelial cells (Zhou et al., 1997b). The invasive cytotrophoblast also produces various proteases, which facilitates its ability to migrate as well as remodel the maternal arterial walls (Blankenship and Enders, 1997b). Interestingly, cytotrophoblast from preeclamptic placenta fail to change integrin expression (Zhou et al., 1993) and fail to express certain endothelial cell adhesion molecules (Zhou et al., 1997a). These results suggest that altered trophoblast adhesion molecule expression is necessary for normal remodeling of the maternal circulation during pregnancy. Recent studies have also shown that $in\ vivo$ inhibition of $\alpha_v\beta_3$ integrin, an integrin that is required for angiogenesis (Brooks et al., 1994), significantly reduced the number of implantation sites in mice (Illera et al., 2000). Thus, appropriate integrin expression/function at the time of implantation as well as later in pregnancy is necessary for successful placentation.

Despite relatively consistent anatomic data regarding vascular formation in the human placenta, there is some conflicting evidence as to when during pregnancy the maternal vasculature becomes perfused (Jaffe et al., 1997). Based primarily on classic histologic studies, the presence of maternal red blood cells in the developing intervillous spaces led to the conclusion that maternal blood flow is established within the first month of conception (Meekins et al., 1997). Doppler flow studies have demonstrated the presence of maternal blood flow within the first trimester of placental development (Kurjak et al., 1993, 1997). Others however suggest that trophoblast plugs obstruct the maternal circulation and prevent blood flow into the intervillous spaces. These plugs do not loosen and allow detectable blood flow until 12 to 13 weeks after conception (Foidart et al., 1992; Jaffe et al., 1997). Recent anatomic results suggest that placental perfusion before 8 weeks is relatively poor due to extensive torturosity of the connections between maternal arteries and the intervillous spaces, and frank arterial connections are not evident until 11 to 12 weeks of pregnancy (Burton et al., 1999). These results are supported by recent evidence of high expression of hypoxia inducible factor-1α (HIF-1α) in human placenta at 5 to 8 weeks of gestation. Placental expression of HIF-1α is absent at 11 to 14 weeks of gestation (Caniggia et al., 2000), a time that corresponds to increased placental oxygenation (Rodesch et al., 1992). Thus, the placenta is relatively hypoxic during the first trimester of development, and premature initiation of maternal blood flow may indeed be characteristic of complicated pregnancies (Jauniaux et al., 1994). The placenta is unique since development of an anatomically complete vasculature may not translate immediately into a physiologically functional circulation.

Although vascular development in the placenta appears to be tightly controlled, the regulatory mechanisms are not completely understood. Knowledge of these mechanisms could have clinical importance in understanding vascular alterations associated with spontaneous abortion and preeclampsia/IUGR, and it may provide insight into the ability of the placenta to undergo vascular adaptations during periods of chronic hypoxia.

ALTERATIONS IN PLACENTAL VASCULARITY IN HUMANS

Spontaneous Abortion

A definitive link between vascularity during early placentation in humans and clinical outcome of the fetus is lacking. Some studies demonstrate an association between deficient chorionic villous vascularity and spontaneous abortion (Meegdes et al., 1988). Vascularized villi composed 89% of the villi in normal placenta following legal abortion, but decreased to 26% in placenta with embryonic death and decreased to only 9% in placenta with blighted ova. In addition, the vascular density was three to four times greater in villi from normal placenta compared to vascularized villi from spontaneously aborted groups (Meegdes et al., 1988). Although it is difficult to determine causality, early villous vascularity may influence the ultimate outcome of pregnancy. Others have found an increase in vascular density in the decidua parietalis, but not the decidua basalis, during spontaneous abortion within the first trimester (Vailhe et al., 1999). However, it is not known if the increased vascularity is causally related to the abortion or is a reflection of the disruption of placentation in failing pregnancies.

Chronic Hypoxia

Fetal capillaries in human placenta are not static since they have been shown to adapt to a variety of hypoxic stresses. Recently, the fetal capillary network has been shown to adapt to hypoxia associated with chronic maternal smoking (Pfarrer et al., 1999a). A denser network and/or a high degree of coiling characterize the terminal villous capillaries in smokers. These adaptive changes are thought to increase the surface area for diffusion (Pfarrer et al., 1999a) and may be responsible for the increased placenta weights in women who smoke during pregnancy (Williams et al., 1997).

Human reproduction at high altitude is usually associated with a reduction in mean birth weight and placental weight or volume (Reshetnikova et al., 1994; Ali et al., 1996; Khalid et al., 1997), which correlates with the level of hypobaric hypoxia experienced by the fetus (Reshetnikova et al., 1994). However, reported changes in placental weight (Kruger and Arias-Stella, 1970; Reshetnikova et al., 1994; Ali et al., 1996; Burton et al., 1996; Khalid et al., 1997) and villous surface area (Jackson et al., 1987; Reshetnikova et al., 1994) are conflicting and probably result from, among other factors, differences in altitude, maternal characteristics, and/or technical aspects between studies. Nevertheless, it is clear that the placenta can undergo vascular adaptations so as to increase its diffusion capacity at altitude (Reshetnikova et al., 1994; Mayhew, 1998). One of the most influential variables on diffusion within the placenta is the mean thickness of the villous membrane (Mayhew et al., 1986). Within intermediate and terminal villi, there is a significant reduction in both the arithmetic and harmonic mean thickness of the villous membrane with increasing altitude (Reshetnikova et al., 1994). These changes are also associated with a significant increase in the fractional volume occupied by fetal capillaries (Reshetnikova et al., 1994; Ali et al., 1996; Burton et al., 1996) and trends toward increasing capillary length and diameter (Reshetnikova et al., 1994). Consequently, the morphometric diffusing capacity of the villous membrane is increased. Decreased diffusion distances within the placenta at high altitude is thought to be due to the peripheralization of villous capillaries (Jackson et al., 1988). Normally, fetal capillaries are relatively unbranched, long vessels that may perfuse several villi in series. Following chronic hypoxia, there is a series of shorter, more branched capillary loops (Ali et al., 1996), and these vascular adaptations may cause a concomitant change in the pattern of villous branching. This increased villous capillary branching and coiling in the human placenta has been described as villous hypercapillarization (Kaufmann et al., 1985). Chronic hypoxic hypoxia also increases capillary bed branching and coiling in sheep (Krebs et al., 1997) and guinea pig (Scheffen et al., 1990) placentae. The resulting increased diffusion capacity of the placental microvasculature is thought to facilitate oxygen supply to the fetus during chronic hypoxia.

Maternal anemia is another hypoxic stress in which the terminal villi must undergo structural adaptations to maintain oxygen-diffusing capacity. Large-scale studies (Godfrey et al., 1991) have shown that maternal anemia is associated with increased placenta weight and placenta weight to birth weight ratio. There are also adaptations of the placental vasculature during maternal anemia. At term, there is a reduction in villous volume and surface areas of anemic placentae that is due in part to diminished growth of the villous tree and a high incidence of placental infarction (Burton et al., 1996). However, diffusion capacity is maintained principally through a reduction in thickness of the villous membrane and increases in the villous capillary volume fraction (Reshetnikova et al., 1995) and surface density (Burton et al., 1996). The increase in capillary volume fraction could be due to larger capillaries and/or vessel proliferation since mean capillary diameter tended to be greater and the ratio of capillary to villous length was significantly greater with anemia (Burton et al., 1996). Recently, others have also found alterations to the normal pattern of placental villous capillary development during the first trimester of pregnancy in anemic women. The early alterations are characterized by increased number of proliferating stromal cells and increased fetal capillary density in the outer part of the villous stroma (Kadyrov et al., 1998).

Interestingly, intermittent exposure to normobaric hypoxia during pregnancy produces adaptive changes in the vascularity of human myometrium (Rakusan et al., 1999). Intermittently breathing a hypoxia gas mixture (12% to 14% O_2) daily during weeks 30 to 32 of pregnancy produced a 27% increase in vascular numerical density and a 15% increase in average vascular area in myometrial tissue samples compared to samples from women kept on room air during that time period of pregnancy. These vascular changes resulted in a 49% increase in total vascular volume density; there was no significant change in myocyte or stromal cell volume density or newborn weights between groups. These results confirm that even intermittent hypoxia can initiate vascular growth in the myometrium of pregnant women.

The exact mechanisms mediating the placental villous and vascular adaptations during hypoxia are not known. Previous studies have shown that the expression of vascular endothelial growth factor (VEGF) is upregulated in placental fibroblasts (Wheeler et al., 1995) and trophoblasts (Shore et al., 1997) during hypoxia. VEGF has been proposed to be responsible for some of the vascular adaptations of anemic hypoxia including the increased diameter and proliferation of capillaries as well as the increased branching and coiling of the capillaries demonstrated in some studies (Kadyrov et al., 1998). Another possibility is that the increased capillary diameter is caused by increased blood flow during hypoxic stress. This increased blood flow to villi would increase sheer stress that may lead to vascular remodeling. Similarly, increased perfusion pressure may increase capillary diameter and contribute to the observed thinning of the villous membrane (Burton et al., 1996). Experimentally, increased perfusion pressure results in significantly greater villous endothelial cell proliferation than perfusion at lower pressures (Karimu and Burton, 1994). Whether the placental vascular adaptations during chronic hypoxia are a direct response to hypoxia, growth factors, or other mechanical factors is not known.

Clearly, uncomplicated pregnancy in the face of hypoxic stress can be partly if not mostly attributed to vascular adaptations that serve to maintain diffusion capacity within the placenta. Although these vascular adaptations are important in maintaining adequate oxygen and nutrient supply to the fetus, the mechanisms mediating these changes are unknown.

Preeclampsia and IUGR

Preeclampsia is multifactorial disease process (Brown, 1995; Vinatier and Monnier, 1995; van Beck and Peeters, 1998) most likely involving immune components (Taylor, 1997), inflammation (Redman et al., 1999), and genetic (Sutherland et al., 1981), cytokine (Conrad and Benyo, 1997), and vascular factors (Taylor et al., 1998; Torry et al., 1998b; Hayman et al., 1999). The involvement of these components culminates in generalized maternal endothelial cell dysfunction that is characteristic of preeclampsia (Taylor, 1997; Roberts, 1998; van Beck and Peeters, 1998; Redman et al., 1999). Classically, vascular pathologies associated with preeclampsia involve inadequate trophoblast invasion and remodeling of the uterine spiral arteries (Starzyk et al., 1999). A key question, yet to be answered, is whether inadequate trophoblast invasiveness and vascular remodeling result from an intrinsic defect in the trophoblast of preeclamptic placentae or whether extrinsic factors in the maternal environment mediate the effects (Pijnenborg, 1998). (A complete review of the pathophysiology and etiology of preeclampsia is beyond the scope of this chapter.)

Although there is variability in both the number of myometrial and decidual arteries invaded as well as the extent of trophoblast invasion in individual arteries (Meekins et al., 1994), it is generally well accepted that shallow trophoblast invasion results in reduced placental perfusion associated with preeclampsia (van Beck and Peeters, 1998). Insufficient trophoblast invasion results in more tortuous arteries with smaller lumina and thicker walls compared to normal placental bed arteries (Starzyk et al., 1997), and these structural changes probably contribute to reduced placental perfusion and placental/fetal hypoxia during the second and third trimesters (Kingdom and Kaufmann, 1999). Indeed, morphometric studies have confirmed various vascular deficiencies in pregnancies culminating in IUGR or small-for-date infants. Most studies describe a reduction in the arterial/arteriolar number (Giles et al., 1985, 1993; McCowan et al., 1987; Bracero et al., 1989; Kreczy et al., 1995) as a main determinant of the increased placental vascular resistance associated with IUGR, but controversy exists as to whether the mechanism responsible for the decrease in the arterial tree is a result of blunted angiogenesis (Bracero et al., 1989; Kreczy et al., 1995) or selective obliteration of arteries/arterioles (Giles et al., 1985; McCowan et al., 1987). Recent studies have challenged these results (Macara et al., 1995) and attribute the placental vascular insufficiencies to reduced capillarity (Teasdale, 1984; Krebs et al., 1996; Todros et al., 1999) and general hypovascularization of the placenta (Hitschold et al., 1993). Collectively, these studies provide an anatomic basis for the elevated resistance to blood flow in placentas from growth-retarded fetuses, and they further highlight the importance of the placental vasculature in normal pregnancies.

MECHANISMS REGULATING VASCULARIZATION OF THE PLACENTA

There is a great number of growth factors, cytokines, matrix proteins, proteases, metabolites, ions, and mechanical factors capable of promoting or inhibiting angiogenesis (Diaz-Flores et al., 1994; Cockerill et al., 1995) many of which are present in the endometrium (Giudice, 1994; Goodger and Rogers, 1995b; Giudice and Irwin, 1999) and placenta (Lala and Hamilton, 1996; Ahmed, 1997; Giudice and Irwin, 1999). It is beyond the scope of this chapter to review the potential angiogenic role of all these factors. Instead, we focus on relatively novel factors that directly, or indirectly through trophoblast interactions, appear to influence vascular development in the placenta.

Transcription Factors/Gene Products

There is a growing number of gene products that have been shown to influence various steps in placentation (Cross et al., 1994; Rinkenberger et al., 1997; Cross, 1998). Many of these, when absent, result in embryonic death secondary to placental failure in mice. Products such as the Mash-2 (Guillemot et al., 1994), Hand1 (Riley et al., 1998), and Ets2 (Yamamoto et al., 1998) transcription factors; estrogen-receptor–related protein-β (Luo et al., 1997); heat-shock proteins (Voss et al., 2000); α_4 integrins (Yang et al., 1995); and vascular cell adhesion molecule-1 (VCAM-1) (Gurtner et al., 1995) have direct influences on trophoblast function. These studies, and others, have clearly demonstrated the requirement for normal trophoblast function in successful placentation. Since trophoblasts form the functional interface between the developing embryo and maternal tissues throughout gestation, it would stand to reason that they play an important role in the regulation of placental vascularization. The intimate relationship between trophoblast and placental vascularity is confirmed by several elegant studies in mice using gene knockout technology.

a. Mek1

Endothelial cell proliferation and migration are important steps in angiogenesis, and one key component of cellular proliferation and migration is the Ras/mitogen-activated protein (MAP) kinase signaling pathway (McCormick, 1993), of which Mek1 is a member. The importance of Mek1 in vascular development of the placenta has recently been shown (Giroux et al., 1999). Disruption of the *Mek1* gene results in embryonic death at 10.5 days. Absence of Mek1 function does not significantly alter expression of classic trophoblast markers in the placenta or early vascular development of the embryo itself. However, histologic assessment of the placenta reveals a lack of embryonic blood vessels in the labyrinth of the placenta. *Mek1*$^{-/-}$ embryos still produce endothelial cells, but they remain restricted to the chorioallantoic region, suggesting a defect in endothelial cell migration. This hypothesis is confirmed using *in vitro* studies in which migration of *Mek1*$^{-/-}$ fibroblasts on fibronectin is severely reduced while migratory capacity is restored following transfection with functional Mek1 into the *Mek1*$^{-/-}$ cells (Giroux et al., 1999).

b. Tfeb

Tfeb, a member of the basic helix-loop-helix leucine zipper family of transcription factors, has also been shown to be important in the vascularization of the placenta (Steingrimsson et al., 1998). Germline mutations in the gene result in death of homozygous embryos between 9.5 and 11.5 days p.c. with specific alterations in the vasculature of the placenta. In normal placenta, high Tfeb expression is confined only to labyrinth trophoblasts; the maternal and embryonic vasculature do not express Tfeb. The lack of trophoblast Tfeb expression in homozygous mutant animals results in a lack of embryonic vessels penetrating into the labyrinth of the placenta. Although embryonic vessels enter the labyrinth initially (i.e., day 8.5 p.c.), they do not invade the placenta during subsequent development. Normal-appearing vasculature is evident in the chorion (suggesting normal vasculogenesis), and maternal sinuses are present, although they are fewer and smaller compared to those of wild-type mice. Since the vascular alterations preceded embryonic death, the authors concluded that inadequate placental vascularization resulted in lethality in Tfeb homozygous mutants. Interestingly, absence of Tfeb expression is also associated with a lack of VEGF expression in the placental labyrinth (Steingrimsson et al., 1998). These results suggest trophoblast expression/function of transcription factors that may activate other genes (like VEGF) that then facilitate successful vascularization of the developing placenta.

c. von Hippel–Lindau (VHL) Tumor Suppressor Gene

VHL tumor suppressor gene product (pVHL) is thought to downregulate transcriptional elongation and therefore significantly influences cell cycle regulation, angiogenesis, and extracellular matrix production (Ohh and Kaelin, 1999). Mutations in the gene are responsible for von Hippel–Lindau disease in humans, which is characterized by the development of well-vascularized tumors (Kaelin et al., 1998).

VHL gene function has also been shown to influence vascular growth/development in the placenta (Gnarra et al., 1997). Murine VHL messenger RNA (mRNA) is detected in the placental labyrinth and pVHL is found in placental trophoblast, allantoic mesoderm, and some embryonic endothelial cells of $VHL^{+/+}$ animals. However, disruption of VHL gene function in mice leads to a lack of embryonic blood vessels in the placental labyrinth and embryonic death at 10.5 to 12.5 days' gestation (Gnarra et al., 1997). Histologically, the embryo itself is generally normal until placental lesions develop. Heterozygous $VHL^{+/-}$ mice show no evident pathologies some 15 months after birth. Quite surprisingly, given that pVHL suppresses VEGF expression (Pal et al., 1997; Kaelin et al., 1998; Krieg et al., 1998), placental VEGF expression is reduced in $VHL^{-/-}$ mice. The authors concluded that loss of VHL expression produces deficient vascular development of the placenta, which results in embryonic lethality, and these effects may be mediated by decreased VEGF expression in the placenta of $VHL^{-/-}$ mice (Gnarra et al., 1997).

d. ARNT

Arylhydrocarbon receptor nuclear translocator (ARNT) protein (or hypoxia-inducible factor-1β) is a member of a family of transcription factors that, following dimerization with proteins such as hypoxia-inducible factor 1α (Wang and Semenza, 1995), regulate cell gene expression during low oxygen concentrations (Bunn and Poyton, 1996). The intimate relationship between hypoxia and induction of angiogenesis makes ARNT an attractive potential regulator of angiogenesis *in vivo*.

The role of ARNT in regulation of placental vascular growth has been described in gene knockout mice (Kozak et al., 1997; Maltepe et al., 1997). In both studies, $ARNT^{-/-}$ mice do not survive past embryonic day 10.5, while $ARNT^{+/-}$ heterozygotes survive to term and beyond (Kozak et al., 1997). In the normal 9.5-day placenta, ARNT is expressed in embryonic trophoblast and maternal decidua (Kozak et al., 1997). Although decidual ARNT expression is evident in the surrogate mother, there is an absence of embryonic trophoblast ARNT expression, and the placenta is smaller and contains minimal vascular profiles compared to the normal placenta (Kozak et al., 1997). In contrast, vascularity of the $ARNT^{-/-}$ yolk sac is similar to that of $ARNT^{+/+}$ embryos. Although developmental defects are also noted in $ARNT^{-/-}$ embryos, lethality was due to inadequate vascularization of the embryonic component of the placenta (Kozak et al., 1997). Essentially similar survival data were obtained in $ARNT^{-/-}$ embryos in another study; however, the most striking phenotypic alterations in the vasculature occurred in the yolk sac of $ARNT^{-/-}$ embryos (Maltepe et al., 1997). The yolk sac of $ARNT^{-/-}$ embryos exhibit apparently normal blood islands (suggesting normal vasculogenesis) but an absence of, or grossly disorganized, microvasculature (suggesting defective angiogenesis). The authors suggest that decreased angiogenesis in ARNT-deficient embryos may be from the lack of hypoxic induction of VEGF expression. Indeed, they and others have shown that hypoxia fails to increase VEGF production in $ARNT^{-/-}$ cells *in vitro* (Forsythe et al., 1996; Wood et al., 1996; Maltepe et al.,

1997). Collectively, these results suggest that local hypoxia during early placentation may trigger the ARNT-induced angiogenic activity required for adequate vascular growth and development in the placenta.

e. Wnt2

The *Wnt* genes compose a relatively large family of genes associated with various physiologic processes from *Drosophila* to humans. Targeted disruption of the *Wnt2* gene in mice induces vascular defects in the placenta. Specifically, many of the mutant placentae demonstrate various degrees of edema, a reduction in the size of the labyrinth and chorioallantoic regions, remodeled maternal vasculature, and in some cases reduced fetal capillary numbers. These defects, however, do not become apparent until after day 14.5 p.c., suggesting that *Wnt2* expression is more important in the maintenance of the fetal placental vasculature (Monkley et al., 1996).

f. c-Ets1

Studies designed to elucidate the mechanisms regulating vascular growth during early placentation in the human are limited. One study investigated the expression of c-*Ets1* proto-oncogene in trophoblast from legal first trimester abortions (Luton et al., 1997). Ets1 proteins are typically transcription activators that facilitate angiogenesis by inducing matrix-degrading proteases *in vitro* (Iwasaka et al., 1996) as well as in various human models of angiogenesis and cell invasion (Wernert et al., 1992). The role of metalloproteases in angiogenesis is well known (Mignatti and Rifkin, 1996) and c-*Ets1* involvement in implantation and placentation in the mouse has been described (Grevin et al., 1993). In the first trimester human placenta, only fetal villous endothelial cells and invading trophoblast expressed c-*Ets1* mRNA; villous cytotrophoblasts, syncytiotrophoblasts, and maternal endothelial cells did not express c-*Ets1* (Luton et al., 1997). Associations between c-Ets1 expression, metalloproteases, and active angiogenesis suggest c-Ets1 could play a role in trophoblast invasion as well as vascularization of the human placenta. Clearly, more mechanistic studies are needed to confirm this hypothesis in humans.

Growth Factors

Hepatocyte Growth Factor/Scatter Factor (HGF/SF)

HGF/SF, in concert with its tyrosine kinase receptor (c-met), are important mediators of epithelial cell proliferation, migration, and differentiation, and its influence in placentation has been described (Stewart, 1996). Gene knockout studies demonstrate that lack of HGF/SF results in severely impaired placentation, hepatocyte differentiation, and death of the embryo between 13.5 and 15.5 days p.c. The placental labyrinth of these animals contains fewer trophoblast and both the maternal and fetal vasculature are poorly developed (Uehara et al., 1995).

Descriptive findings in humans also suggest that HGF/SF/c-met may modulate placentation. Most studies find HGF/SF localization in the villous core tissue (Clark et al., 1996b; Kauma et al., 1997), with the perivascular stromal cells demonstrating intense hybridization signals (Kilby et al., 1996; Somerset et al., 1998). HGF/SF immunoreactivity can also be found on vascular endothelial cells in the term placenta (Kilby et al., 1996). In addition, most studies describe c-met protein and/or mRNA expression on cytotrophoblast (Furugori et al., 1997; Kauma et al., 1997) and vascular endothelium (Kilby et al., 1996; Somerset et al., 1998). Quantitative analysis of HGF/SF mRNA expression in human placenta reveals that expression peaks during the second trimester (Furugori et al., 1997; Somerset et al., 1998) and then decreases throughout the third trimester (Somerset et al., 1998). Interestingly, second trimester human placental explants produce approximately twofold more HGF/SF in culture than third trimester explants (Kauma et al., 1997), suggesting an intrinsic difference in the temporal regulation of human placental HGF/SF.

A key role for HGF/SF in placentation is supported by studies assessing its expression in complicated pregnancies. IUGR is associated with a significant reduction in placental HGF/SF mRNA expression (Somerset et al., 1998), although this is not reflected by decreased serum levels of the protein (Clark et al., 1998a). Placental expression of HGF/SF is also significantly reduced in preeclamptic placentae (Furugori et al., 1997).

In addition to its direct effect on trophoblast function, HGF/SF also directly influences endothelial cell migration, proliferation, invasiveness, and tube formation (Rosen et al., 1997). HGF/SF induces VEGF gene transcription in various epithelial-derived cells (Gille et al., 1998; Wojta et al., 1999) and increases expression of VEGF receptor on endothelial cells (Wojta et al., 1999). Others have shown an additive effect of HGF/SF and VEGF on endothelial cell proliferation and a synergistic effect on endothelial cell migration *in vitro* (Van Belle et al., 1998). Importantly, exogenous HGF/SF also promotes angiogenesis *in vivo* (Van Belle et al., 1998). Collectively, these studies suggest that HGF/SF may have a significant role in regulating placental development and, given its direct angiogenic potential, may directly influence vascularization of the placenta. Further studies assessing the role of this pleiotropic growth factor in placental vascular development are warranted.

Basic Fibroblast Growth Factor (bFGF)

The fibroblast growth factor family is composed of a large number of pleuripotent, heparin-binding peptides. Basic fibroblast growth factor (bFGF), which was isolated from human placental tissue (Moscatelli et al., 1986, 1988), is a known endothelial cell mitogen and is angiogenic *in vivo*. Several descriptive studies provide rationale for the role of bFGF in vascular development in the human reproductive tract. For instance, bFGF and its receptor (FGF-R1) are thought to play a critical role in the regeneration of human endometrium following menstruation. Hybridization signals for bFGF mRNA are found in stromal cells and glandular epithelial cells throughout the menstrual cycle with quantitatively greater bFGF and FGF receptor (FGF-R1) expression in proliferative endometrium (Sangha et al., 1997).

A role for bFGF in vascular development of the placenta is also likely. bFGF mRNA expression is associated with syncytiotrophoblasts and cytotrophoblasts of the first trimester human placenta (Shams and Ahmed, 1994). In the term placenta, syncytiotrophoblast (Shams and Ahmed, 1994; Arany and Hill, 1998), vascular endothelium (Arany and Hill, 1998) and vascular smooth muscle cells (Ferriani et al., 1994) express bFGF mRNA. Interestingly, bFGF gene expression is greater in the first trimester than in the term placenta, suggesting a developmental control of its expression (Di Blasio et al., 1997). In the human placenta, bFGF protein is primarily found associated with stromal cells and endothelial cells of villi (Arany and Hill, 1998). In rats, bFGF mRNA expression increases and coincides temporally with active angiogenesis within the mesometrial decidua, and the antimesometrial decidua expresses low level of bFGF transcripts (Srivastava et al., 1998). bFGF protein distribution also demonstrates temporal and spatial changes during the periimplantation period in the rat (Wordinger et al., 1994). *In vitro* studies have confirmed that trophoblasts produce and release biologically active bFGF into culture media (Hamai et al., 1998). The presence of FGF-R1 on endothelial cells and some trophoblasts suggests that bFGF may act in a paracrine fashion to promote vascular growth in the placenta as well as to function in an autocrine manner to promote trophoblast growth (Ferriani et al., 1994).

Collectively, these findings suggest that bFGF could be involved in the vascularization process within the developing placenta. However, further studies are needed to determine if bFGF is required for adequate vascular development of the placentation. Although we have chosen to concentrate only on the distribution of bFGF within the placenta, the localization and potential role of other heparin-binding growth factors in the placenta has been reviewed (Ahmed, 1997).

Vascular Endothelial Growth Factor (VEGF)

The molecular and biologic properties of VEGF are well characterized (Ferrara, 1995; Neufeld et al., 1999). VEGF is an indispensable angiogenic factor, which is confirmed by (1) gene knockout studies that show expression of both VEGF alleles is required for vasculogenesis/angiogenesis during development (Carmeliet et al., 1996; Ferrara et al., 1996), (2) functional inhibition by neutralizing VEGF antibodies (Kim et al., 1992; Asano et al., 1995; Borgstrom et al., 1996) or VEGF antisense expression (Saleh et al., 1996) limits tumor growth *in vivo*, (3) signaling-defective VEGF receptor mutants (Millauer et al., 1994, 1996) also inhibit tumor associated angiogenesis, and (4) disruption of flt-1 (Fong et al., 1995) or the rodent homologue of human kinase insert domain-contains receptor KDR/*Flk-1* (Shalaby et al., 1995) gene expression during embryogenesis results in an embryonic lethal phenotype. These results demonstrate that VEGF plays a critical role in normal and pathologic vascular growth, and therefore may represent an attractive regulator of vascular growth in the placenta (Torry and Torry, 1997; Smith et al., 2000) and other reproductive tissues (Lebovic et al., 1999).

Immunohistologic studies localize VEGF protein in human trophoblasts during pregnancy (Ahmed et al., 1995; Clark et al., 1996a, 1998c; Shiraishi et al., 1996; Vuorela et al., 1997). In most studies (Jackson et al., 1994; Ahmed et al., 1995; Clark et al., 1996a), but not all (Shiraishi et al., 1996), VEGF reactivity was noted in villous cytotrophoblasts in first trimester placentae. Syncytiotrophoblasts at this stage expressed relatively low levels of VEGF (Ahmed et al., 1995; Clark et al., 1996a), and the pattern was not uniform (Shiraishi et al., 1996). In general, the VEGF protein expression pattern remains relatively stable in second and third trimester samples (Ahmed et al., 1995; Clark et al., 1996a; Shiraishi et al., 1996), while Western blot analyses of VEGF shows an increase of about two- to three-fold from first trimester to term samples (Jackson et al., 1994).

Studies in the rabbit (Das et al., 1997) and porcine (Winther et al., 1999) placenta have also documented VEGF expression in trophoblast throughout pregnancy, and VEGF receptor expression increases during the periimplantation period in golden hamsters (Yi et al., 1999). In the mouse, temporal and spatial associations between $VEGF_{164}$ expression, flk-1, and neuropilin-1 receptor expression and angiogenesis/vascular permeability changes during implantation have been shown (Halder et al., 2000). Neuropilin-1 is a co-receptor for $VEGF_{165}$ (Soker et al., 1998) that facilitates the binding and physiologic activity of $VEGF_{165}$ and KDR/Flk-1 receptor (Neufeld et al., 1999; Petrova et al., 1999). Importantly, $VEGF_{164}$ was the most prominent VEGF isoform expressed during implantation in the mouse (Halder et al., 2000). The expression pattern of neuropilin-1 and its influence on vascularization of the human placenta, if any, have not been described.

Initial *in situ* hybridization studies (Sharkey et al., 1993) showed weak VEGF mRNA expression in both cytotrophoblasts and syncytiotrophoblasts in first trimester human placentae and no expression in extravillous trophoblast. At term, most trophoblast subpopulations are weakly positive to negative (Charnock-Jones et al., 1994). We have shown that isolated cytotrophoblasts and *in vitro* differentiated syncytiotrophoblasts from term human placentae express low levels of VEGF mRNA and that there was no significant difference in expression following differentiation (Shore et al., 1997). Assessment of VEGF mRNA expression in placental biopsies during the normal third trimester shows that expression decreases as term approaches (Cooper et al., 1996). The placenta does not appear to express VEGF-B or VEGF-C (Clark et al., 1998c).

There is little information concerning the regulation of VEGF expression in the placenta. VEGF expression has been shown to be inducible in many cells by hypoxia, and expression returns to normal levels upon reoxygenation (Shweiki et al., 1995). We found hypoxia significantly upregulates VEGF mRNA expression in isolated term trophoblasts (Shore et al., 1997). Cytokines and growth factors can also influence VEGF expression: interleukin-1 (IL-1) (Li et al., 1995), epidermal growth factor (EGF) (Goldman et al., 1993), transforming growth factor-β (TGF-β) (Pepper et al., 1993) and platelet-derived growth factor (PDGF) (Finkenzeller et al., 1992; Brogi et al., 1994) are able to induce VEGF expression in a variety of cells. Clearly, regulation of placental VEGF expression may vary according to the presence and/or concentration of other cytokines in the tissue.

Past investigations of VEGF expression in reproductive tissues and its up-regulation by hypoxia have provided some impetus for its role in preeclampsia. Unfortunately, the literature addressing maternal blood VEGF levels during preeclampsia is conflicting. Some studies have shown that maternal serum VEGF levels increase in preeclampsia (Baker et al., 1995; Sharkey et al., 1996; Kupferminc et al., 1997), while others have shown that levels decrease (Lyall et al., 1997a; Reuvekamp et al., 1999). Irrespective, interpretation of maternal serum VEGF studies is cautioned because of the diverse potential origins of VEGF expression (see Torry and Torry, 1997). Studies utilizing umbilical vein blood samples, which provide a more restricted representation of VEGF expression within the placenta, show that VEGF levels do not differ between normal and preeclamptic pregnancies (Baker et al., 1995; Lyall et al., 1997b). Indeed, VEGF mRNA expression is significantly lower in preeclamptic compared to normal placentae (Cooper et al., 1996).

Soluble Flt-1 Receptor

The direct influence of VEGF on placental vascularization is also complicated by the presence of the soluble Flt-1 (sFlt-1) receptor. This VEGF binding protein is a splice variant of Flt-1 gene expression that functions to antagonize VEGF biologic activity (Kendall and Thomas, 1993; Kendall et al., 1996). High concentrations of sFlt-1 (or sFlt-1–like VEGF-binding protein) are found in sera and amniotic fluid during pregnancy (Banks et al., 1998; Clark et al., 1998b; Hornig et al., 1999, 2000; Vuorela-Vepsalainen et al., 1999; Vuorela et al., 2000). Human endothelial cells produce sFlt-1 in culture (Hornig et al., 2000) and human trophoblasts contain mRNA for sFlt-1 *in vivo* and secrete the protein *in vitro* (Clark et al., 1998b). Hypoxia has been shown to increase sFlt-1 production in cultured term placental villi extracts (Hornig et al., 2000). In studies on mice, sFlt-1 expression in spongiotrophoblasts increased from day 11 to 17 of pregnancy, and exogenous VEGF administration over this time period enhanced fibrin deposition and the number of resorption sites in the mice (He et al., 1999). Thus, sFlt-1 expression in the placenta may help regulate VEGF's biologic effects *in vivo*, and this regulation appears to be necessary for successful placental development and function. Indeed, the significantly higher levels of sFlt-1 in amniotic fluid from preeclamptic pregnancies (Vuorela et al., 2000) supports an association between aberrant placentation and dysregulation of VEGF biologic activity. Unfortunately, the mechanisms regulating sFlt-1 expression and its definitive role in altering vascular development of the placenta in humans are not fully understood.

Angiopoietins

Other modulators of angiogenesis, such as the angiopoietins, have recently been described. Angiopoietin-1 is an endogenous agonist for endothelial cell specific Tie-2 receptor, while angiopoietin-2 is an endogenous antagonist for the same receptor, and deletion of either angiopoietin-1 or angiopoietin-2 is lethal to mice during development (reviewed in Davis and Yancopoulos, 1999). Postnatally, overexpression of angiopoietin-1 promotes vascular expansion (Suri et al., 1998; Thurston et al., 1999), and angiopoietin-2 is expressed preferentially at sites of vascular remodeling (Maisonpierre et al., 1997). Both appear to facilitate the angiogenic effect of VEGF *in vivo*, with angiopoietin-1 promoting vessel maturation and angiopoietin-2 promoting neovascularization (Asahara et al., 1998). Unfortunately, relatively little is known regarding the regulation and role of the angiopoietins in the placenta. Angiopoietin-2 mRNA is expressed in human syncytiotrophoblasts, and its receptor, Tie-2, is expressed in fetal and maternal endothelial cells and endovascular trophoblasts (Goldman-Wohl et al., 2000). Abundant angiopoietin-2 expression, restricted to the first trimester (Goldman-Wohl et al., 2000), provides an intriguing hypothesis into trophoblast influence on neovascularization and/or remodeling of existing vasculature within the human placenta. Indeed, the spatial and temporal differences in angiopoietin expression throughout gestation and the decrease in angiopoietin-2 protein expression in placenta with IUGR (Dunk et al., 2000) provide compelling rational for their importance in regulating placental vascularity. Although similar temporal relationships between angiopoietin expression and angiogenesis in other reproductive tissues have been described (Goede et al., 1998), more mechanistic studies are needed to precisely determine the role angiopoietins have, either directly or in concert with VEGF, in regulating vascularization of the human placenta.

In summary, these results suggest that VEGF protein is produced by trophoblasts throughout pregnancy, and VEGF expression in early gestation may play a role in uterine vessel remodeling and angiogenesis that occurs at this time and may also influence development of the villous vasculature. The presence of flt-1 receptors on trophoblasts (Shore et al., 1997) suggests that trophoblast-derived VEGF could also act in an autocrine manner to influence trophoblast function. In addition, local expression of angiopoietins and the presence of soluble Flt-1 receptor could further modulate the biologic activity of placenta-derived VEGF. Clearly more research is needed to fully understand the influence of VEGF and the angiopoietins on vascular development of the placenta.

Placenta Growth Factor (PlGF)

PlGF was originally cloned from a human placental cDNA library (Maglione et al., 1991). The primary sequence of PlGF shows significant homology to the PDGF-like domain of VEGF. Like VEGF, alternative splicing of PlGF mRNA accounts for at least three variant isoforms (Hauser and Weich, 1993; Maglione et al., 1993; Cao et al., 1997) and we have proposed a fourth PlGF isoform to be expressed in normal human trophoblasts (Yang et al., 2000). PlGF expression is primarily restricted to the placenta (Maglione et al., 1991; Hauser and Weich, 1993; Khaliq et al., 1996; Cao et al., 1997; Vuorela et al., 1997), and trophoblasts represent the major cellular sources of PlGF (Achen et al., 1997; Shore et al., 1997; Clark et al., 1998c).

Little is known regarding the regulation of trophoblast PlGF expression. In normal cultured trophoblast and JEG-3 choriocarcinoma cells, basal levels of PlGF expression is significantly higher than that of VEGF, and we have found no significant difference in PlGF gene expression following *in vitro* differentiation of normal cytotrophoblasts to syncytiotrophoblasts (Shore et al., 1997). In contrast to VEGF, PlGF expression in normal trophoblasts (Shore et al., 1997) as well as choriocarcinoma cells (Gleadle et al., 1995) is significantly downregulated by hypoxia. The molecular mechanisms mediating hypoxic downregulation of PlGF expression are not known, and, as with VEGF, the influence of cytokines and growth factors that could govern expression of PlGF in trophoblasts remain to be elucidated.

Based on structural homology to VEGF and high-affinity binding to the flt-1 receptor, it is thought that PlGF is an angiogenic factor. Indeed, several reports have shown PlGF to have mitogenic activity on endothelial cells (Maglione et al., 1991; Hauser and Weich, 1993; Ziche et al., 1997). PlGF is able to induce all aspects of *in vivo* angiogenesis, and its effects on endothelial cell migration and proliferation are similar to those of more classic angiogenic factors (Ziche et al., 1997). The addition of rhPlGF to endothelial cell cultures induces activation of the ERK1 and 2 mitogen-activated protein kinase (MAPK) signal transduction pathways. The magnitude and kinetics of these responses are similar to those induced by the known endothelial cell mitogens, VEGF and epidermal growth factor (Desai et al., 1999). These molecular results are consistent with PlGF being able to promote mitosis in endothelial cells. However, others have shown purified PlGF has little or no mitogenic activity for human umbilical vein or bovine adrenal cortex–derived endothelial cells (Park et al., 1994) and little or no chemoattractant properties on endothelial cells *in vitro* (Cao et al., 1996a). Although there are several potential explanations for this discrepancy (Torry et al., 1999), PlGF does potentiate the mitogenic activity and permeability effects of very low concentrations of VEGF on endothelial cells (Park et al., 1994). The basic residues of PlGF-2, which lack direct effects on endothelial cell proliferation *in vitro*, increase the release of endogenous bFGF and VEGF from their heparin-binding sites in the extracellular matrix. The increased bioavailability of these potent angiogenic molecules in turn increased endothelial cell proliferation *in vitro* (Barillari et al., 1998). These results support the hypothesis that PlGF could also indirectly influence placental vascularity by facilitating the release of sequestered endogenous growth factors.

Experiments addressing the *in vivo* angiogenic properties of PlGF are not much clearer. Some have found that exogenous PlGF had little or no direct angiogenic effect in mice (Cao et al., 1996b) or the chick chorioallantoic assay (Kurz et al., 1998). However, these results are in direct conflict with a report showing that PlGF-1 induces a strong vascular growth response in rabbit cornea and chick chorioallantoic membrane assays (Ziche et al., 1997).

Although the direct effects of PlGF on angiogenesis may be controversial, there is little reason to doubt that PlGF can indirectly modulate vascular responses in tissues demonstrating high PlGF expression. Thus, trophoblast-derived PlGF could, via paracrine methods, influence vascular growth, stability, permeability, and remodeling within the placental bed, and aberrant expression of PlGF could be associated with vascular insufficiencies that culminate in reduced placental size and poorer fetal outcomes. Indeed, the restricted expression patterns of PlGF and the ability of hypoxia to reduce trophoblast expression of PlGF *in vitro* (Shore et al., 1997) prompted our hypothesis that pregnancies complicated by abnormal perfusion of the placental bed may exhibit altered expression of PlGF *in vivo*. Accordingly, we first determined the serum PlGF levels in normal pregnant women and then measured serum PlGF levels in gestational age–matched normal and preeclamptic women (Torry et al., 1998b). In normal pregnancies, maternal serum levels of PlGF increase significantly from late first trimester to late second trimester and then decline steadily from peak levels at ~28 to 30 weeks to term (Fig. 11.2). These fluctuations in serum PlGF levels reflect the changes noted in PlGF gene expression in placental tissue at these same time points (Fig. 11.3). Interestingly, average serum PlGF levels in preeclamptic women were significantly lower than those in normal pregnant women (Fig. 11.4). In particular, at 30 to 32 weeks' gestation, when PlGF levels in normal pregnant women are near maximal, levels in preeclamptic women are approximately five times lower. Importantly, the low levels of PlGF noted in the preeclamptic patients in this study were not due to a decrease in placental mass, indicating that trophoblast production of PlGF is severely altered in preeclampsia (Torry et al., 1998b). In a subsequent prospective longitudinal study we found that maternal serum levels of PlGF are significantly lower by 12 to 16 weeks' gestation in women who eventually developed preeclampsia as compared to those with uneventful course of pregnancy (Tidwell

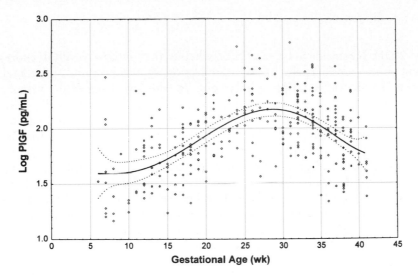

FIGURE 11.2. Serum placenta growth factor (PlGF) levels during normal pregnancy. Systemic PlGF levels were determined by antigen capture enzyme-linked immunosorbent assay (ELISA) in 308 normal pregnant women and plotted as a function of gestational age. Each data point represents the mean of a single determination done in duplicate. Solid line represents the best-fit polynomial line for the distribution. Dotted lines represent the 95% confidence levels for nonparametric data. All serum PlGF levels were converted to log(10) values (Torry et al., 1998b).

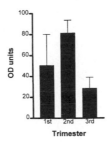

FIGURE 11.3. Placental PlGF messenger RNA (mRNA) expression during gestation. Total RNA (7 g) were separated on denaturing 1% agarose gels containing 2.2 M formaldehyde, transferred to nylon membranes, and immobilized by ultraviolet (UV) cross-linking. Membranes were prehybridized and expression of PlGF determined by Northern blot analyses using an *in vitro* transcribed PlGF antisense riboprobe from pPlGF-1. Membranes were washed and exposed to x-ray film at −80°C. Laser densitometry readings were performed to quantitate the PlGF signals. OD-Optical Density.

FIGURE 11.4. Serum PlGF levels in normal and preeclamptic pregnant women. Mean serum PlGF values ± standard error of the mean (SEM) for preeclamptic patients (*n* = 30) and gestational age–matched controls (*n* = 30) are shown. Serum PlGF levels were determined by antigen capture ELISA (Torry et al., 1998b).

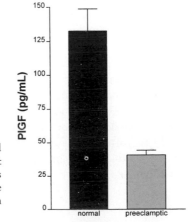

et al., 2001). It remains to be elucidated whether the low level of PlGF expression during preeclampsia is the cause or reflection of the presumed placental bed hypoxia that develops during preeclampsia. In addition to regulating vascular development during pregnancy, there is substantial evidence supporting roles for PlGF in regulating proliferation in normal first trimester trophoblasts (Athanassiades and Lala, 1998), and we (Desai et al., 1999) have shown PlGF protects term trophoblasts from apoptosis.

The potential of PlGF to affect maternal vasculature and trophoblast function is extended to include fetal vascular development in the placenta. Although PlGF mRNA expression has been localized exclusively to villous trophoblast, protein localization was also noted in the fetal stem vessels, implying that trophoblast-derived PlGF may act in a paracrine fashion to influence placental blood vessels (Vuorela et al., 1997). Recent studies from our lab, as well as others, now provide evidence that trophoblast-derived PlGF enters the fetal circulation. Although abundant PlGF mRNA expression is restricted exclusively to the trophoblast of the mouse and there is no PlGF expression evident within the embryo proper (Achen et al., 1997), our studies in humans show that PlGF protein is present in umbilical cord serum and that titers tightly correlate with maternal PlGF titers (Torry et al., 1998b). These studies confirm that trophoblast PlGF enters the fetal circulation. Vascularization of the murine and human placenta originates from the allantois, and studies show that the PlGF receptor, flt-1, is highly expressed in the allantoic bud region (Breier et al., 1995). Expression of PlGF in trophoblast has led to the conclusion that PlGF may direct embryonic vascular growth in the placenta (Achen et al., 1997). Thus, a unifying hypothesis (Torry et al., 1999) is that decreased trophoblast PlGF expression may compromise endothelial cell and trophoblast function, which further contribute to the vascular pathology noted in obstetrical complications as preeclampsia (Roberts et al., 1989) and this may also account for the abnormal appearance of villous vascular structures previously noted in perfusion compromised conditions like IUGR (Krebs et al., 1996).

SUMMARY

Substantial vascular growth and remodeling is necessary in the development of a functional placenta, and vascular insufficiencies of pregnancy are thought to contribute to many obstetrical complications. For instance, aberrant vascular function and/or perfusion at the maternal-fetal interface is associated with recurrent spontaneous abortion (Torry et al., 1998a), pregnancy induced hypertension (Roberts et al., 1989; Conrad and Benyo, 1997), and IUGR (Krebs et al., 1996). The etiologies of such complications remain speculative and are undoubtedly multifactorial; nonetheless, involvement of the vasculature may represent a unifying focus point. Any maternal/fetal condition that disrupts the delicate balance of positive and negative regulators of vascular growth could have severe consequences on the overall growth and function of the placenta. It is likely therefore that knowledge of the regulation of vascular growth in the placenta could provide insights into novel therapies for placental-related vascular insufficiencies.

ADDENDUM

Since submission of this manuscript, additional reviews of placental angiogenesis have been published (Reynolds and Redmer, Biol Reprod 64:1033 1040, 2001; Sherer and Abulafia, Placenta 22:1–13, 2001).

REFERENCES

Abrahamsohn, P., Lundkvist, O., Nilsson, O. (1983). Ultrastructure of the endometrial blood vessels during implantation of the rat blastocyst. Cell Tissue Res 229:269–280.

Abulafia, O., Sherer, D.M. (1999). Angiogenesis of the endometrium. Obstet Gynecol 94:148–153.

Achen, M.G., Gad, J.M., Stacker, S.A., Wilks, A.F. (1997). Placenta growth factor and vascular endothelial growth factor are co-expressed during early embryonic development. Growth Factors 15:69–80.

Ahmed, A. (1997). Heparin-binding angiogenic growth factors in pregnancy. Trophoblast Res 10:215–258.

Ahmed, A., Li, X.F., Dunk, C., Whittle, M.J., Rushton, D.I., Rollason, T. (1995). Colocalisation of vascular endothelial growth factor and its Flt-1 receptor in human placenta. Growth Factors 12:235–243.

Ali, K.Z., Burton, G.J., Morad, N., Ali, M.E. (1996). Does hypercapillarization influence the branching pattern of terminal villi in the human placenta at high altitude? Placenta 17:677–682.

Arany, E., Hill, D.J. (1998). Fibroblast growth factor-2 and fibroblast growth factor receptor-1 mRNA expression and peptide localization in placentae from normal and diabetic pregnancies. Placenta 19:133–142.

Asahara, T., Chen, D.H., Takahashi, T., et al. (1998). Tie2 receptor ligands, angiopoietin-1 and angiopoietin-2, modulate VEGF-induced postnatal neovascularization. Circ Res 83:233–240.

Asan, E., Kaymaz, F.F., Cakar, A.N., Dagdeviren, A., Beksac, M.S. (1999). Vasculogenesis in early human placental villi: an ultrastructural study. Anat Anz 181:549–554.

Asano, M., Yukita, A., Matsumoto, T., Kondo, S., Suzuki, H. (1995). Inhibition of tumor growth and metastasis by an immunoneutralizing monoclonal antibody to human vascular endothelial growth factor/vascular permeability factor121. Cancer Res 55: 5296–5301.

Athanassiades, A., Lala, P.K. (1998). Role of placenta growth factor (PlGF) in human extravillous trophoblast proliferation, migration and invasiveness. Placenta 19:465–473.

Baker, P.N., Krasnow, J., Roberts, J.M., Yeo, K.T. (1995). Elevated serum levels of vascular endothelial growth factor in patients with preeclampsia. Obstet Gynecol 86:815–821.

Banks, R.E., Forbes, M.A., Searles, J., et al. (1998). Evidence for the existence of a novel pregnancy-associated soluble variant of the vascular endothelial growth factor receptor, Flt-1. Mol Hum Reprod 4:377–386.

Barillari, G., Albonici, L., Franzese, O., et al. (1998). The basic residues of placenta growth factor type 2 retrieve sequestered angiogenic factors into a soluble form: implications for tumor angiogenesis. Am J Pathol 152:1161–1166.

Battegay, E.J. (1995). Angiogenesis: mechanistic insights, neovascular diseases, and therapeutic prospects. J Mol Med 73:333–346.

Blankenship, T.N., Enders, A.C. (1997a). Expression of platelet-endothelial cell adhesion molecule-1 (PECAM) by macaque trophoblast cells during invasion of the spiral arteries. Anat Rec 247:413–419.

Blankenship, T.N., Enders, A.C. (1997b). Trophoblast cell-mediated modifications to uterine spiral arteries during early gestation in the macaque. Acta Anat (Basel) 158: 227–236.

Borgstrom, P., Hillan, K.J., Sriramarao, P., Ferrara, N. (1996). Complete inhibition of angiogenesis and growth of microtumors by antivascular endothelial growth-factor neutralizing antibody—novel concepts of angiostatic therapy from intravital videomicroscopy. Cancer Res 56:4032–4039.

Bracero, L.A., Beneck, D., Kirshenbaum, N., Peiffer, M., Stalter, P., Schulman, H. (1989). Doppler velocimetry and placental disease. Am J Obstet Gynecol 161:388–393.

Breier, G., Clauss, M., Risau, W. (1995). Coordinate expression of vascular endothelial growth factor receptor-1 (flt-1) and its ligand suggests a paracrine regulation of murine vascular development. Dev Dyn 204:228–239.

Brogi, E., Wu, T., Namiki, A., Isner, J.M. (1994). Indirect angiogenic cytokines upregulate VEGF and bFGF gene expression in vascular smooth muscle cells, whereas hypoxia upregulates VEGF expression only. Circulation 90:649–652.

Brooks, P.C., Clark, R.A., Cheresh, D.A. (1994). Requirement of vascular integrin alpha v beta 3 for angiogenesis. Science 264:569–571.

Brown, M.A. (1995). The physiology of pre-eclampsia. Clin Exp Pharmacol Physiol 22: 781–791.

Bunn, H.F., Poyton, R.O. (1996). Oxygen sensing and molecular adaptation to hypoxia. Physiol Rev 76:839–885.

Burrows, T.D., King, A., Loke, Y.W. (1996). Trophoblast migration during human placental implantation. Hum Reprod Update 2:307–321.

Burton, G.J., Jauniaux, E., Watson, A.L. (1999). Maternal arterial connections to the placental intervillous space during the first trimester of human pregnancy: the Boyd collection revisited. Am J Obstet Gynecol 181:718–724.

Burton, G.J., Reshetnikova, O.S., Milovanov, A.P., Teleshova, O.V. (1996). Stereological evaluation of vascular adaptations in human placental villi to differing forms of hypoxic stress. Placenta 17:49–55.

Burton, G.J., Tham, S.W. (1992). Formation of vasculo-syncytial membranes in the human placenta. J Dev Physiol 18:43–47.

Caniggia, I., Mostachfi, H., Winter, J., et al. (2000). Hypoxia-inducible factor-1 mediates the biological effects of oxygen on human trophoblast differentiation through TGFbeta(3). J Clin Invest 105:577–587.

Cao, Y., Chen, H., Zhou, L., et al. (1996a). Heterodimers of placenta growth factor/vascular endothelial growth factor. Endothelial activity, tumor cell expression, and high affinity binding to Flk-1/KDR. J Biol Chem 271:3154–3162.

Cao, Y., Ji, W.R., Qi, P., Rosin, A., Cao, Y. (1997). Placenta growth factor: identification and characterization of a novel isoform generated by RNA alternative splicing. Biochem Biophys Res Commun 235:493–498.

Cao, Y.H., Linden, P., Shima, D., Browne, F., Folkman, J. (1996b). In vivo angiogenic activity and hypoxia induction of heterodimers of placenta growth factor vascular endothelial growth factor. J Clin Invest 98:2507–2511.

Carmeliet, P., Collen, D. (1997). Molecular analysis of blood vessel formation and disease. Am J Physiol 273:H2091–2104.

Carmeliet, P., Ferreira, V., Breier, G., et al. (1996). Abnormal blood vessel development and lethality in embryos lacking a single VEGF allele. Nature 380:435–439.

Charnock-Jones, D.S., Sharkey, A.M., Boocock, C.A., et al. (1994). Vascular endothelial growth factor receptor localization and activation in human trophoblast and choriocarcinoma cells. Biol Reprod 51:524–530.

Christofferson, R.H., Nilsson, B.O. (1988). Morphology of the endometrial microvasculature during early placentation in the rat. Cell Tissue Res 253:209–220.

Clark, D.E., Salvig, J.D., Smith, S.K., Charnock-Jones, D.S. (1998a). Hepatocyte growth factor levels during normal and intra-uterine growth-restricted pregnancies. Placenta 19:671–673.

Clark, D.E., Smith, S.K., He, Y., et al. (1998b). A vascular endothelial growth factor antagonist is produced by the human placenta and released into the maternal circulation. Biol Reprod 59:1540–1548.

Clark, D.E., Smith, S.K., Licence, D., Evans, A.L., Charnock-Jones, D.S. (1998c). Comparison of expression patterns for placenta growth factor, vascular endothelial growth factor (VEGF), VEGF-B and VEGF-C in the human placenta throughout gestation. J Endocrinol 159:459–467.

Clark, D.E., Smith, S.K., Sharkey, A.M., Charnockjones, D.S. (1996a). Localization of VEGF and expression of its receptors Flt and Kdr in human placenta throughout pregnancy. Hum Reprod 11:1090–1098.

Clark, D.E., Smith, S.K., Sharkey, A.M., Sowter, H.M., Charnock-Jones, D.S. (1996b). Hepatocyte growth factor/scatter factor and its receptor c-met: localisation and expression in the human placenta throughout pregnancy. J Endocrinol 151:459–467.

Cockerill, G.W., Gamble, J.R., Vadas, M.A. (1995). Angiogenesis: models and modulators. Int Rev Cytol 159:113–160.

Conrad, K.P., Benyo, D.F. (1997). Placental cytokines and the pathogenesis of preeclampsia. Am J Reprod Immunol 37:240–249.

Cooper, J.C., Sharkey, A.M., CharnockJones, D.S., Palmer, C.R., Smith, S.K. (1996). VEGF mRNA levels in placentae from pregnancies complicated by pre-eclampsia. Br J Obstet Gynaecol 103:1191–1196.

Craven, C.M., Morgan, T., Ward, K. (1998). Decidual spiral artery remodelling begins before cellular interaction with cytotrophoblasts. Placenta 19:241–252.

Cross, J.C. (1998). Formation of the placenta and extraembryonic membranes. Ann NY Acad Sci 857:23–32.

Cross, J.C., Werb, Z., Fisher, S.J. (1994). Implantation and the placenta: key pieces of the development puzzle. Science 266:1508–1518.

Damsky, C.H., Fisher, S.J. (1998). Trophoblast pseudo-vasculogenesis: faking it with endothelial adhesion receptors. Curr Opin Cell Biol 10:660–666.

Dantzer, V., Leiser, R. (1994). Initial vascularization in the pig placenta: I. Demonstration of nonglandular areas by histology and corrosion casts. Anat Rec 238:177–190.

Das, S.K., Chakraborty, I., Wang, J., Dey, S.K., Hoffman, L.H. (1997). Expression of vascular endothelial growth factor (VEGF) and VEGF-receptor messenger ribonucleic acids in the peri-implantation rabbit uterus. Biol Reprod 56:1390–1399.

Davis, S., Yancopoulos, G.D. (1999). The angiopoietins: Yin and Yang in angiogenesis. Curr Top Microbiol Immunol 237:173–185.

Demir, R., Kaufmann, P., Castellucci, M., Erbengi, T., Kotowski, A. (1989). Fetal vasculogenesis and angiogenesis in human placental villi. Acta Anat (Basel) 136:190–203.

Desai, J., Holt-Shore, V., Torry, R.J., Caudle, M.R., Torry, D.S. (1999). Signal transduction and biological function of placenta growth factor in primary human trophoblast. Biol Reprod 60:887–892.

Di Blasio, A.M., Carniti, C., Vigano, P., Florio, P., Petraglia, F., Vignali, M. (1997). Basic fibroblast growth factor messenger ribonucleic acid levels in human placentas from normal and pathological pregnancies. Mol Hum Reprod 3:1119–1123.

Diaz-Flores, L., Gutierrez, R., Varela, H. (1994). Angiogenesis: an update. Histol Histopathol 9:807–843.

Dunk, C., Shams, M., Nijjar, S., et al. (2000). Angiopoietin-1 and angiopoietin-2 activate trophoblast Tie-2 to promote growth and migration during placental development. Am J Pathol 156:2185–2199.

Enders, A.C. (1995). Transition from lacunar to villous stage of implantation in the macaque, including establishment of the trophoblastic shell. Acta Anat (Basel) 152: 151–169.

Enders, A.C., Blankenship, T.N. (1997). Modification of endometrial arteries during invasion by cytotrophoblast cells in the pregnant macaque. Acta Anat (Basel) 159: 169–193.

Enders, A.C., Hendrickx, A.G., Schlafke, S. (1983). Implantation in the rhesus monkey: initial penetration of endometrium. Am J Anat 167:275–298.

Enders, A.C., Lantz, K.C., Schlafke, S. (1996). Preference of invasive cytotrophoblast for maternal vessels in early implantation in the macaque. Acta Anat (Basel) 155:145–162.

Engerman, R.L., Pfaffenbach, D., Davis, M.D. (1967). Cell turnover of capillaries. Lab Invest 17:738–743.

Ferrara, N. (1995). The role of vascular endothelial growth factor in pathological angiogenesis. Breast Cancer Res Treat 36:127–137.

Ferrara, N., Carver-Moore, K., Chen, H., et al. (1996). Heterozygous embryonic lethality induced by targeted inactivation of the VEGF gene. Nature 380:439–442.

Ferriani, R.A., Ahmed, A., Sharkey, A., Smith, S.K. (1994). Colocalization of acidic and basic fibroblast growth factor (FGF) in human placenta and the cellular effects of bFGF in trophoblast cell line JEG-3. Growth Factors 10:259–268.

Findlay, J.K. (1986). Angiogenesis in reproductive tissues. J Endocrinol 111:357–366.

Finkenzeller, G., Marme, D., Weich, H.A., Hug, H. (1992). Platelet-derived growth factor-induced transcription of the vascular endothelial growth factor gene is mediated by protein kinase C. Cancer Res 52:4821–4823.

Flamme, I., Frolich, T., Risau, W. (1997). Molecular mechanisms of vasculogenesis and embryonic angiogenesis. J Cell Physiol 173:206–210.

Foidart, J.M., Hustin, J., Dubois, M., Schaaps, J.P. (1992). The human placenta becomes haemochorial at the 13th week of pregnancy. Int J Dev Biol 36:451–453.

Folkman, J. (1995). Clinical applications of research on angiogenesis. N Engl J Med 333:1757–1763.

Fong, G.H., Rossant, J., Gertsenstein, M., Breitman, M.L. (1995). Role of the Flt-1 receptor tyrosine kinase in regulating the assembly of vascular endothelium. Nature 376: 66–70.

Forsythe, J.A., Jiang, B.H., Iyer, N.V., et al. (1996). Activation of vascular endothelial growth factor gene transcription by hypoxia-inducible factor 1. Mol Cell Biol 16: 4604–4613.

Furugori, K., Kurauchi, O., Itakura, A., et al. (1997). Levels of hepatocyte growth factor and its messenger ribonucleic acid in uncomplicated pregnancies and those complicated by preeclampsia. J Clin Endocrinol Metab 82:2726–2730.

Gargett, C.E., Lederman, F.L., Lau, T.M., Taylor, N.H., Rogers, P.A. (1999). Lack of correlation between vascular endothelial growth factor production and endothelial cell proliferation in the human endometrium. Hum Reprod 14:2080–2088.

Giles, W., Trudinger, B., Cook, C., Connelly, A. (1993). Placental microvascular changes in twin pregnancies with abnormal umbilical artery waveforms. Obstet Gynecol 81: 556–559.

Giles, W.B., Trudinger, B.J., Baird, P.J. (1985). Fetal umbilical artery flow velocity waveforms and placental resistance: pathological correlation. Br J Obstet Gynaecol 92: 31–38.

Gille, J., Khalik, M., Konig, V., Kaufmann, R. (1998). Hepatocyte growth factor/scatter factor (HGF/SF) induces vascular permeability factor (VPF/VEGF) expression by cultured keratinocytes. J Invest Dermatol 111:1160–1165.

Giroux, S., Tremblay, M., Bernard, D., et al. (1999). Embryonic death of Mek1-deficient mice reveals a role for this kinase in angiogenesis in the labyrinthine region of the placenta. Curr Biol 9:369–372.

Giudice, L. (1994). Growth factors and growth modulators in human uterine endometrium: their potential relevance to reproductive medicine. Fertil Steril 61:1–17.

Giudice, L.C., Irwin, J.C. (1999). Roles of the insulinlike growth factor family in non-pregnant human endometrium and at the decidual: trophoblast interface. Semin Reprod Endocrinol 17:13–21.

Gleadle, J.M., Ebert, B.L., Firth, J.D., Ratcliffe, P.J. (1995). Regulation of angiogenic growth factor expression by hypoxia, transition metals, and chelating agents. Am J Physiol 268: C1362–1368.

Gnarra, J.R., Ward, J.M., Porter, F.D., et al. (1997). Defective placental vasculogenesis causes embryonic lethality in VHL-deficient mice. Proc Natl Acad Sci USA 94: 9102–9107.

Godfrey, K.M., Redman, C.W., Barker, D.J., Osmond, C. (1991). The effect of maternal anaemia and iron deficiency on the ratio of fetal weight to placental weight [see comments]. Br J Obstet Gynaecol 98:886–891.

Goede, V., Schmidt, T., Kimmina, S., Kozian, D., Augustin, H.G. (1998). Analysis of blood vessel maturation processes during cyclic ovarian angiogenesis. Lab Invest 78:1385–1394.

Goldman, C.K., Kim, J., Wong, W.L., King, V., Brock, T., Gillespie, G.Y. (1993). Epidermal growth factor stimulates vascular endothelial growth factor production by human malignant glioma cells: a model of glioblastoma multiforme pathophysiology. Mol Biol Cell 4:121–133.

Goldman-Wohl, D.S., Ariel, I., Greenfield, C., Lavy, Y., Yagel, S. (2000). Tie-2 and angiopoietin-2 expression at the fetal-maternal interface: a receptor ligand model for vascular remodelling. Mol Hum Reprod 6:81–87.

Goodger, A.M., Rogers, P.A. (1993). Uterine endothelial cell proliferation before and after embryo implantation in rats. J Reprod Fertil 99:451–457.

Goodger, A.M., Rogers, P.A. (1994). Endometrial endothelial cell proliferation during the menstrual cycle. Hum Reprod 9:399–405.

Goodger, A.M., Rogers, P.A. (1995a). Blood vessel growth and endothelial cell density in rat endometrium. J Reprod Fertil 105:259–261.

Goodger, A.M., Rogers, P.A. (1995b). Blood vessel growth in the endometrium. Microcirculation 2:329–343.

Gordon, J.D., Shifren, J.L., Foulk, R.A., Taylor, R.N., Jaffe, R.B. (1995). Angiogenesis in the human female reproductive tract. Obstet Gynecol Surv 50:688–697.

Grevin, D., Chen, J.H., Raes, M.B., Stehelin, D., Vandenbunder, B., Desbiens, X. (1993). Involvement of the proto-oncogene c-ets 1 and the urokinase plasminogen activator during mouse implantation and placentation. Int J Dev Biol 37:519–529.

Guillemot, F., Nagy, A., Auerbach, A., Rossant, J., Joyner, A.L. (1994). Essential role of Mash-2 in extraembryonic development. Nature 371:333–336.

Gurtner, G.C., Davis, V., Li, H., McCoy, M.J., Sharpe, A., Cybulsky, M.I. (1995). Targeted disruption of the murine VCAM1 gene: essential role of VCAM-1 in chorioallantoic fusion and placentation. Genes Dev 9:1–14.

Halder, J.B., Zhao, X., Soker, S., et al. (2000). Differential expression of VEGF isoforms and VEGF(164)-specific receptor neuropilin-1 in the mouse uterus suggests a role for VEGF(164) in vascular permeability and angiogenesis during implantation. Genesis 26: 213–224.

Hamai, Y., Fujii, T., Yamashita, T., Kozuma, S., Okai, T., Taketani, Y. (1998). Evidence for basic fibroblast growth factor as a crucial angiogenic growth factor, released from human trophoblasts during early gestation. Placenta 19:149–155.

Hauser, S., Weich, H.A. (1993). A heparin-binding form of placenta growth factor (PlGF-2) is expressed in human umbilical vein endothelial cells and in placenta. Growth Factors 9:259–268.

Hayman, R., Brockelsby, J., Kenny, L., Baker, P. (1999). Preeclampsia: the endothelium, circulating factor(s) and vascular endothelial growth factor. J Soc Gynecol Invest 6:3–10.

He, Y., Smith, S.K., Day, K.A., Clark, D.E., Licence, D.R., Charnock-Jones, D.S. (1999). Alternative splicing of vascular endothelial growth factor (VEGF)-R1 (FLT-1) pre-mRNA is important for the regulation of VEGF activity. Mol Endocrinol 13:537–545.

Hitschold, T., Weiss, E., Beck, T., Hunterfering, H., Berle, P. (1993). Low target birth weight or growth retardation? Umbilical Doppler flow velocity waveforms and histometric analysis of fetoplacental vascular tree. Am J Obstet Gynecol 168:1260–1264.

Hornig, C., Barleon, B., Ahmad, S., Vuorela, P., Ahmed, A., Weich, H.A. (2000). Release and complex formation of soluble VEGFR-1 from endothelial cells and biological fluids. Lab Invest 80:443–454.

Hornig, C., Behn, T., Bartsch, W., Yayon, A., Weich, H.A. (1999). Detection and quantification of complexed and free soluble human vascular endothelial growth factor receptor-1 (sVEGFR-1) by ELISA. J Immunol Methods 226:169–177.

Illera, M.J., Cullinan, E., Gui, Y., Yuan, L., Beyler, S.A., Lessey, B.A. (2000). Blockade of the alpha(v)beta(3) integrin adversely affects implantation in the mouse. Biol Reprod 62:1285–1290.

Ingber, D., Fujita, T., Kishimoto, S., et al. (1990). Synthetic analogues of fumagillin that inhibit angiogenesis and suppress tumour growth. Nature 348:555–557.

Isner, J.M., Asahara, T. (1999). Angiogenesis and vasculogenesis as therapeutic strategies for postnatal neovascularization. J Clin Invest 103:1231–1236.

Iwasaka, C., Tanaka, K., Abe, M., Sato, Y. (1996). Ets-1 regulates angiogenesis by inducing the expression of urokinase-type plasminogen activator and matrix metalloproteinase-1 and the migration of vascular endothelial cells. J Cell Physiol 169:522–531.

Jackson, M.R., Carney, E.W., Lye, S.J., Ritchie, J.W. (1994). Localization of two angiogenic growth factors (PDECGF and VEGF) in human placentae throughout gestation. Placenta 15:341–353.

Jackson, M.R., Mayhew, T.M., Boyd, P.A. (1992). Quantitative description of the elaboration and maturation of villi from 10 weeks of gestation to term. Placenta 13:357–370.

Jackson, M.R., Mayhew, T.M., Haas, J.D. (1987). Morphometric studies on villi in human term placentae and the effects of altitude, ethnic grouping and sex of newborn. Placenta 8:487–495.

Jackson, M.R., Mayhew, T.M., Haas, J.D. (1988). On the factors which contribute to thinning of the villous membrane in human placentae at high altitude. II. An increase in the degree of peripheralization of fetal capillaries. Placenta 9:9–18.

Jaffe, R. (1998). First trimester utero-placental circulation: maternal-fetal interaction. J Perinat Med 26:168–174.

Jaffe, R., Jauniaux, E., Hustin, J. (1997). Maternal circulation in the first-trimester human placenta—myth or reality? Am J Obstet Gynecol 176:695–705.

Jaffe, R.B. (2000). Importance of angiogenesis in reproductive physiology. Semin Perinatol 24:79–81.

Jauniaux, E., Burton, G.J., Moscoso, G.J., Hustin, J. (1991). Development of the early human placenta: a morphometric study. Placenta 12:269–276.

Jauniaux, E., Zaidi, J., Jurkovic, D., Campbell, S., Hustin, J. (1994). Comparison of colour Doppler features and pathological findings in complicated early pregnancy. Hum Reprod 9:2432–2437.

Jirkovska, M., Kubinova, L., Krekule, I., Hach, P. (1998). Spatial arrangement of fetal placental capillaries in terminal villi: a study using confocal microscopy. Anat Embryol (Berl) 197:263–272.

Kadyrov, M., Kosanke, G., Kingdom, J., Kaufmann, P. (1998). Increased fetoplacental angiogenesis during first trimester in anaemic women. Lancet 352:1747–1749.

Kaelin, W.G., Iliopoulos, O., Lonergan, K.M., Ohh, M. (1998). Functions of the von Hippel-Lindau tumour suppressor protein. J Intern Med 243:535–539.

Kaiserman-Abramof, I.R., Padykula, H.A. (1989). Angiogenesis in the postovulatory primate endometrium: the coiled arteriolar system. Anat Rec 224:479–489.

Kamat, B., Brown, L., Manseau, E., Senger, D., Dvorak, H. (1995). Expression of vascular permeability factor/vascular endothelial growth factor by human granulosa and theca lutein cells. Role in corpus luteum development. Am J Pathol 146:157–165.

Karimu, A.L., Burton, G.J. (1994). Significance of changes in fetal perfusion pressure to factors controlling angiogenesis in the human term placenta. J Reprod Fertil 102: 447–450.

Kaufmann, P., Bruns, U., Leiser, R., Luckhardt, M., Winterhager, E. (1985). The fetal vascularisation of term human placental villi. II. Intermediate and terminal villi. Anat Embryol (Berl) 173:203–214.

Kauma, S., Hayes, N., Weatherford, S. (1997). The differential expression of hepatocyte growth factor and met in human placenta [see comments]. J Clin Endocrinol Metab 82: 949–954.

Kendall, R.L., Thomas, K.A. (1993). Inhibition of vascular endothelial cell growth factor activity by an endogenously encoded soluble receptor. Proc Natl Acad Sci USA 90: 10705–10709.

Kendall, R.L., Wang, G., Thomas, K.A. (1996). Identification of a natural soluble form of the vascular endothelial growth factor receptor, FLT-1, and its heterodimerization with KDR. Biochem Biophys Res Commun 226:324–328.

Khalid, M.E., Ali, M.E., Ali, K.Z. (1997). Full-term birth weight and placental morphology at high and low altitude. Int J Gynaecol Obstet 57:259–265.

Khaliq, A., Li, X.F., Shams, M., et al. (1996). Localisation of placenta growth factor (PlGF) in human term placenta. Growth Factors 13:243–250.

Khong, T.Y., De Wolf, F., Robertson, W.B., Brosens, I. (1986). Inadequate maternal vascular response to placentation in pregnancies complicated by pre-eclampsia and by small-for-gestational age infants. Br J Obstet Gynaecol 93:1049–1059.

Kilby, M.D., Afford, S., Li, X.F., Strain, A.J., Ahmed, A., Whittle, M.J. (1996). Localisation of hepatocyte growth factor and its receptor (c-met) protein and mRNA in human term placenta. Growth Factors 13:133–139.

Kim, K.J., Li, B., Houck, K., Winer, J., Ferrara, N. (1992). The vascular endothelial growth factor proteins: identification of biologically relevant regions by neutralizing monoclonal antibodies. Growth Factors 7:53–64.

Kingdom, J.C., Kaufmann, P. (1999). Oxygen and placental vascular development. Adv Exp Med Biol 474:259–275.

Klauber, N., Rohan, R.M., Flynn, E., D'Amato, R.J. (1997). Critical components of the female reproductive pathway are suppressed by the angiogenesis inhibitor AGM-1470. Nat Med 3:443–446.

Kozak, K.R., Abbott, B., Hankinson, O. (1997). ARNT-deficient mice and placental differentiation. Dev Biol 191:297–305.

Krebs, C., Longo, L.D., Leiser, R. (1997). Term ovine placental vasculature: comparison of sea level and high altitude conditions by corrosion cast and histomorphometry. Placenta 18:43–51.

Krebs, C., Macara, L.M., Leiser, R., Bowman, A.W., Greer, I.A., Kingdom, J.C. (1996). Intrauterine growth restriction with absent end-diastolic flow velocity in the umbilical artery is associated with maldevelopment of the placental terminal villous tree. Am J Obstet Gynecol 175:1534–1542.

Kreczy, A., Fusi, L., Wigglesworth, J.S. (1995). Correlation between umbilical arterial flow and placental morphology. Int J Gynecol Pathol 14:306–309.

Krieg, M., Marti, H.H., Plate, K.H. (1998). Coexpression of erythropoietin and vascular endothelial growth factor in nervous system tumors associated with von Hippel-Lindau tumor suppressor gene loss of function. Blood 92:3388–3393.

Kruger, H., Arias-Stella, J. (1970). The placenta and the newborn infant at high altitudes. Am J Obstet Gynecol 106:586–591.

Kupferminc, M.J., Daniel, Y., Englender, T., et al. (1997). Vascular endothelial growth factor is increased in patients with preeclampsia. Am J Reprod Immunol 38:302–306.

Kurjak, A., Kupesic, S., Hafner, T., Kos, M., Kostovic-Knezevic, L., Grbesa, D. (1997). Conflicting data on intervillous circulation in early pregnancy. J Perinat Med 25:225–236.

Kurjak, A., Predanic, M., Kupesic-Urek, S. (1993). Transvaginal color Doppler in the assessment of placental blood flow. Eur J Obstet Gynecol Reprod Biol 49:29–32.

Kurz, H., Wilting, J., Sandau, K., Christ, B. (1998). Automated evaluation of angiogenic effects mediated by VEGF and PlGF homo- and heterodimers. Microvasc Res 55:92–102.

Lala, P.K., Hamilton, G.S. (1996). Growth factors, proteases and protease inhibitors in the maternal-fetal dialogue. Placenta 17:545–555.

Lebovic, D.I., Mueller, M.D., Taylor, R.N. (1999). Vascular endothelial growth factor in reproductive biology. Curr Opin Obstet Gynecol 11:255–260.

Leiser, R., Beier, H. (1988). Morphological studies of lacunae formation in early rabbit placenta. In: Kaufmann P., Miller, R., eds. Trophoblast research: placental vascularization and blood flow, vol. 3, pp. 97–110. New York: Plenum Press.

Leiser, R., Kaufmann, P. (1994). Placental structure: in a comparative aspect. Exp Clin Endocrinol 102:122–134.

Leiser, R., Krebs, C., Klisch, K., et al. (1997). Fetal villosity and microvasculature of the bovine placentome in the second half of gestation. J Anat 191:517–527.

Li, J., Perrella, M.A., Tsai, J.C., et al. (1995). Induction of vascular endothelial growth factor gene expression by interleukin-1 beta in rat aortic smooth muscle cells. J Biol Chem 270:308–312.

Luo, J., Sladek, R., Bader, J.A., Matthyssen, A., Rossant, J., Giguere, V. (1997). Placental abnormalities in mouse embryos lacking the orphan nuclear receptor ERR-beta. Nature 388:778–782.

Luton, D., Sibony, O., Oury, J.F., Blot, P., Dieterlen-Lievre, F., Pardanaud, L. (1997). The c-ets1 protooncogene is expressed in human trophoblast during the first trimester of pregnancy. Early Hum Dev 47:147–156.

Lyall, F., Greer, I.A., Boswell, F., Fleming, R. (1997a). Suppression of serum vascular endothelial growth factor immunoreactivity in normal pregnancy and in pre-eclampsia. Br J Obstet Gynaecol 104:223–228.

Lyall, F., Young, A., Boswell, F., Kingdom, J.C., Greer, I.A. (1997b). Placental expression of vascular endothelial growth factor in placentae from pregnancies complicated by preeclampsia and intrauterine growth restriction does not support placental hypoxia at delivery. Placenta 18:269–276.

Macara, L., Kingdom, J.C., Kohnen, G., Bowman, A.W., Greer, I.A., Kaufmann, P. (1995). Elaboration of stem villous vessels in growth restricted pregnancies with abnormal umbilical artery Doppler waveforms [see comments]. Br J Obstet Gynaecol 102:807–812.

Maglione, D., Guerriero, V., Viglietto, G., Delli-Bovi, P., Persico, M.G. (1991). Isolation of a human placenta cDNA coding for a protein related to the vascular permeability factor. Proc Natl Acad Sci USA 88:9267–9271.

Maglione, D., Guerriero, V., Viglietto, G., et al. (1993). Two alternative mRNAs coding for the angiogenic factor, placenta growth factor (PlGF), are transcribed from a single gene of chromosome 14. Oncogene 8:925–931.

Maisonpierre, P.C., Suri, C., Jones, P.F., et al. (1997). Angiopoietin-2, a natural antagonist for Tie2 that disrupts in vivo angiogenesis. Science 277:55–60.

Maltepe, E., Schmidt, J.V., Baunoch, D., Bradfield, C.A., Simon, M.C. (1997). Abnormal angiogenesis and responses to glucose and oxygen deprivation in mice lacking the protein ARNT. Nature 386:403–407.

Mayhew, T.M. (1998). Thinning of the intervascular tissue layers of the human placenta is an adaptive response to passive diffusion in vivo and may help to predict the origins of fetal hypoxia. Eur J Obstet Gynecol Reprod Biol 81:101–109.

Mayhew, T.M., Jackson, M.R., Haas, J.D. (1986). Microscopical morphology of the human placenta and its effects on oxygen diffusion: a morphometric model. Placenta 7:121–131.

McCormick, F. (1993). Signal transduction. How receptors turn Ras on. Nature 363:15–16.

McCowan, L.M., Mullen, B.M., Ritchie, K. (1987). Umbilical artery flow velocity waveforms and the placental vascular bed. Am J Obstet Gynecol 157:900–902.

Meegdes, B.H., Ingenhoes, R., Peeters, L.L., Exalto, N. (1988). Early pregnancy wastage: relationship between chorionic vascularization and embryonic development. Fertil Steril 49:216–220.

Meekins, J.W., Luckas, M.J., Pijnenborg, R., McFadyen, I.R. (1997). Histological study of decidual spiral arteries and the presence of maternal erythrocytes in the intervillous space during the first trimester of normal human pregnancy. Placenta 18:459–464.

Meekins, J.W., Pijnenborg, R., Hanssens, M., McFadyen, I.R., van Asshe, A. (1994). A study of placental bed spiral arteries and trophoblast invasion in normal and severe preeclamptic pregnancies. Br J Obstet Gynaecol 101:669–674.

Mignatti, P., Rifkin, D.B. (1996). Plasminogen activators and matrix metalloproteinases in angiogenesis. Enzyme Protein 49:117–137.

Millauer, B., Longhi, M.P., Plate, K.H., et al. (1996). Dominant-negative inhibition of Flk-1 suppresses the growth of many tumor types in vivo. Cancer Res 56:1615–1620.

Millauer, B., Shawver, L.K., Plate, K.H., Risau, W., Ullrich, A. (1994). Glioblastoma growth inhibited in vivo by a dominant-negative Flk-1 mutant. Nature 367:576–579.

Monkley, S.J., Delaney, S.J., Pennisi, D.J., Christiansen, J.H., Wainwright, B.J. (1996). Targeted disruption of the Wnt2 gene results in placentation defects. Development 122:3343–3353.

Moscatelli, D., Joseph-Silverstein, J., Presta, M., Rifkin, D.B. (1988). Multiple forms of an angiogenesis factor: basic fibroblast growth factor. Biochimie 70:83–87.

Moscatelli, D., Presta, M., Rifkin, D.B. (1986). Purification of a factor from human placenta that stimulates capillary endothelial cell protease production, DNA synthesis, and migration. Proc Natl Acad Sci USA 83:2091–2095.

Neufeld, G., Cohen, T., Gengrinovitch, S., Poltorak, Z. (1999). Vascular endothelial growth factor (VEGF) and its receptors. FASEB J 13:9–22.

Ohh, M., Kaelin, W.G., Jr. (1999). The von Hippel-Lindau tumour suppressor protein: new perspectives. Mol Med Today 5:257–263.

Pal, S., Claffey, K.P., Dvorak, H.F., Mukhopadhyay, D. (1997). The von Hippel-Lindau gene product inhibits vascular permeability factor/vascular endothelial growth factor expression in renal cell carcinoma by blocking protein kinase C pathways. J Biol Chem 272:27509–27512.

Park, J.E., Chen, H.H., Winer, J., Houck, K.A., Ferrara, N. (1994). Placenta growth factor. Potentiation of vascular endothelial growth factor bioactivity, in vitro and in vivo, and high affinity binding to Flt-1 but not to Flk-1/KDR. J Biol Chem 269:25646–25654.

Peek, M., Landgren, B.M., Johannisson, E. (1992). The endometrial capillaries during the normal menstrual cycle: a morphometric study. Hum Reprod 7:906–911.

Pepper, M.S., Vassalli, J.D., Orci, L., Montesano, R. (1993). Biphasic effect of transforming growth factor-beta 1 on in vitro angiogenesis. Exp Cell Res 204:356–363.

Petrova, T.V., Makinen, T., Alitalo, K. (1999). Signaling via vascular endothelial growth factor receptors. Exp Cell Res 253:117–130.

Pfarrer, C., Macara, L., Leiser, R., Kingdom, J. (1999a). Adaptive angiogenesis in placentas of heavy smokers. Lancet 354:303.

Pfarrer, C., Winther, H., Leiser, R., Dantzer, V. (1999b). The development of the endotheliochorial mink placenta: light microscopy and scanning electron microscopical morphometry of maternal vascular casts. Anat Embryol (Berl) 199:63–74.

Pijnenborg, R. (1998). The origin and future of placental bed research. Eur J Obstet Gynecol Reprod Biol 81:185–190.

Rakusan, K., Ehrenburg, I.V., Gulyaeva, N.V., Tkatchouk, E.N. (1999). The effect of intermittent normobaric hypoxia on vascularization of human myometrium. Microvasc Res
58:200–203.

Ramsey, E., Donner, M. (1980). Placental vasculature and circulation. Philadelphia: W.B.
Saunders.

Redman, C.W., Sacks, G.P., Sargent, I.L. (1999). Preeclampsia: an excessive maternal inflammatory response to pregnancy. Am J Obstet Gynecol 180:499–506.

Rees, M.C.P., Bicknell, R. (1998). Angiogenesis in the endometrium. Angiogenesis 2:29–35.

Reshetnikova, O.S., Burton, G.J., Milovanov, A.P. (1994). Effects of hypobaric hypoxia on
the fetoplacental unit: the morphometric diffusing capacity of the villous membrane at
high altitude. Am J Obstet Gynecol 171:1560–1565.

Reshetnikova, O.S., Burton, G.J., Teleshova, O.V. (1995). Placental histomorphometry and
morphometric diffusing capacity of the villous membrane in pregnancies complicated by
maternal iron-deficiency anemia. Am J Obstet Gynecol 173:724–727.

Reuvekamp, A., Velsing-Aarts, F.V., Poulina, I.E., Capello, J.J., Duits, A.J. (1999). Selective
deficit of angiogenic growth factors characterises pregnancies complicated by preeclampsia. Br J Obstet Gynaecol 106:1019–1022.

Reynolds, L.P., Killilea, S.D., Redmer, D.A. (1992). Angiogenesis in the female reproductive system. FASEB J 6:886–892.

Riley, P., Anson-Cartwright, L., Cross, J.C. (1998). The Hand1 bHLH transcription factor
is essential for placentation and cardiac morphogenesis. Nat Genet 18:271–275.

Rinkenberger, J.L., Cross, J.C., Werb, Z. (1997). Molecular genetics of implantation in the
mouse. Dev Genet 21:6–20.

Risau, W. (1997). Mechanisms of angiogenesis. Nature 386:671–674.

Roberts, J.M. (1998). Endothelial dysfunction in preeclampsia. Semin Reprod Endocrinol
16:5–15.

Roberts, J.M., Taylor, R.N., Musci, T.J., Rodgers, G.M., Hubel, C.A., McLaughlin, M.K.
(1989). Preeclampsia: an endothelial cell disorder. Am J Obstet Gynecol 161:1200–1204.

Rodesch, F., Simon, P., Donner, C., Jauniaux, E. (1992). Oxygen measurements in endometrial and trophoblastic tissues during early pregnancy. Obstet Gynecol 80:283–285.

Rogers, P.A. (1992). Early endometrial microvascular response during implantation in the
rat. Reprod Fertil Dev 4:261–264.

Rogers, P.A., Abberton, K.M., Susil, B. (1992). Endothelial cell migratory signal produced
by human endometrium during the menstrual cycle. Hum Reprod 7:1061–1066.

Rogers, P.A., Lederman, F., Taylor, N. (1998). Endometrial microvascular growth in normal
and dysfunctional states. Hum Reprod Update 4:503–508.

Rogers, P.A., Macpherson, A.M., Beaton, L.A. (1988). Vascular response in a non-uterine
site to implantation-stage embryos in the rat and guinea-pig: in vivo and ultrastructural
studies. Cell Tissue Res 254:217–224.

Rosen, E.M., Lamszus, K., Laterra, J., Polverini, P.J., Rubin, J.S., Goldberg, I.D. (1997).
HGF/SF in angiogenesis. Ciba Found Symp 212:215–226; discussion 227–229.

Saleh, M., Stacker, S.A., Wilks, A.F. (1996). Inhibition of growth of C6 glioma cells in vivo
by expression of antisense vascular endothelial growth factor sequence. Cancer Res 56:
393–401.

Sangha, R.K., Li, X.F., Shams, M., Ahmed, A. (1997). Fibroblast growth factor receptor-1
is a critical component for endometrial remodeling: localization and expression of basic
fibroblast growth factor and FGF-R1 in human endometrium during the menstrual cycle
and decreased FGF-R1 expression in menorrhagia. Lab Invest 77:389–402.

Scheffen, I., Kaufmann, P., Philippens, L., Leiser, R., Geisen, C., Mottaghy, K. (1990). Alterations of the fetal capillary bed in the guinea pig placenta following long-term hypoxia.
Adv Exp Med Biol 277:779–790.

Shalaby, F., Rossant, J., Yamaguchi, T.P., et al. (1995). Failure of blood-island formation and
vasculogenesis in Flk-1-deficient mice. Nature 376:62–66.

Shams, M., Ahmed, A. (1994). Localization of mRNA for basic fibroblast growth factor in human placenta. Growth Factors 11:105–111.

Sharkey, A.M., Charnock-Jones, D.S., Boocock, C.A., Brown, K.D., Smith, S.K. (1993). Expression of mRNA for vascular endothelial growth factor in human placenta. J Reprod Fertil 99:609–615.

Sharkey, A.M., Cooper, J.C., Balmforth, J.R., et al. (1996). Maternal plasma levels of vascular endothelial growth factor in normotensive pregnancies and in pregnancies complicated by preeclampsia. Eur J Clin Invest 26:1182–1185.

Shaw, S.T., Jr., Macaulay, L.K., Hohman, W.R. (1979). Vessel density in endometrium of women with and without intrauterine contraceptive devices: a morphometric evaluation. Am J Obstet Gynecol 135:202–206.

Shiraishi, S., Nakagawa, K., Kinukawa, N., Nakano, H., Sueishi, K. (1996). Immunohistochemical localization of vascular endothelial growth-factor in the human placenta. Placenta 17:2–3.

Shore, V.H., Wang, T.H., Wang, C.L., Torry, R.J., Caudle, M.R., Torry, D.S. (1997). Vascular endothelial growth factor, placenta growth factor and their receptors in isolated human trophoblast. Placenta 18:657–665.

Shweiki, D., Neeman, M., Itin, A., Keshet, E. (1995). Induction of vascular endothelial growth factor expression by hypoxia and by glucose deficiency in multicell spheroids: implications for tumor angiogenesis. Proc Natl Acad Sci USA 92:768–772.

Smith, S.K. (1998). Angiogenesis, vascular endothelial growth factor and the endometrium. Human Reprod Update 4:509 519.

Smith, S.K., He, Y., Clark, D.E., Charnock-Jones, D.S. (2000). Angiogenic growth factor expression in placenta. Semin Perinatol 24:82–86.

Soker, S., Takashima, S., Miao, H.Q., Neufeld, G., Klagsbrun, M. (1998). Neuropilin-1 is expressed by endothelial and tumor cells as an isoform-specific receptor for vascular endothelial growth factor. Cell 92:735–745.

Somerset, D.A., Li, X.F., Afford, S., et al. (1998). Ontogeny of hepatocyte growth factor (HGF) and its receptor (c-met) in human placenta: reduced HGF expression in intrauterine growth restriction. Am J Pathol 153:1139–1147.

Srivastava, R.K., Gu, Y., Ayloo, S., Zilberstein, M., Gibori, G. (1998). Developmental expression and regulation of basic fibroblast growth factor and vascular endothelial growth factor in rat decidua and in a decidual cell line. J Mol Endocrinol 21:355 362.

Starzyk, K.A., Pijnenborg, R., Salafia, C.M. (1999). Decidual and vascular pathophysiology in pregnancy compromise. Semin Reprod Endocrinol 17:63–72.

Starzyk, K.A., Salafia, C.M., Pezzullo, J.C., et al. (1997). Quantitative differences in arterial morphometry define the placental bed in preeclampsia. Hum Pathol 28:353–358.

Steingrimsson, E., Tessarollo, L., Reid, S.W., Jenkins, N.A., Copeland, N.G. (1998). The bHLH-Zip transcription factor Tfeb is essential for placental vascularization. Development 125:4607–4616.

Stewart, F. (1996). Roles of mesenchymal-epithelial interactions and hepatocyte growth factor-scatter factor (HGF-SF) in placental development. Rev Reprod 1:144–148.

Stoz, F., Schuhmann, R.A., Schebesta, B. (1988). The development of the placental villus during normal pregnancy: morphometric data base. Arch Gynecol Obstet 244:23–32.

Suri, C., McClain, J., Thurston, G., et al. (1998). Increased vascularization in mice overexpressing angiopoietin-1. Science 282:468–471.

Sutherland, A., Cooper, D.W., Howie, P.W., Liston, W.A., MacGillivray, I. (1981). The indicence of severe pre-eclampsia amongst mothers and mothers-in-law of preeclamptics and controls. Br J Obstet Gynaecol 88:785–791.

Tanaka, C., Kuwabara, Y., Sakai, T. (1999). Structural identification and characterization of arteries and veins in the placental stem villi. Anat Embryol (Berl) 199:407–418.

Tawia, S.A., Rogers, P.A. (1992). In vivo microscopy of the subepithelial capillary plexus of the endometrium of rats during embryo implantation. J Reprod Fertil 96:673–680.

Taylor, R.N. (1997). Review: immunobiology of preeclampsia. Am J Reprod Immunol 37: 79–86.

Taylor, R.N., de Groot, C.J., Cho, Y.K., Lim, K.H. (1998). Circulating factors as markers and mediators of endothelial cell dysfunction in preeclampsia. Semin Reprod Endocrinol 16:17–31.

te Velde, E.A., Exalto, N., Hesseling, P., van der Linden, H.C. (1997). First trimester development of human chorionic villous vascularization studied with CD34 immunohistochemistry. Hum Reprod 12:1577–1581.

Teasdale, F. (1984). Idiopathic intrauterine growth retardation: histomorphometry of the human placenta. Placenta 5:83–92.

Thurston, G., Suri, C., Smith, K., et al. (1999). Leakage-resistant blood vessels in mice transgenically overexpressing angiopoietin-1. Science 286:2511–2514.

Tidwell, S.C., Ho, H.-N., Chiu, W.-H., Torry, R.J., Torry, D.S. (2001). Low maternal serum levels of placenta growth factor as an antecedent of clinical preeclampsia. Am J Obstet Gynecol 184:1267–1272.

Todros, T., Sciarrone, A., Piccoli, E., Guiot, C., Kaufmann, P., Kingdom, J. (1999). Umbilical Doppler waveforms and placental villous angiogenesis in pregnancies complicated by fetal growth restriction. Obstet Gynecol 93:499–503.

Torry, D.S., Ahn, H., Barnes, E.L., Torry, R.J. (1999). Placenta growth factor: potential role in pregnancy. Am J Reprod Immunol 41:79–85.

Torry, D.S., Holt, V.J., Keenan, J.A., Harris, G., Caudle, M.R., Torry, R.J. (1996). Vascular endothelial growth factor expression in cycling human endometrium. Fertil Steril 66: 72–80.

Torry, D.S., Labarrere, C.A., McIntyre, J.A. (1998a). Uteroplacental vascular involvement in recurrent spontaneous abortion. Curr Opin Obstet Gynecol 10:379–382.

Torry, D.S., Torry, R.J. (1997). Angiogenesis and the expression of vascular endothelial growth factor in endometrium and placenta. Am J Reprod Immunol 37:21–29.

Torry, D.S., Wang, H.S., Wang, T.H., Caudle, M.R., Torry, R.J. (1998b). Preeclampsia is associated with reduced serum levels of placenta growth factor. Am J Obstet Gynecol 179:1539–1544.

Torry, R.J., Rongish, B.J. (1992). Angiogenesis in the uterus: potential regulation and relation to tumor angiogenesis. Am J Reprod Immunol 27:171–179.

Uehara, Y., Minowa, O., Mori, C., et al. (1995). Placental defect and embryonic lethality in mice lacking hepatocyte growth factor/scatter factor. Nature 373:702–705.

Vailhe, B., Dietl, J., Kapp, M., Toth, B., Arck, P. (1999). Increased blood vessel density in decidua parietalis is associated with spontaneous human first trimester abortion. Hum Reprod 14:1628–1634.

van Beck, E., Peeters, L.L. (1998). Pathogenesis of preeclampsia: a comprehensive model. Obstet Gynecol Surv 53:233–239.

Van Belle, E., Witzenbichler, B., Chen, D., et al. (1998). Potentiated angiogenic effect of scatter factor/hepatocyte growth factor via induction of vascular endothelial growth factor: the case for paracrine amplification of angiogenesis. Circulation 97:381–390.

Vinatier, D., Monnier, J.C. (1995). Pre-eclampsia: physiology and immunological aspects. Eur J Obstet Gynecol Reprod Biol 61:85–97.

Voss, A.K., Thomas, T., Gruss, P. (2000). Mice lacking HSP90beta fail to develop a placental labyrinth. Development 127:1–11.

Vuorela, P., Hatva, E., Lymboussaki, A., et al. (1997). Expression of vascular endothelial growth factor and placenta growth factor in human placenta. Biol Reprod 56:489–494.

Vuorela, P., Helske, S., Hornig, C., Alitalo, K., Weich, H., Halmesmaki, E. (2000). Amniotic fluid—soluble vascular endothelial growth factor receptor-1 in preeclampsia. Obstet Gynecol 95:353–357.

Vuorela-Vepsalainen, P., Alfthan, H., Orpana, A., Alitalo, K., Stenman, U.H., Halmesmaki, E. (1999). Vascular endothelial growth factor is bound in amniotic fluid and maternal serum. Hum Reprod 14:1346–1351.

Wang, G.L., Semenza, G.L. (1995). Purification and characterization of hypoxia-inducible factor 1. J Biol Chem 270:1230–1237.

Welsh, A.O., Enders, A.C. (1991). Chorioallantoic placenta formation in the rat: II. Angiogenesis and maternal blood circulation in the mesometrial region of the implantation chamber prior to placenta formation. Am J Anat 192:347–365.

Wernert, N., Raes, M.B., Lassalle, P., et al. (1992). c-ets1 proto-oncogene is a transcription factor expressed in endothelial cells during tumor vascularization and other forms of angiogenesis in humans. Am J Pathol 140:119–127.

Wheeler, T., Elcock, C.L., Anthony, F.W. (1995). Angiogenesis and the placental environment. Placenta 16:289–296.

Wilcox, A.J., Weinberg, C.R., O'Connor, J.F., et al. (1988). Incidence of early loss of pregnancy. N Engl J Med 319:189–194.

Williams, L.A., Evans, S.F., Newnham, J.P. (1997). Prospective cohort study of factors influencing the relative weights of the placenta and the newborn infant. BMJ 314:1864–1868.

Winther, H., Ahmed, A., Dantzer, V. (1999). Immunohistochemical localization of vascular endothelial growth factor (VEGF) and its two specific receptors, Flt-1 and KDR, in the porcine placenta and non-pregnant uterus. Placenta 20:35–43.

Wojta, J., Kaun, C., Breuss, J.M., et al. (1999). Hepatocyte growth factor increases expression of vascular endothelial growth factor and plasminogen activator inhibitor-1 in human keratinocytes and the vascular endothelial growth factor receptor flk-1 in human endothelial cells. Lab Invest 79:427–438.

Wood, S.M., Gleadle, J.M., Pugh, C.W., Hankinson, O., Ratcliffe, P.J. (1996). The role of the aryl hydrocarbon receptor nuclear translocator (ARNT) in hypoxic induction of gene expression. Studies in ARNT-deficient cells. J Biol Chem 271:15117–15123.

Wordinger, R.J., Smith, K.J., Bell, C., Chang, I.F. (1994). The immunolocalization of basic fibroblast growth factor in the mouse uterus during the initial stages of embryo implantation. Growth Factors 11:175–186.

Yamamoto, H., Flannery, M.L., Kupriyanov, S., et al. (1998). Defective trophoblast function in mice with a targeted mutation of Ets2. Genes Dev 12:1315–1326.

Yang, J.T., Rayburn, H., Hynes, R.O. (1995). Cell adhesion events mediated by alpha 4 integrins are essential in placental and cardiac development. Development 121:549–560.

Yang, W., Ann, H., Desai, J.B., Torry, R.J., Torry, D.S. (2000). Characterization of a novel isoform of placenta growth factor expressed in human trophoblast and endothelial cells. Am J Reprod and Immunol 43(6):332.

Yi, X.J., Jiang, H.Y., Lee, K.K., O, W.S., Tang, P.L., Chow, P.H. (1999). Expression of vascular endothelial growth factor (VEGF) and its receptors during embryonic implantation in the golden hamster (Mesocricetus auratus). Cell Tissue Res 296:339–349.

Zhou, Y., Damsky, C.H., Chiu, K., Roberts, J.M., Fisher, S.J. (1993). Preeclampsia is associated with abnormal expression of adhesion molecules by invasive cytotrophoblasts. J Clin Invest 91:950–960.

Zhou, Y., Damsky, C.H., Fisher, S.J. (1997a). Preeclampsia is associated with failure of human cytotrophoblasts to mimic a vascular adhesion phenotype. One cause of defective endovascular invasion in this syndrome? J Clin Invest 99:2152–2164.

Zhou, Y., Fisher, S.J., Janatpour, M., et al. (1997b). Human cytotrophoblasts adopt a vascular phenotype as they differentiate. A strategy for successful endovascular invasion? J Clin Invest 99:2139–2151.

Ziche, M., Maglione, D., Ribatti, D., et al. (1997). Placenta growth factor-1 is chemotactic, mitogenic, and angiogenic. Lab Invest 76:517–531.

Index